MW00713397

People-Plant Relationships: Setting Research Priorities

Joel Flagler
Raymond P. Poincelot
Editors

CRC Press is an imprint of the
Taylor & Francis Group, an informa business

Published by

Food Products Press, 10 Alice Street, Binghamton, NY 13904-1580, USA.

Food Products Press is an imprint of The Haworth Press, Inc., 10 Alice Street, Binghamton, NY 13904-1580, USA.

Reprinted 2009 by CRC Press

The Haworth Press, Inc., 10 Alice Street, Binghamton, NY 13904-1580 USA

Library of Congress Cataloging-in-Publication Data

People-plant relationships: setting research priorities / Joel Flagler, Raymond P. Poincelot, editors.
p. cm.
Includes bibliographical references and index.
ISBN 1-56022-050-3 (alk. paper)
1. Human-plant relationships. 2. Human-plant relationships–Research. I. Flagler, Joel. II. Poincelot, Raymond P., 1944- .
QK46.5.H85P46 1994
304.2'7–dc20 94-3031
CIP

People-Plant Relationships: Setting Research Priorities

CONTENTS

RESEARCH IMPLEMENTATION

ABOUT THE EDITORS

Joel Flagler, MFS, is an Associate Professor at Rutgers University-Cook College. He serves as the Cooperative Extension Agent in Agriculture and Resource Management for Bergen County, NJ, and is liaison to the Bergen County Department of Parks. Professor Flagler is also a Registered Horticultural Therapist. He has published articles in many journals, given talks nationwide, and for six years has written a weekly gardening column for *The Bergen Record*. Currently serving his second elected term on the Board of Directors of the American Horticultural Therapy Association, Mr. Flagler designed the horticultural therapy curriculum for the New York Botanical Garden in the Bronx. He has received regional and national recognition for his work with correctional youth, training them for possible horticultural careers, and has been the recipient of many professional awards and honors.

Raymond P. Poincelot, PhD, is currently a Professor of Biology at Fairfield University. In addition, he serves as the University Safety Officer/Chemical Hygiene Officer and was a former Chair of Biology. Professor Poincelot is also a Senior Editor for Food Products Press, where he is the Editor of the *Journal of Sustainable Agriculture*, *Journal of Home & Consumer Horticulture*, and two book series dealing with similar topics. Dr. Poincelot has authored four books dealing with horticulture and sustainable agriculture and over 70 research and professional publications. His research involves the use of humic/seaweed extracts and their effects upon plant propagation and growth. He teaches courses concerned with plant biology, horticulture and the environment.

About the Contributors

Harry Betros is Program Assistant for Rutgers University Cooperative Extension.

Margaret Burchett is affiliated with the Urban Horticulture Unit at the School of Biological and Biomedical Sciences, University of Technology, Sydney, Australia.

Alejandro Ching is Research Horticulturist and Director of the Alternative Crops Research Center, Northwest Missouri State University, Maryville, Missouri.

Dina Chuensanguansat is a Graduate Student in the Department of Geography at the University of Hawaii, Honolulu, Hawaii.

John Dotter is Master Gardener and Garden Programs Supervisor for the Recreation, Parks and Community Services Department of San Jose, California.

Nina L. Etkin is Associate Professor in the Department of Anthropology, and Associate Researcher in the Social Science Research Institute of the University of Hawaii-Manoa, Honolulu, Hawaii.

Joel Flagler is Assistant Professor at Rutgers University-Cook College. He serves as the Agricultural/Resource Management Agent for Bergen County, New Jersey. Mr. Flagler is the A.S.H.S. delegate to the People-Plant Council and is on the Board of Directors of A.H.T.A. He chaired and coordinated the 1992 People-Plant Symposium and the HIH workshop at A.S.H.S., Hawaii. He is also a Registered Horticultural Therapist.

Karen Green was employed, from 1992-1994, by the Horticultural Society of New York teaching the Appleseed Program. She was also employed by the Harlem School District.

Dan Greenlee is Rehabilitation Facilities Representative for Crop-king Inc., Medina, Ohio.

Gert Gröning, Hochschule der Künste Berlin, Fachbereich Architektur, Hardenbergstraße 33, 1000 Berlin 12, Germany.

William K. Hallman is a Psychologist and member of the faculty of the Department of Human Ecology, Cook College, Rutgers, The State University of New Jersey.

William T. Hlubik is Assistant Professor/Agricultural and Resource Management Agent for Rutgers University Cooperative Extension, New Brunswick, New Jersey.

Toshihiko Ikagawa is affiliated with the Department of Geography at the University of Hawaii at Manoa, Honolulu, Hawaii.

Duane Jewell is Agricultural Economist and Head of the Agriculture Department, Northwest Missouri State University, Maryville, Missouri.

Maxine Jewel Kaplan has worked at the Westchester County Medical Center, Valhalla, New York, since 1981. She has programs both at the Ruth Taylor Rehabilitation and Geriatric Institute and the Psychiatric Institute, which are part of the medical center's facilities.

Jean Stephans Kavanagh is affiliated with the Department of Landscape Architecture at Texas Tech University in Lubbock, Texas.

Terry Keller is Director of the NYBG Bronx Green-Up Program, Bronx, New York.

Charles A. Lewis is a retired Research Fellow in Horticulture, Morton Arboretum, Lisle, Illinois.

Clare S. Liptak is County Agricultural Agent with Rutgers Cooperative Extension of Somerset County, Somerville, New Jersey.

Virginia I. Lohr is Associate Professor in the Department of Horticulture and Landscape Architecture at Washington State University, Pullman, Washington.

Pete Madsen is Information Officer in the Department of Horticulture at Virginia Tech University, Blacksburg, Virginia.

Lorisa M. W. Mock is Head Gardener at an entertainment farm in suburban Philadelphia.

Geraldine Moreno-Black is affiliated with the Department of Anthropology at the University of Oregon, Eugene, Oregon.

Thomas A. Musiak is a member of the faculty and Chair of the Department of Landscape Architecture at Texas Tech University, Lubbock, Texas.

Russ Parsons, is affiliated with the College of Architecture, Texas A&M University, College Station, Texas.

Ishwarbhai C. Patel is County Agricultural Agent, Rutgers Urban Gardening, Newark, New Jersey.

Diane Relf is Associate Professor in the Department of Horticulture at Virginia Tech University, Blacksburg, Virginia.

Linda L. Remy is Senior Research Analyst with the Institute for Health Policy Studies at the University of California, San Francisco, San Francisco, California.

Jay Stone Rice is a Program Consultant with the San Francisco Sheriff's Department, San Rafael, California.

Judith Schwartz is Adjunct Faculty Member in Biology at Keene State College, Keene, New Hampshire. She is a Partner in Town and Country Gardens, Marlow, New Hampshire.

Rhonda Roland Shearer is an Artist and Sculptor. She resides in New York, New York.

Bonnie Sherk is Creator, Developer, and Planner of unique artistic ecological, and educational environments with integrated programs and curricula, called LIFE FRAMES™. She is Creator and Executive Director of A LIVING LIBRARY™, designed to connect communities, schools and institutions around the globe.

Candice A. Shoemaker is Assistant Professor of Horticulture at Berry College, Mt. Berry, Georgia.

Deborah C. Smith-Fiola is an Entomologist and County Agricultural Agent with the Rutgers Cooperative Extension of Ocean County, New Jersey.

Martha C. Straus is currently Supervising Director of Horticultural Therapy at Friends Hospital. She has served on the Board of the American Horticultural Therapy Association and as its Secretary and Vice-President.

Louis G. Tassinary is affiliated with the College of Architecture, Texas A&M University, College Station, Texas.

Roger E. Ulrich is Research Professor at Western Michigan University, Kalamazoo, Michigan.

Roger S. Ulrich is affiliated with the College of Architecture, Texas A&M University, College Station, Texas.

Lyndon L. Wester is Associate Professor in the Department of Geography at the University of Hawaii, Honolulu, Hawaii.

Joachim Wolschke-Bulmahn is Director of Studies in Landscape Architecture at Dumbarton Oaks/Trustees for Harvard University, Washington, DC.

Ronald Wood is affiliated with the Urban Horticulture Unit at the School of Biological and Biomedical Sciences, University of Technology, Sydney, Australia.

Madelaine H. Zadik is a Registered Horticultural Therapist (HTR). She has served on the board of the American Horticultural Therapy Association and as President of the New York chapter of AHTA.

James W. Zampini is President of Lake County Nursery, Inc., Perry, Ohio.

Preface

This collection of papers is from the symposium, "People-Plant Relationships: Setting Research Priorities." The sponsors were: Rutgers University Cooperative Extension, American Society of Horticultural Science, American Association of Botanical Gardens and Arboreta, American Horticultural Therapy Association and the People-Plant Council. The symposium was held in East Rutherford, NJ. The Symposium Organizing Committee consisted of Joel Flagler (Chair and Conference Coordinator) from Rutgers University, Virginia Lohr from Washington State University, Candice Shoemaker from Berry College, Diane Relf from Virginia Polytechnic Institute and State University, Terry Keller from the New York Botanical Garden and Ishwarbhai Patel from Rutgers University. The objectives and focus of the symposium were:

> To identify research priorities which will lead to improved understanding of the relationships between people and plants and develop a research agenda.
>
> To identify research methodologies and mechanisms by which horticulturists can work with social scientists and others to more fully understand and utilize people-plant relationships.
>
> To expand a network for researchers, academicians, funders and industry specialists interested in people-plant relationships.

The topics addressed at the meeting included plants and human culture, plants and the community, plants and the individual, horticultural therapy and research implementation. I am pleased to share this special collection with a guest editor, Joel Flagler. Joel was responsible for the on-site organization of the workshop and handled the gathering and reviewing of the papers that were presented. I thank Joel for all his help and the pleasure of working with him. I hope you enjoy the reading of these papers as much as I did.

Raymond P. Poincelot

Foreword

An understanding of the psychological, physiological and social responses of people to plants in their environment can be a valuable tool in improving the physical and mental health of individuals and communities. An increasingly significant body of research has been accumulated quantifying peoples' preference, for specific types of configurations of vegetation both in urban and forested areas. Studies give us some insight into peoples' perceptions of plants and their values. Other research examines the impact a view of plants has on individuals' emotional states through self-reporting and through physiological measurements. As the data accumulates, we are given a tantalizing indication of how much more there is to be understood about the human relationship to plants and our need for plants in our environment. There is a feeling of excitement among researchers and educators that we may be on the frontier of recognizing and utilizing the natural environment for psychological, physiological, and social health, much as the knowledge of the plant's role in medicine and nutrition has contributed to physical health.

By bringing together social scientists and researchers from the arts and humanities with plant scientists, this symposium highlights the breadth of disciplines that can contribute to this knowledge base. At the same time, it focuses on the need to establish research priorities to enable us to work together toward achievable goals and it provides a forum for identifying these priorities.

Diane Relf
The People-Plant Council

Introduction

The underlying hypothesis for this second People-Plant Symposium can be stated as follows–*An understanding of the psychological, physiological, and social responses of people to the plants in their environments can play a significant role in improved physical and mental health for individuals and communities.*

This important conference was spawned by the first Symposium in April, 1990, in Arlington, Virginia. Many important results came from that landmark event, including the formation of the People-Plant Council, a network for the dissemination of research results.

This symposium was officially titled *People-Plant Relationships: Setting Research Priorities.* Two major objectives were to identify the methodologies and mechanisms by which research can proceed, and to set a research agenda. An additional focus was to expand a network for researchers, academicians, funders and industry specialists interested in people-plant relationships.

People-plant issues have never been more popular nor more important. Educators, researchers and industry leaders are taking a much closer look than ever before at the ways in which people and plants interact. New committees and working groups have been established at universities and major associations to focus on the human issues in horticulture. Plant growers, florists, landscapers and nurserymen, likewise, are beginning to embrace the people-plant connection.

This climate of cooperation was clearly demonstrated by the groups that have collaborated to bring you this event. The Symposium was sponsored by Rutgers University Cooperative Extension, The American Society for Horticultural Science, The American Association of Botanical Gardens and Arboreta, The American Horticultural Therapy Association, and The People-Plant Council.

A pioneer in the people-plant movement, Charles Lewis of the Morton Arboretum, suggests a new vantage point for looking at plants. From it we can begin to see plants as much more than mere botanical entities. People react and respond to plants, in places where we live, work and recreate. By viewing plants from a human perspective we acknowledge the many ways in which the presence of vegetation positively affects human beings.

The enhanced understanding of people-plant interactions can be put to good use. In communities and inner city environments gardens can be a source of neighborhood pride. Abandoned lots and crime infested areas are being reclaimed and re-vegetated. Significant social benefits are being observed and documented. A set of parallel benefits is seen in medical and rehabilitative settings where horticulture is put to use as a therapeutic modality. It has been demonstrated that gardening and planting activities can help in the treatment of individuals with mental, physical, and emotional disabilities.

In his welcome address Dr. Daryl Lund, Executive Dean of Agriculture and Natural Resources at Rutgers University-Cook College, pointed to the innovative nature of people-plant research. The investigation into the human issues serves to broaden the scope of what horticulture is and can be.

Within the broad scope of people-plant relationships we have identified five topic areas which provide a framework for the symposium proceedings. They are: Plants and Human Culture; Plants and the Community; Plants and the Individual; Horticultural Therapy: A Specific Application; and Research Implementation.

I wish to acknowledge all those who presented their work at the Symposium. Researchers came from Canada, Europe, Australia, Hawaii and all over the continental United States. Horticulturists were well represented from industry, universities, botanical gardens, and horticultural therapy programs. Also, professionals from many related disciplines participated including geographers, biologists, anthropologists, urban foresters, psychologists, landscape architects and artists.

Special thanks go to the individuals on the organizing committee who also served as Symposium moderators. They are Diane Relf, Virginia Polytech; Candice Shoemaker, Berry College, Georgia; Virginia Lohr, Washington State University; and Terry Keller, New York Botanical Garden. Their dedication and contribution helped steer this important event toward success.

Joel Flagler
Rutgers University Cooperative Extension

PLANTS AND HUMAN CULTURE

Chapter 1

Plants and Human Culture

Candice A. Shoemaker

SUMMARY. Historically and presently, plants serve an intimate role in human culture. They influence language, art and literature; performing arts and modern mass media; politics and world events. Plants serve as symbols in many celebrations and rituals such as holidays, weddings, and funerals. Understanding the past and current roles plants have in human culture can provide insights into society's values and can ensure a continued relationship between plants and people. This paper is a commentary on the relationship between plants and human culture and a review of current research on this topic.

The green plant is fundamental to all other life. Humankind could perish from this planet and plants of all kinds would continue to grow and thrive. In contrast, the disappearance of plants would be accompanied by the disappearance of all animal life, including humankind.

We are very aware of our need for plants for basic survival. The oxygen we breathe, the nutrients we consume, the fuels we burn, many of the most important materials we use, are all related to plant life. Plants provide all of our food, either directly or indirectly. They are also utilized as a source of construction materials for our homes and work sites. Raw plant materials are used in the manufacture of fabrics and paper and such synthetics as plastics and rayon. We have come to depend upon many of the complex substances that plants produce–dyes, tannins, waxes, resins, flavorings, medicines, and drugs. In general, plants are necessary to satisfy our basic instinct for survival by providing food, fiber, and shelter.

Early humans, from the beginning, depended upon botanical knowledge for existence (Janick 1992). They understood the life cycles of plants, knew the seasons of the year, and when and where the natural plant food resources could be harvested in greatest abundance with the least effort. They spun fibers, wove cloth, and made string, cord, baskets, canoes, shields, spears, bows and arrows, and a variety of household utensils. They were familiar with a variety of drug and medicinal plants. As they used plants and plant material for survival and to make life more comfortable, they also used plants in ceremonies and rituals, in pictures, and to make musical instruments.

The development of civilization and the development of agriculture coincide. The development of civilization, or culture, was possible due to the "invention" of agriculture. As crop production became more and more efficient, less time was needed to provide for food and shelter and more time was provided for activities such as contemplation, poetry, and art.

The discovery of agriculture is remarkable for two reasons (Janick 1992). One is its universality, that is, agriculture is an independent discovery throughout many parts of the world. For example, we find each great ancient civilization based on grain–wheat in Europe and the near East, rice in Asia, maize in the Americas, and sorghum in Africa. The second remarkable aspect of agriculture is the ability of each population in selecting out the desirable species and domesticating them.

The idea of universality can also be related to the use of plants for things other than survival. Flowers have been used for centuries and still are used extensively in ceremonies for expression of joy, affection, welcome, gratitude, sympathy, celebration, grief, friendship, marital union, or spiritual contemplation. For example, in this country many people associate gladiolus with funerals, the rose with Valentine's Day, and mistletoe with Christmas.

McDonald and Bruce (1992) showed that we associate horticultural elements with holidays. For example, when asked to describe a Christmas scene, 82 percent of the respondents included a horticultural element in the description. Also, subjects rated Christmas descriptions that included a horticultural element as more meaningful and enjoyable than descriptions that excluded horticultural elements. Many of our holidays–both secular and religious–are rich in plant symbolism. We toast the New Year with champagne, bake a cherry pie for President's Day, and give roses on Valentine's Day. What is Easter without an Easter Lily, Halloween without a pumpkin and Christmas without a Christmas tree?

Plants influence our language, art, and literature, as well as the performing arts, and modem mass media. Through language, we refer to plants everyday. For example, as researchers we want to "get to the root of the problem," sometimes get "stumped," we may be told by a "branch research station" that it is a "thorny problem," and finally realize we "can't compare apples and oranges." For more examples of "plant talk" refer to Bryant's (1992) review.

Poetry and literature are filled with reference to plants. In Shakespeare's writing, over 200 plants are mentioned. Shakespeare displays an intimate knowledge of plant growth, propagation, grafting, pruning, manuring, weeding, ripening, and decay (Janick 1992). The horticultural imagery and allusions have become part of our culture and fuse plant lore and literature.

What has the industrial revolution done to our relationship with plants? Are we close to being cut off from direct contact with plant life? Our athletic fields have been stripped of grass and replaced with astroturf, our bouquets are silk flowers, and open spaces are being eaten up by buildings and roads.

Our arrival on the moon provided a perspective of the Earth we never had before–as a very small part of space but beautiful and precious. We needed to start taking care of it. The ecology movement has grown and with it more and more people are becoming aware of this very important relationship between people and plants.

When considering plants and human culture, we must study past and current relationships between plants and people to ensure a direct relationship in the future.

Studying the role plants have had in the development of our culture may help us understand our current position. We can look at what the role of plants was in art and what it is today. How were plants used? What plants were used? Why? What were they symbolizing? These same questions can be asked regarding the many rituals, ceremonies, and celebrations we have that have plants in them. For example, Rosenfield (1992) studied the role of gardens in civic virtue in the Italian Renaissance. He concluded that the Renaissance garden afforded its visitors the temporary repose thought to restore mental vitality, it lent a sense of tranquility, and demonstrated the rewards of detachment and the ability to shift one's perspective. Today, research is being conducted on these very ideas (Kaplan 1992). Specifically, Kaplan is studying what makes an environment serve a restorative function, something that appears to be similar to the purpose of the Renaissance garden.

This age of high technology, too much stress, over-working, and fast pace, can be exhausting. If research shows the restorative value of nature, what company wouldn't include plants as a critical component of their business? The corporate world has begun to see the value of plants in the work place. How many office buildings do you walk in today that are not plantscaped? Corporate America has also seen the value of well-developed grounds and public access to the grounds. The attitude is that gardens and well-groomed, park-like settings enhance the corporate image (Parker 1992). Gardens in the work place can be a means to accomplish profit, serve human needs, and be an element of social responsibility.

Within the arts, plants have frequently been used in paintings as potent vehicles for symbolism (Cremone and Doherty 1992). Artists have often utilized flowers to convey religious, moral, or social lessons. Such floral symbolism included the lily illustrating purity, the carnation representing fidelity, and the tulip indicating greed. Religious doctrines were also given botanical symbols: wheat became a metaphor for life, jasmine symbolized Divine love, and the passion-flower recalled the instruments of the Passion of Christ.

Plants are often at the center of political debate and controversy. Tobacco smoking is the topic of heated, social debate today. Thirty years ago smoking was an acceptable social behavior in all public places and today it is illegal in most public buildings. "Clearcutting" and other forestry practices have their vocal critics and active opponents. The hole in the ozone and the "greenhouse effect" suggest that "clearcutting" and the rampant development of land in this century may be dramatically affecting the fine balance of nature. The results are changes in human behavior–recycling, renewing, reusing are the mantra of the 1990's. Clearly, the human/plant relationship can be a matter of serious, social consequence.

Obviously, plants have been a part of human culture since the beginning of time and will more than likely continue to be. Plants are a part of the social world as well as the physical world. Therefore, it is not surprising that they influence human behavior and that human culture is imprinted with a botanical reality.

REFERENCES

Bryant, C. 1992. 'Plant talk': Some notes on the Botanical Bent of American Language. *HortTechnology* 3(1).

Cremone, J. & Doherty, R. 1992. *Vita Brevis*: Moral symbolism from nature. In *The Role of Horticulture in Human Well-Being and Social Development*, edited by P.D. Relf. Timber Press.

Janick, J. 1992. Horticulture and human culture. In *The Role of Horticulture in Human Well-Being and Social Development*, edited by P.D. Relf. Timber Press.

Kaplan, S. 1992. The restorative environment: nature and human experience. In *The Role of Horticulture in Human Well-Being and Social Development*, edited by P.D. Relf. Timber Press.

McDonald, B. & Bruce, A.J. 1992. Can you have a merry Christmas without a tree? In *The Role of Horticulture in Human Well-Being and Social Development*, edited by P.D. Relf. Timber Press.

Parker, D. 1992. The corporate garden. In *The Role of Horticulture in Human Well-Being and Social Development*. Timber Press.

Rosenfield, L. 1992. Gardens and civic virtue in the Italian Renaissance. In *The Role of Horticulture in Human Well-Being and Social Development*, edited by P.D. Relf. Timber Press.

Chapter 2

American Women and Their Gardens: A Study in Health, Happiness, and Power, 1600-1900

Judith Schwartz

SUMMARY. In looking at historical perspectives of the people-plant relationship, accepted histories must be re-read with a fresh outlook, as this subject is not usually documented as such. Women's history must also be pulled from between the lines of already-written volumes in which their stories are largely omitted, although alluded to occasionally. The special relationship that women share with plants, horticulture, agriculture, botany, etc., can be traced from mythological and biblical stories and allusions through the centuries. The history of women in America, from 1600 to 1900, coincides with the development of "western" horticulture in this country, specifically in terms of personal and public health, happiness and social position (i.e., power)–which are primary manifestations of the people-plant relationship. The pursuit of this research is often like the proverbial "looking for a needle in a haystack," although undoc-

This paper is an annotated version of a presentation given at the Meadowlands Sheraton in East Rutherford, NJ, as a part of the People-Plant Symposium, "Setting Research Priorities," sponsored by Rutgers University, the People-Plant Council, the American Society for Horticultural Science, the American Association of Botanical Gardens and Arboreta, and the American Horticultural Therapy Association. Research assistance was provided by Ava Miedzinski, 1722 Sunset Boulevard., Houston, Texas; editorial assistance was provided by Nancy Gitchell, 563 West Street, Keene, New Hampshire.

umented and sentimental appraisals of women's relationship to plants abound, especially in histories of the early years of our nation. The virtually non-existent literacy rate among Colonial women makes early, first-hand accounts very rare. Prescriptive literature written by men does, however, make clear many of the goodwife's skills and responsibilities. As literacy increased from 1600 to 1900, women's letters and diaries, as well as published garden notes and even botanical field guides, attest to the special relationship that women have shared with plants and horticulture from pre-history to the present.

INTRODUCTION

The relationship of women to plants, to horticulture, and to agriculture is as ancient as the story of Eve and the apple. Probably older. Myth and history, as presented through the voice and pen of *men*, abound with both positive and negative images of Earth goddesses and heavenly goddesses, potion-brewing witches as well as herbal healers; women in positions of socio-economic-political, as well as emotional, strength vs. women in positions of sentimental frailty and uselessness; women in positions of good and in positions of evil.

In my professional work as a garden designer, I have noticed that women form a major part of my client base. Some work physically in their gardens, providing food and flowers for their families and their homes; others use their gardens more as places for spiritual and emotional comfort–places of beauty and quiet, places to rest and to retreat from the cares of the world, places to visit with and to converse with friends and family; still others use their gardens as showcases for status and prestige in the community, for symbolizing their wealth or position in society. My involvement with horticultural therapy has, moreover, made me sensitive to looking at not only why women garden, but how they garden as well.

During the course of belatedly completing my bachelor's degree in the fall of 1990, I found myself in an upper-level history class called "Women in America: 1600-1900." The course objective was to learn methods of extracting women's stories from the general histories that have been presented to us–mostly by men–throughout the ages, and to learn to search for firsthand documents in which women have been able to tell their stories for themselves and for each other. While the subject matter was not presented from an aggressively political feminist viewpoint, in exploring "how women's lives in America changed from the seventeenth to the nineteenth centuries," readings and lecture themes looked at "women's power, the gender division of labor, women in private vs. public spheres, women in relationships and as individuals, and women's political, economic, cultural and social lives" (Ford, 1990). All of these issues, I discovered, could be brought to bear on the subject of how and why women garden. The fact that my own research was currently constrained by the historical parameters of the course (i.e., American women 1600-1900) in no way diminishes the fact that women the world over, throughout the ages, and in all societies and cultures, have been intimately and importantly involved in horticulture, in agriculture, in purely scientific botanical study, as well as in an enormous variety of domestic and leisure arts using plant products and motifs.

When I first began my exploration of this subject, research librarians at several public libraries, two historical societies, and at two colleges (one of them an Ivy-league women's school, founded with a focus on botanical studies!) informed me that there was no material to be found that would in any way relate to my inquiry. If I had taken their responses at face value, I would not currently have over fifty pages of research and analysis, as well as an extensive bibliography to reference, which I add to daily.

As the American Horticultural Therapy Association and the People-Plant Council have discovered in developing their own bibliographies of documents having any connection with the people-plant relationship, research of this kind must be done in a reticulated manner. There are no clearly highlighted references to follow, like the Yellow Brick Road which led Dorothy through the poppy fields to the Land of Oz. And, as it was difficult enough to ferret out the people-plant relationship itself, adding the dimension of women's place and purpose to the package increased the challenge tenfold. Accepted histories must be re-read with a fresh outlook and the slightest reference or most innocent-looking document must be re-analyzed through the filter of the feminist perspective.

As I reviewed the history of women in America in the seventeenth, eighteenth, and nineteenth centuries, I looked for the relationship that these women had specifically with *gardens* and with *gardening*. (The bigger issue of women and plants was touched on, but would have turned a semester's project into a lifetime study.) Questions concerning gender roles led me to ask whether or not Colonial and Victorian women actually physically worked in their gardens. If so, did they do so out of necessity, out of pleasure–or both? If not, why not? Did social custom not "allow" or encourage them to, or did they simply not have the interest? Questions revolving around the adjunctive therapeutic value of horticulture surfaced repeatedly. For what physical, psychological, and emotional reasons did women participate in gardening activities? And, what kinds of personal, family, and community "powers" were derived from women's horticultural endeavors? How were these questions answered and documented? I am still looking for answers and there may be additional questions as well. Some of what I have found, I will share with you.

HEALTH, HAPPINESS AND POWER

When Europeans crossed the Atlantic to "settle" the North American wilderness, breaking garden ground and sowing ocean-borne seeds took second place to *no other activity.* In her book, *Old Time Gardens,* Alice Morse Earle (1901) tells us that "The first entry in the Plymouth Records . . . is the assignment of 'Meresteads and Garden-Plotes,' not meresteads alone, which were farm lands, but home gardens" (p. 3). Rudy Favretti, an authority on Colonial garden reproduction, states that "These home-lot gardens were tended by the women of the household," while men were involved in larger-scale agricultural crop production on the outskirts of the settlements (pp. 18-20). But he is writing in 1977 (reprinted 1990) and does not directly quote his sources.[1]

Undocumented and sentimental appraisals of women's relationship to plants abound, especially in histories of the early years of our nation. Except for garden historian Ann Leighton, who discovered that the available primary sources were men's records, these texts, without crediting their own references, created a sympathetic picture of the women to whom early American gardens apparently meant much. Alice Morse Earle supplies my favorite:

> . . . what must that sweet air from the land have been to the sea-weary Puritan women on shipboard, laden to them with its promise of a garden! for I doubt not every woman bore with her across seas some little package of seeds and bulbs from her English home garden, and perhaps a tiny slip or plant of some endeared flower; watered each day, I fear, with many tears, as well as from the surprisingly scant water supply which we know was on board that ship. (p. 2)[2]

To be sure, colonial housewives came to this country well-versed in the skills and activities necessary to realizing productive growth from their ocean-borne propagules. Prescriptive literature, written by men for the benefit of the fairer sex, had long instructed them in the means and methods of providing food, flowers, and medicines for their families and friends. As early as 1577, a certain poetical Thomas Tusser was ordering English goodwives about their gardening business, with advice for every month of the year. An oft-quoted example:

> In March and in April, from morning to night;
> In sowing and setting good huswives delight.
> To have in their garden, or some other plot:
> To trim up their house, and to furnish their pot.[3]

The versatility of her garden must have made the colonial housewife a vital part of the early American family. Strewing herbs to keep her house fresh-smelling and vermin-free, seasoning the food for variety and palatability, making bright colors from dye plants, decorating the outside and inside of her house, and having to provide various remedies–all these tasks were essential to the health and well-being of her family and friends.

One in particular, the distillation and dispensation of medicinal remedies, was an important source of power for the seventeenth-century housewife. Unfortunately, there are no documented written records of her cures. Even Cotton Mather, although he encouraged women in medicine and "urges gentlewomen to keep their gardens full of helpful plants and their closets ready stocked with 'several harmless and useful (and especially external) remedies for the help of their poor neighbors,'" does not abide with women taking notes (Leighton, 1970; reprinted 1986, p. 120).

The Euro-American home gardener of the 1600's then grew cabbages and peas, beans and greens, herbs, and–yes–a few flowers. The important food and medicinal plants grown in Colonial home gardens were a necessary and integral part of the domestic and community economy, and the distillation and dispensation of medicinal remedies was an important source of personal and public power for the seventeenth-century housewife. Additionally, Early American home gardeners emotionally relied on their gardens for comfort in a strange and sometimes harsh land, and for relief from their homesickness for the formal and cottage gardens they had left behind. To conclude her poem entitled *A Puritan Lady's Garden*, the nineteenth-century American poetess, Sarah N. Cleghorn, says:

> When ships for England cleared the bay,
> If long beside these reefs of foam
> She stood, and watched them sail away,
> It was her garden first enticed her
> To turn, and call this country "home."[4]

Eighteenth-century American home gardeners were still growing many of the same indispensable herbs and vegetables as did their foremothers, but as "civilization" created feelings of safety, security, and stability, garden boundaries expanded and gardens, especially for the wealthy, became places of social gathering and true leisure activity.[5] Hudson River and Southern gardens, among others, were being laid out in formal design and were tended, not by family members, but by hired staffs of trained gardeners.[6] Urban dwellers were beginning to rely more and more on trade than on individual home gardens to support burgeoning populations, and outlying farmsteads were the providers of seasonal food supplies. The city-bred gentlewoman's requirement of growing vital herbs and medicines for physical well-being also became limited, and then totally eclipsed, by the patriarchal society which *continues* to put male physicians in the place of midwives and wise women. It is still the gentlewomen of the 1700's, however, who conspire to create gardens of beauty and peaceful serenity–places for contemplation, social gathering, and family activity, thus ensuring the emotional health and happiness of their families. And home gardens were still domestic necessities on ever-expanding frontiers, as well as in less well-to-do country homes, where summer vegetables and herbal medicines continued to thrive under the expert care of the woman of the house.

As class structure and economic opportunity became functioning variables, individual personality traits and interests could parallel these factors. Colonial women were beginning to have choices about actually working in their gardens, versus merely supervising work and reaping the life-sustaining physical and emotional benefits. Nonetheless, in the agriculturally based economy of Colonial America, gender roles and sociopolitical ideologies do not seem to have *prevented* women from gardening when necessity or desire arose.[7]

Moreover, the eighteenth century in American gardening provides many more primary sources relative to female participation in the people-plant equation. As the literacy rate improved, an increasing number of women left journals or diaries, or wrote "memoirs,"[8] and a fair amount of correspondence exists in various archival collections throughout the country, especially in the New England and Southern states.[9]

In her book, *A Midwife's Tale*, in which she analyzes the life of Martha Ballard, based on her diary from 1785 through 1812, Pulitzer Prize winning historian Laurel Thatcher Ulrich (1990) devotes an entire chapter to her subject's garden work. In this, she shows how the reliable and repetitive natural cycles of gardening provided the Maine midwife with a sense of personal power and orderliness in the face of family, community, and global distress. Nonetheless, as Thatcher writes, "Historians have written a great deal about field agriculture in early America but not enough about the intricate horticulture that belonged to women, the intense labor of cultivation and preservation that allowed one season to stretch almost to another" (p. 324). She also documents the midwife's occasional, but increasingly more frequent, interactions and conflicts with the male physicians who were encroaching on her community, and she devotes an entire Appendix to "Medicinal Ingredients Mentioned in Martha Ballard's Diary." Many of these were plant-based remedies, but Ulrich is not always able to distinguish between wild-gathered, native, or naturalized plants, versus plants specifically cultivated for their healing properties.

In addition to their gardening work in the 1700's, a few independent women actually delved professionally into the study of botany and horticulture and published manuscript works or prescriptive literature for the general gardening public.[10] It should also be noted that the gardens of several Southern women were repositories and trading points for the male botanical explorers of the time.[11]

In looking at gardening trends in the nineteenth century, we need to understand that we are dealing with some rather more complicated issues. With the Industrial Revolution in England, cheap glass turned the botanical and gardening world on its head. Not only could travelling botanists send home their live specimens in terrarium-like glass cases, but gardeners were able, inside enormous glass houses, to grow ever-more-tender plants from South America, Africa, and the Orient. Victorian England was obsessed with collecting, propagating, hybridizing, and "bedding out" plants. This national fever cut across all social classes and sexual spheres. *English* women had always gardened and it was now considered very genteel to tend one's plants in one's own glass bell jar, larger cold-frame, or even larger glass-covered conservatory.

In nineteenth-century America, life was still not quite so cavalier, although the wealthier class, always influenced by English gardening, continued to expand their estate gardens, and landscape architecture and design began to make its mark on the American countryside (Leighton, 1987, p. 120). With the arrival of industrial technology in the 1800's, we see very interesting trends in the struggle to re-define woman's role in the home, in the public eye, in society as a whole. As agriculturally based society waned, not only were gender roles becoming more separated, but families found themselves economically less dependent upon their own survival-oriented gardens. This naturally would undermine woman's important role in providing for her family. The so-called "Doctrine of Separate Spheres," along with a growing sense that women were somehow more pure and virtuous than men, made them responsible for the morals of their menfolk and, therefore, of the nation. (Although women continued to simultaneously be seen as temptresses and *un*doers of the morals of men!) This created new and conflicting agendas for action–and non-action.

In one of woman's most important roles, that of growing, preparing, and prescribing of herbal remedies, we see some changes over time. Ann Leighton (1987), in *American Gardens in the Nineteenth Century: "For Comfort and Affluence,"* sums it up: "By the nineteenth century, medicine, especially of the home variety, had become a modest, but lasting, amateur science. Useful medicinal plants were listed in books about gardening and in seedsmen's catalogues. But they were planted to flourish with the vegetables, leaving 'pleasure' gardens to follow the styles of the changing times" (p. 16). Nevertheless, Elizabeth Gemming (1968), in *Child Life in Old New England*, asserts that still ". . . Mothers all know how to make tea from coltsfoot and flaxseed and honey to soothe their children's coughs" (pp. 113-14).

Herbology aside, Suzanne Lebsock (1985), in *The Free Women of Petersburg*, says that several Petersburg women ". . . wrote with special fondness and verve about their gardens." Along with her examples, she further states that ". . . no code of gentility prevented . . . women from engaging in physical labor." And she concludes that the very physical act of gardening not only ". . . was much of the pleasure [but] more than any other household task, gardening brought women some peace."[12]

Because they were no longer needed to grow food and medicine for their families and friends, many nineteenth-century urban women may have lost some of what we are calling personal and public "power." They certainly were not experiencing that sense of peace that their still-gardening sisters were enjoying in more rural settings. Moreover, with the increasingly eccentric Victorian view on femininity, espoused by "professional" males of the time, came the idea that women were, in fact, too frail and sentimental to be overtaxing their physical and emotional strengths by working in the out-of-doors. Fortunately, garden writers of the day, as well as moral and health reformers, recommended that women take up their garden tools to improve not only their own personal health and happiness, but the face of the neighborhood as well–the latter falling under the category of patriotism and civic duty (Leighton, 1987).[13]

The "Republican Mother," in charge of her children's education and moral upbringing, was encouraged in her knowledge of "watered-down" botanical identification and taxonomy. Poetry and sentiment were used to make the demanding intellectual exercises of botany easier for women to understand and relate to. The Victorian concept of "The Language of

Flowers" arises from this approach, problems in communication prevailing only when senders and receivers of messages used different lexicon. These floral dictionaries were, in fact, mostly written by women of the time and unfortunately were not very reliable (Leighton, 1987, p. 88).

By the middle of the century, prescriptive literature recommending gardening to ladies was getting more serious. In 1827, both the Pennsylvania and Massachusetts Horticultural Societies had formed, along with others in several more states (but most of these were not long-lived). By 1829, the Pennsylvania Society had three ladies in its membership. In 1842, ladies were admitted to the Massachusetts Horticultural Society annual meeting, after a hard debate between them and alcoholic refreshment (Ibid. pp. 106-9).

In the 1850's, the Massachusetts Horticultural Society Committee on Ornamental Gardening visited the garden of a certain Mr. Fay in Chelsea and ". . . found Mrs. Fay 'arranging, planting, and cultivating with her own hand the flower treasures of the earth' (subsequent lamentation that 'so few of the ladies of our land imitate her example')" (Ibid. p. 111).

A great leader in nineteenth-century American horticulture and landscape gardening, Andrew Jackson Downing, also lamented the lack of American ladies participating in beautifying their homes and improving their health and education meanwhile. He published many of his own gardening texts and periodicals and was instrumental in publishing American editions of several English volumes, notably Mrs. Jane Loudon's *Gardening for Ladies* and *Ladies' Companion to the Flower Garden.* Other garden writers as well, insisted ". . . on the extreme importance of floriculture, especially in relation to the happiness of the family and the improvement of the home landscape" (Ibid. p. 90-99).

Moral reformers, along with landscape and garden writers, encouraged women to get fresh air and exercise. In 1859, Henry Ward Beecher expounded on the health-giving virtues of floriculture for women and children alike.[14] Ten years later, Catharine Beecher and her sister, Harriet Beecher Stowe, also advocated the advancement of horticulture for the benefit of children, especially young girls.[15] The Beecher sisters temper their argument by asking that women and children confine themselves to the planting and maintenance of gardens, leaving the physically more demanding work of preparation to the father. In addition to healthful habits, also mentioned by Henry Ward Beecher, Catharine Beecher and Harriet Beecher Stowe insist that "benevolent and social feelings could also be cultivated, by influencing children to share their fruits and flowers with friends and neighbors, as well as to distribute roots and seeds to those who have not the means of procuring them" (Lacy, ed., 1988, p. 27).

Women did, in fact, participate somewhat in this type of horticultural reform. As keepers of the nation's standards of beauty and morality, three ladies of the Flower Committee of the Massachusetts Horticultural Society started a children's horticulture program in 1878. Through various Sunday schools, the women instructed children of the working class, beginning with the planting of window boxes (Leighton, 1987, p. 111).

Alice Morse Earle (1901) tells of a horticulture program at the County Jail in Fitchburg, Massachusetts. She says that ". . . laid out by the wife of the warden, aided by the manual labor of convicted prisoners . . . [was] . . . a Box-edged garden of much beauty and large extent." The warden's wife hoped that ". . . working among flowers would have a benefitting and softening influence on these criminals."[16]

As the century progressed, American women took back gardening in a serious way, and by 1890, the first Ladies Garden Club was organized in Athens, Georgia (Hadfield, 1964, p. 52). This gave women a forum for exploring gardens and gardening on their own terms. Rose gardens, rock gardens, woodland gardens, and one-color borders, gave women well-defined parameters within which to experience horticulture according to their own limitations or interests. Herb gardens abounded. The controlling factors of size and special interest gave women gardens that were freed up from the constraints and regulations and "expertise" of professional male landscape designers (Leighton, 1987).[17]

By the end of the nineteenth century, women were elbow-deep in the study of botany, the practice of gardening and–they were writing about it, almost nonstop. They recognized the need to familiarize themselves with botanical identification and nomenclature, and also the limitations for them of both the more formal texts of the day, as well as the more poetic, but unreliable ones.[18]

Celia Thaxter, writing from Appledore Island off the coast of New Hampshire, was published posthumously in 1894. She is no meek observer of flower form and habit, but a gardener in every sense of the word. In her "Prefatory" to *An Island Garden*, she says, "Ever since I could remember anything, flowers have been like dear friends to me, comforters, inspirers, powers to uplift and cheer . . . I began a little garden when not more than five years old." Her book goes through the gardening season, from early spring until leave-taking time in autumn. Like other ladies of her age, she did have help with initial ground-breaking in her garden, but in her writings we hear the voice of the truly dedicated gardener: "In the first week of April the ground is spaded for me; after that no hands touch it save my own throughout the whole season. Day after day it is so pleasant working in the bright cool spring air" (pp. 18-19).

A few weeks later, she says, "... Some morning in the first of May I sit in the sunshine and soft air, transplanting my young Pansies and Gilly-flowers into the garden beds." And then she gives us proof of the pleasures of working with the soil itself and even with the tools that she uses:

> So deeply is the gardener's instinct implanted in my soul,
> I really love the tools with which I work, the iron fork,
> the spade, the hoe, the rake, the trowel, and the watering-pot
> are pleasant objects in my eyes ...
>
> ... I like to take the hoe in my hands and break to pieces the
> clods of earth left by the overturning spade, to work
> into the soil the dark, velvet-smooth, scentless barnyard
> manure, which is to furnish the best of food for my flowers;
> it is a pleasure to handle the light rake, drawing it evenly
> through the soil and combing out every stick and stone and
> straw and lump, till the ground is as smooth and fine as meal ...
> then is the ground ready for the sowing of seeds. (pp. 24-5)

Celia Thaxter writes, in its own time, about every aspect of growing her gardens. Sowing seeds; fighting weeds; observations of the growth of foliage and flowers; design and arrangement; gathering seed pods in autumn; as well as a constant awareness of and comment on the bird-life in her garden–flow through her essays in a timeless manner. We have waited a long time for such expression–from the pen of a woman.

CONCLUSION

It is evident that gardening of some sort or other has, throughout the centuries, been of great significance in the lives of women, both for their own emotional support, as well as by putting them in some position of power in family and community life. In seventeenth-century America, women's garden-derived power came from the physical dependency on food and medicine. The eighteenth century saw a transition from survival garden to a domestic and social center of activity and entertainment. This transition also saw the importance of women's gardens in providing, for botanists, outdoor laboratories for collection, propagation, hybridization and study of new species and cultivars. By the nineteenth century, as Republican Mothers and as social reformers, aside from simply beautifying their domestic surrounds for their personal health and happiness, women had the emotional and moral character of the nation in their gardening hands.

I feel strongly that more research is needed on this subject, as well as on the broader relationship of women to plants, and on the incorporation of plant materials and symbols in domestic and leisure arts, as well as in myth, magic, and ritual. Women have participated in gardening and agriculture for physical, emotional, psychological, and spiritual reasons, throughout every age and culture. Dedicated anthropologists, sociologists, and historians must look more deeply *for and into* primary sources so that they are able to extract the stories of women and their connection to the living world, from the letters, diaries, and records of all the ages that *her* story has been told by him.

NOTES

1. Rudy Favretti, Professor Emeritus of Landscape Architecture at the University of Connecticut, has specialized in publishing monographs and small books which detail historic garden reproduction, including design considerations and appropriate plant lists. His wife, Joy, is a botanist and librarian, and has co-authored the volume cited. We can assume from their credentials that they have arrived at reasonable conclusions from their own research, yet it is frustrating, to the meticulous historian, that references are not cited thoroughly and in a formal manner.

In an earlier volume, *Colonial Gardens* (1972), co-authored with Gordon P. DeWolf of Harvard's Arnold Arboretum, Favretti concludes with an interesting bibliography. He also clearly explains that historic garden preservation or restoration does require thorough research into the site and its inhabitants. Probate inventories, old deeds, diaries, journals, and letters are good sources of pertinent primary information. "Personal letters written from husband to wife, sister to sister, brother to brother, reveal much because it was the custom of the day to speak of plants in bloom in the garden, the change of season and its effect on the garden, what was harvested, and much more" (p. 30). Town histories, news articles, and advertisements can also reveal a lot, and can be found in the archives of horticultural and historical societies. Unfortunately, he does not directly cite any of these in his publications.

Additionally, and most pertinent to the interdisciplinary nature of people-plant studies, however, Favretti states: "The author relies heavily on paintings for information on the design of gardens. These works often suggest a fence style or garden arrangement typical of a particular town or region . . . " (p. 31). I would also speculate that they indicate something of gender roles and societal position, as well, and he does illustrate some of his works with interesting examples of this nature.

2. The problem of "common knowledge" assumptions about women and gardens provoked me from the start of this project. Samples of undocumented entries from a variety of gardening and history texts follow:

"Our mothers and grandmothers came honestly by their love of gardens. They inherited this affection from their Puritan, Quaker or Dutch forebears, perhaps from the days when the famous hanging gardens of Babylon were made for a woman. Bacon says: 'A garden is the purest of human pleasures, it is the greatest refreshment to the spirits of man.'* A garden was certainly the greatest refreshment to the spirits of woman in the early colonial days, and the purest of her pleasures–too often her only pleasure" (Earle, 1898; reprinted 1975, pp. 431-2).

> *God Almighty first planted a garden. And indeed, it is the purest of human pleasures. It is the greatest refreshment to the spirits of man; without which, buildings and palaces are but gross handiworks: and a man shall ever see, that when ages grow to civility and elegancy, men come to build stately, sooner than to garden finely: as if gardening were the greater perfection. SIR FRANCIS BACON (1561-1626)

(Spencer, 1935, p. 49)

"There was so little beauty in the lives of colonial women–perhaps one cherished heirloom, . . . [One] source of beauty was the flower garden that bloomed by the doorstep, carefully tended in moments stolen from weeding the cornpatch. The Huguenot women who came from France are said to have brought seeds of hollyhocks and primroses with them in their apron pockets. Dutch women unpacked bulbs of the gay tulips that would transplant a bit of Holland to New Amsterdam. English women soothed their homesickness with the familiar flowers of home–roses and lilies of the valley, foxglove, and Canterbury bells. A frontier bride, riding on a pillion behind her new husband through a frightening wilderness, remembered with comfort the little packet of seeds her mother had tucked in among her household items" (Speare, 1963, pp. 81-2).

"Amongst their few belongings, the first settlers of America took with them treasured seeds and roots of their favorite herbs . . . To the Colonial goodwife fell the responsibility of the kitchen and herb garden . . . During her chore-filled days, probably her greatest pleasure was tending her fragrant and flavorsome herbs" (Seuffert, 1987).

3. Ann Leighton (1970; reprinted 1986) says, "In the cheerful verses of Thomas Tusser, described as one of England's first didactic poets, we see the housewife's lot described. Coming from East Anglia in the reign of Henry VIII, educated at Eton and Cambridge, Tusser "devised" his book, *Five Hundred Points of Good Husbandry*, published first in 1557. . . . In the chapter on "September's Husbandry" which begins his order of months, Tusser in mid-poem addresses the housewife:

> Wife, into thy garden, and set me a plot,
> With strawberry roots, of the best to be got;
> Such growing abroad, among thorns in the wood,
> Well chosen and picked, prove excellent good.
>
> The barberry, respies and gooseberry too,

Look now to be planted as other things do.
The gooseberry, respies and roses, all three,
With strawberries under them, trimly agree.

[In December]

Hide strawberries, wife,
To save their life,
Know, border and all,
Now cover ye shall.

If frost do continue, take this for a law
The strawberries look to be covered with straw,
Laid overly trim upon crochets and bows,
And after uncovered as weather allows.

The gillyflowers also, the skillful do know
Both look to be covered, in frost and in snow.
The knot and the border, and rosemary gay,
Do crave the like succor, for dying away.
(pp. 164-5)

Sara Midda, in *In and Out the Garden* (1981), quotes springtime and summertime duties from a 1580 edition:

In March and in April, from morning to night;
In sowing and setting good huswives delight.
To have in their garden, or some other plot:
To trim up their house, and to furnish their pot.

Good huswives in summer will save their own seedes
Against the next yere–as occasion needes.
One seede for another, to make an exchange
With fellowlie neighbourhood seemeth not strange.

And Ann Leighton (1970; 1986) has the list of remedies a gentlewoman should have on hand:

Good huswives provide, ere a sickness do come,
Of sundry good things in her house to have some.
Good aqua composita and vinegar tart,
Rose water and treacle to comfort the heart.

Cold herbs in her garden, for agues that burn,
That over strong heat, to good temper may turn,
White endive and succory, with spinage enough,
All such with good pot herbs, should follow the plough.

Get water of fumitory, liver to cool,
And other the like, or else go like a fool.
Conserves of barberry, quinces and such
With sirops that easeth the sickly so much. (p. 127)

4. A PURITAN LADY'S GARDEN

This fairy pleasance in the brake–
This maze run wild of flower and vine–
Our fathers planted for the sake
Of eyes that longed for English gardens
Amid the virgin wastes of pine.

Here by the broken, moldering wall,
Where still the tiger-lilies ride,
Once grew the crown imperial,
The tall blue larkspur, white Queen Margaret,
Prince's feather, and mourning bride.

Beyond their pale, a humbler throng,
Grew bouncing Bet and columbine;
The mountain fringe ran all along
The thick-set hedge of cinnamon roses,
And overhung the eglantine.

And Sunday flowers were here as well–
Adam-and-Eve, within their hood,
The stately Canterbury bell,
And, oft in churches breathing fragrance,
The sweet and pungent southernwood.

When ships for England cleared the bay,
If long beside these reefs of foam
She stood, and watched them sail away,
It was her garden, first enticed her
To turn, and call this country 'home.'

SARAH N. CLEGHORN (1876 - ?)

(Spencer, 1935, p. 230)

5. "In New York, among the Dutch," says Carl Holliday (1982), "community life had the pleasant familiarity of one large family" (p. 209). He quotes Mrs. Anne Grant (see Note 8), whose *Memoirs of An American Lady* (1808) paints the following picture:

> Every house had its garden, well and a little green behind; before every door a tree was planted . . . every one planting the kind that best pleased with him, or which he thought would afford the most agreeable shade to the open portion by his door, which was surrounded by seats, and ascended by a few steps. It was in these that each domestic group was seated in summer evenings to enjoy the balmy twilight or the serenely clear moon light. (p. 33)

Alice Morse Earle (1898; reprinted 1975) also talks about the gardens of the wealthy Dutch. She says that "In New York, before the Revolution, were many beautiful gardens, such as that of Madam Alexander on Broad Street, where in their proper season grew 'paus bloemen of all hues, laylocks and tall May roses and snowballs intermixed with choice vegetables and herbs all bounded and hemmed in by huge rows of neatly clipped box edgings.'" (This quote turns up in other literature fairly frequently–never cited, however.) She continues her descriptions of luxury and leisure by saying, "We have a pretty picture also, in the letters of Catharine Rutherford, of an entire company gathering rose-leaves in June in Madam Clark's garden, and setting the rose-still at work to turn their sweet scented spoils into rose-water" (pp. 438-40).

6. Alice Morse Earle (1898; 1975) found "in an early account" that by 1682, gardens in South Carolina were " . . . supplied with such European Plants and Herbs as are necessary for the Kitchen, and they begin to be beautiful and adorned with such Flowers as to the Smell or Eye are pleasing or agreeable" (p. 438). And that by 1750, " . . . many exquisite gardens could be seen in Charleston, and they were the pride of Southern colonial dames. Those of Mrs. Lamboll, Mrs. Hopton, and Mrs. Logan were the largest . . . Mrs. Laurens had another splendid garden. These Southern ladies *and their gardeners* (my emphasis) constantly sent specimens to England and received others in return" (p. 439).

Ann Leighton (1976; reprinted 1986), quoting from Anne Grant's *Memoirs of An American Lady*, says that the wealthy Schuylers of upstate New York "'. . . had gardeners and their gardens were laid out in the European manner'" (p. 372).

What does the phrase ". . . and their gardeners . . . " mean to these lady gardeners? Are they working in their gardens, or just supervising, and enjoying the fruits of others' labours? If we look at snatches of correspondence from some of our country's businessmen and founding fathers, ". . . we find Sir Harry Frankland order Daffodils and Tulips for the garden *he made for Agnes Surriage*; and it is said that the first lilacs ever seen in Hopkinton were planted *by him for her*" (Earle, 1901, [my emphasis]). Carl Holliday, in *Woman's Life in Colonial Days* (1982), says of Benjamin Franklin, "It does the soul good and warms the heart towards old Benjamin to see him stopping in the midst of his labors for America to write his wife: 'I send you some curious Beans for your Garden.'" (p. 138; quoted from Smyth: *Writings of Franklin.* Vol III, p. 325; in London, sometime between 1758 and 1765). Did Mrs. Franklin take an "active" interest in her garden? Or does she supervise a crew of men to keep her garden green and growing?

7. Alice Morse Earle (1901) finds Richard Stockton from Princeton, New Jersey, writing to his wife while he was in England in 1766, "I am making you a charming collection of bulbous roots which shall be sent over as soon as the prospect of freezing on your coast is over. The first of April, I believe, will be time enough for you to put them in your sweet little flower garden, which you so fondly cultivate . . . " (p. 30).

And Mrs. Anne Grant again speaks to us about the upstate New York Dutch, as she says, "Not only the training of children, but of plants, such as needed peculiar care or skill to rear them, was the female province . . . I have so often beheld, both in town and country, a respectable mistress of a family going out to her garden, in an April morning, with her great calash, her little painted basket of seeds, and her rake over her shoulder to her garden labours . . . A woman in very easy circumstances and abundantly gentle in form and manner would sow and plant and rake incessantly" (*Memoirs*, p. 29) (Holliday, 1982, p. 211).

8. The oft-quoted Mrs. Anne Grant was born in Glasgow, Scotland on February 21, 1755. In 1758, she and her mother came to Claverack on the Hudson, where her father, Duncan MacVicar, was an officer in a Highland regiment. She was taught to read by her mother, and learned to speak Dutch as well. In 1762, having caught the attention of the wealthy Madame Schuyler, she began several years of residence with the Schuyler family in Albany, New York. Anne Mac Vicar, however, left the New World in 1768, at the age of thirteen, returning to Scotland with her family. In 1779 she married the Reverend James Grant, a military chaplain, who died in 1801, leaving her with eight children to support. According to *Appleton's Cyclopaedia of American Biography* (Wilson and Fiske, eds., 1887), her poems were collected in an octavo volume in 1803, and were followed in 1806 by her "Letters from the Mountains." "Her best-known work, begun at the age of fifty-two, and issued in London in 1808, is entitled 'Memoirs of An American Lady.' It is a charming picture of New York colonial life, and one that was greatly admired by Sir Walter Scott and Robert Southey . . . A second edition of the memoir of Mrs. Schuyler appeared in 1809, and was reprinted the same year in Boston and New York. Other editions were issued in the latter city in 1836 and 1846, while a third edition was published in London in 1817. The previous American editions being out of print, another appeared in 1876, accompanied by a fine steel portrait of Mrs. Grant, and a memoir written by her godson." Mrs. Grant died in Edinburgh November 7, 1838 (p. 707).

9. The letters of Thomas Jefferson, that most famous of American gardeners, speak obsessively of the tasks and plants and plans for his gardens at Monticello. Hugh Matthews, who recently published his own monthly *Notes of a New England Gardener*, felt that "In letters to his daughters we can gain some insight into whether they shared [Thomas Jefferson's] love of gardening" (Nov. 1988, Vol. 1, No. 11).

In discussion of Southern gardens, Alice Morse Earle (1898; 1975) says that ". . . the letters of the day, especially those of Eliza Lucas Pinckney, ever interested in floriculture and arboriculture, show a constant exchange with English flower-lovers" (p. 439).

10. Ann Leighton (1976; 1986) speaks at some length about a Jane Colden, who " . . . instructed by her father . . . learned the Linnean system in the wilds of upper New York" and whose efforts can be appraised today as those of our first woman botanist" (p. 95).

Martha Logan, a prominent Southern gardener, had earned respect and esteem not only from the reputation of her gardens, but also from the publication of her frequently-mentioned garden calendar, " . . . the first horticultural book published in America" (Leighton, 1976; 1986, p. 211 and Earle, 1898; 1975, p. 439).

11. John Bartram, the foremost American botanist in Philadelphia, was very much indulged by prominent lady gardeners in the South. Martha Logan and Mrs. Lamboll were among these (Leighton, 1976; 1986, pp. 128-9).

12. I found Suzanne Lebsock's research (1985) particularly exciting and well-documented. Along with Laurel Thatcher Ulrich, Lebsock and other contemporary professors of women's history *are* exploring women's relationship to plants, to gardens, and to agriculture, but usually as part of a larger look at women's studies in general. Meanwhile, they are getting at the letters, diaries, and memoirs, and they are thoroughly analyzing what they are finding therein. Lebsock states:

> Petersburg women wrote with special fondness and verve about their gardens. Mary Cumming fixed her ambition almost as soon as she stepped off the boat from Ireland in 1811. "I mean to turn gardener," she announced, and when spring came she filled her letters home with reports on the status of her young vegetables. "Do you know I have become a great gardener of late? I have got a variety of seeds sown long since, and a great many are coming up. My peas will be ready for rodding in a day or two, my cabbages are doing very well, and this week I intend getting my melons and cucumbers sown in a large square."
>
> In the beginning, Mary Cumming walked a half mile each way to her garden plot, and she did most of the work herself. This changed after her first harvest, however; her baby daughter died, her health failed her, and she sometimes went for months too weak and too depressed even to write home. She did return to her garden, though, expanded her acreage, hired a gardener, and turned entrepreneur. ". . . I hope to make a great deal of money by the produce of our garden, for we cannot use one quarter of the vegetables and fruit which we raise, so that we send a quantity to market every morning. I generally receive from three shillings to four-and-six a day, which is my money. When Mrs. Bell lived here she once told me she made forty dollars by her asparagus alone."
>
> Mrs. Bell's asparagus was a telling commodity. No doctrine of proper spheres prevented women–even upper-class women–from turning their domestic pursuits into profit-making pursuits in the local market. As Mary Cumming told it, "Almost everyone who has a garden raises vegetables for market and some make very large sums of money." And no code of gentility prevented the same women from engaging in physical labor. Mistresses did at times put slaves to work in their gardens, but the mistresses themselves planted and picked and weeded too. In May 1857, Lizzie Partin

assessed the state of her garden: "No thing extra," she called it, "but very good considering the force to attend it"–herself, that is, and one woman slave. Mildred W. Campbell's enthusiastic gardening had earned her some strained muscles (she was seventy years old and had been cutting down trees), so she hired a man to spend half a day spading up ground. But she still thought that she and her elderly woman slave, Charlotte, together could do all the planting. In physical exertion, it seems, was much of the pleasure. More than any other household task, gardening brought women some peace. (Lebsock, 1985, pp. 150-1)

13. "During the previous centuries, the concept of gardening for beauty and national benefit did not have much urgency," writes Ann Leighton. However, ". . . at the beginning of the nineteenth century, D. David Hosack began to promote the use of gardening for the general good and for the enhancement of domestic surroundings" (Leighton, 1987, p. 120). There were others who saw gardening and related botanical pursuits as very appropriate and necessary for the health of women and children, and therefore, of the nation. Thomas Bewick, writing in 1822 said, "If I could influence the fair sex, there is one thing to which I would draw their attention; and that is Horticulture . . . as this delightful and healthy employment–which has been long enough in the rude hands of men–would entice them into the open air, stimulate them to exertion and draw them away from their sedentary modes of life, mewed up in close rooms, where they are confined like nuns. This would contribute greatly to their amusement and exhilarate their spirits" (Hadfield, 1964. pp. 51-2).

Moral reformers, along with landscape and garden writers, encouraged women to get fresh air and exercise. A certain Walter Elder, writing from Philadelphia in 1848, says that ". . . Gardens are important, as they are 'reforming and moralising to the young' and 'exalt the national character.' . . . Elder feels that a well-kept cottage garden in itself is both 'patriotic' and 'christian'" (Leighton, 1987, p. 81).

14. In 1859, Henry Ward Beecher expounded on the virtues of floriculture for women and children alike. An essay entitled "A Plea for Health and Floriculture," which has been extracted from *Plain and Pleasant Talk About Fruits, Flowers and Farming*, is worth quoting at some length. The flavor of the writing and its message is typical of the times:

Every one knows to what an extent women are afflicted with nervous disorders, neuralgic affections as they are softly termed. Is it equally well known that formerly when women partook from childhood, of out-of-doors labors, were confined less to heated rooms and exciting studies, they had, comparatively, few disorders of this nature.

. . . If our children were early made little enthusiasts for the garden, when they were old they would not depart from it.

. . . A love of flowers would beget early rising, industry, habits of close observation, and of reading. It would incline the mind to notice natural

> phenomena, and to reason upon them. It would occupy the gentle enthusiasm; maintain simplicity of taste; and in connection with personal instruction, unfold in the heart an enlarged, unstraitened, ardent piety. (Lacy, Allen, ed., 1988)

15. In their essay "Gardening and the Education of Women," from their book, *The American Woman's Home* (1869), the tactics of Catharine Beecher and Harriet Beecher Stowe are similar to those of their brother:

> . . . the cultivation of flowers and fruits . . . especially for the daughters of a family, is greatly promotive of health and amusement . . .
>
> No father, who wishes to have his daughters grow up to be healthful women, can take a surer method to secure this end. Let him set apart a portion of his ground for fruits and flowers, and see that the soil is well prepared and dug over, and all the rest may be committed to the care of the children . . . Then . . . every man who has even half an acre could secure a small Eden around his premises. (Lacy, ed., 1988, p. 26)

16. The warden's wife writes, "They all enjoyed being out of doors with their pipes, whether among the flowers or the vegetables; and no attempt at escape was ever made by any of them while in the comparative freedom of the flower-garden." This same woman ". . . planted and marked distinctly . . . over seven hundred groups of annuals and hardy perennials, hoping the men would care to learn the names of the flowers, and through that knowledge, and their practise in the care . . . be able to obtain positions as under-gardeners when their terms of imprisonment expires" (Earle, 1901, pp. 101-2).

17. Ann Leighton asserts that ". . . the proliferation of what we can call specialized gardening . . . may have sprung in large part from the impulses of women wishing to be free to garden on their own, for their own pleasure and within their own powers of maintenance" (1987, pp. 285-7).

18. In 1893, Mrs. William Starr Dana published her popular book, *How to Know the Wildflowers*. In her preface to the first edition, she says:

> The pleasure of a walk in the woods and fields is enhanced a hundredfold by some little knowledge of the flowers which we meet at every turn. Their names alone serve as a clew to their entire histories . . . But if we have never studied botany it has been no easy matter to learn these names . . . While it is more than probable that any attempt to attain our end by means of some "Key," which positively bristles with technical terms and outlandish titles, has only led us to replace the volume in despair.
>
> . . . we have ventured to hope that such a book as this will not be altogether unwelcome . . .
>
> . . . Surely Sir John Lubbock is right in maintaining that "those who love nature can never be dull," provided the love be expressed by an intelligent interest rather than by a purely sentimental rapture. (pp. xiii-iv)

She then proceeds with her own key by offering the wildflowers in color groupings and by introducing each plant by its best-known common name(s). This is followed by correct Linnaean taxonomy and a lovely descriptive text, full of her own observations as well as historical and literary allusion.

BIBLIOGRAPHY

I. Literature Cited

Betts, Edwin Morris (1944). *Thomas Jefferson's Garden Book.* Philadelphia: Memoirs of the American Philosophical Society. Excerpted in Matthews, Hugh (1988). *Notes of a New England Gardener.* Vol. 1, No. 11, November, Barrington, NH.

Dana, Mrs. William Starr (1893). *How to Know the Wildflowers.* New York: Scribners. Reprinted 1989. Boston: Houghton Mifflin.

Earle, Alice Morse (1898). *Homelife in Colonial Days.* New York: Macmillan. Reprinted 1975. Williamstown, MA: Corner House Publications.

______ (1901). *Old Time Gardens.* New York: Macmillan.

Favretti, Rudy and Joy (1977). *For Every House a Garden: A Guide for Reproducing Period Gardens.* Chester, CT: Pequot Press. Reprinted 1990. Hanover and London: University Press of New England.

Favretti, Rudy and DeWolf, Gordon P. (1972). *Colonial Gardens.* Barre, MA: Barre Publishers.

Ford, Linda G. (1990). *Women in America: 1600-1900.* Course Outline, Fall 1990.

Gemming, Elizabeth (1968). *Huckleberry Hill: Child Life in Old New England.*

Hadfield, Miles and John (1964). *Gardens of Delight.* Boston and Toronto: Little, Brown and Company.

Holliday, Carl (1982). *Woman's Life in Colonial Days.* Williamstown, MA: Corner House Pub.

Lacy, Allen, ed. (1988). *The American Gardener: A Sampler.* New York: Farrar Straus Giroux.

Lebsock, Suzanne (1985). *The Free Women of Petersburg: Status and Culture in a Southern Town, 1784-1860.* New York and London: W.W. Norton and Company.

Leighton, Ann (1970). *Early American Gardens "For Meate or Medicine."* Boston: Houghton Mifflin. Reprinted 1986. Amherst: University of Massachusetts Press.

______ (1976). *American Gardens in the Eighteenth Century "For Use or For Delight."* Boston: Houghton Mifflin. Reprinted 1986. Amherst: University of Massachusetts Press.

______ (1987). *American Gardens in the Nineteenth Century "For Comfort and Affluence."* Amherst: University of Massachusetts Press.

Midda, Sara (1981). *In and Out of the Garden.* New York: Workman Pub.

Seuffert, Keven Ostrander (1987). *"Herb Lore" from the Peter Matteson Tavern.* Bennington, VT: Bennington Museum Press.

Speare, Elizabeth George (1963). *Life in Colonial America.* New York: Random House.
Spencer, Sylvia (1935). *Up from the Earth: A Collection of Garden Poems 1300 BC-AD 1935.* Boston: Houghton Mifflin.
Thaxter, Celia (1894). *An Island Garden.* Boston: Houghton Mifflin. Reprinted 1988.
Ulrich, Laurel Thatcher (1990). *A Midwife's Tale: The Life of Martha Ballard, Based on Her Diary, 1785-1812.* New York: Vintage Books.
Wilson, James Grant and Fiske, John, eds. (1887). *Appleton's Cyclopaedia of American Biography,* Vol. II. New York: D. Appleton and Company.

II. Other References

American Silk Grower and Agriculturist. Devoted to the Culture of Silk and the Ordinary Pursuits of the Farmer. Weekly Periodical. (Vol. 1, No. 1, Mon. June 6, 1836–No. 13).
American Society of Landscape Architects (1932). *Colonial Gardens: The Landscape Architecture of George Washington's Time.* Washington, DC: Issued by the United States George Washington Bicentennial Commission, March.
Benson, Albert Emerson (1929). *History of the Massachusetts Horticultural Society.* Printed for the MHS.
Berkin, Carol Ruth and Norton, Mary Beth, eds. (1979). *Women of America: A History.* Boston: Houghton Mifflin.
Berrall, Julia S. (1953). *A History of Flower Arrangement.* London and New York: The Studio Publications, Inc. in Association with Thomas V. Crowell Company.
Boland, Maureen and Bridget (1976). *Old Wives' Lore for Gardeners.* New York: Farrar, Straus and Giroux.
Carter, Tom (1985). *The Victorian Garden.* Salem, NH: Salem House.
Denman, Cherry (1990). *The Little Green Book.* New York: Stewart, Tabori and Chang.
Ely, Helena Rutherford (1903). *A Woman's Hardy Garden.* New York: Macmillan.
Favretti, Rudy J. (1962). *Early New England Gardens.* Sturbridge, MA: Old Sturbridge Village Booklet Series.
______ (1964). *New England Colonial Gardens.* Chester, CT: Pequot Press.
Geyser, Samuel Wood (1945). *Horticulture and Horticulturists in Early Texas.*
Glasscock, Jean, general ed. (1975). *Wellesley College 1875-1975: A Century of Women.* Wellesley College.
Hill, Thomas (1987). *The Gardener's Labyrinth.* Orig. 1577. Reprint based on 1652 ed. Mabey, Richard, ed. New York: Oxford University Press.
Hooker, Worthington, MD. (1873). *The Child's Book of Nature.* New York: Harper and Brothers.
Jones, Julia and Deer, Barbara (1989). *The Country Diary of Garden Lore.* New York: Summit Books, Simon and Schuster.
MacNicol, Mary (1967). *Flower Cookery: The Art of Cooking with Flowers.* New York: Fleet Press Corp.

Martin, Laura C. (1987). *Garden Flower Folklore*. Chester, CT: The Globe Pequot Press.

McCormick, Harriet Hammond (1923). *Landscape Art: Past and Present*. Written by Mrs. McCormick for a meeting of the Friday Club in Chicago in February 1899. Again read at the annual meeting of the American Park and Outdoor Art Association held in Chicago in June 1900. New York and London: Charles Scribner's Sons.

Phipps, Frances (1972). *Colonial Kitchens: Their Furnishings and Their Gardens*. New York: Hawthorne Books.

Prentiss, J. and J.W., pub. (1831). *Catalogue of Books in the Keene Circulating Library*, Keene, NH, August.

Prentiss, John, pub. *The Gentlemen's and Ladies' Diary and ALMANAC of 1803*. Keene, NH.

Rumrill, Alan (1990). "*Palm-Leaf Hats: Outwork in Rural New Hampshire*," Newsletter of the Historical Society of Cheshire County, Vol. 6, No. 4, February. Keene: Historical Society of Cheshire County.

Scourse, Nicolette (1983). *The Victorians and Their Flowers*. Portland, Oregon: Timber Press.

Slosson, Elvena (1951). *Pioneer American Gardening*. New York: Coward-McCann.

Stabler, Lois K., ed. *Very Poor and of a Lo Make: The Journal of Abner Sanger (1774-1782: Keene; 1791-1794: Dublin)*.

Ulrich, Laurel Thatcher (1980). *Good Wives: Image and Reality in the Lives of Women in Northern New England 1650-1750*. New York and Toronto: Oxford University Press.

Wright, Richardson (1928). *Forgotten Ladies: Nine Portraits from the American Family Album*. Philadelphia and London: JB Lippincott.

______ (1949). *Gardener's Tribute*. Philadelphia and New York: JB Lippincott.

Chapter 3

Are We Afraid of Plants? Exploring Patriarchal Society's Devaluing of Plants, Women and Nature

Rhonda Roland Shearer

SUMMARY. Centuries of Western patriarchal values influence how we view and treat nature, women and plants. Ecofeminists have been exploring the deep connections between women and nature. Rhonda Roland Shearer's research shows that the societal devaluing of women and nature extends also to plants. She uses trends in art history as illustrative guide-posts of society's values. She probes how our patriarchal hierarchy has been influencing artists' preference for "masculine subjects" (history painting and monuments) as opposed to flowers and plants, which are historically devalued as a "woman painter's subject." Looking at the philosophical underpinnings of our patriarchal hierarchy, she discusses the deep-rooted fears we have of nature and women conditioned by centuries of Orthodox religion, science and philosophy. By examining society's attitudes toward women, plants and nature, Shearer believes an important transformation can take place. Shearer explores how the new fractal geometry, being a geometry of nature, can be used to put women and nature on an equal basis with men and technology. Before women and nature can be free from violence, all of society (both men and women) must alter their values so that both masculine and feminine characteristics have equal status.

> Woman is nature, hence detestable. (Charles Baudelaire, Les Fleurs Du Mal, 1857)

> Men and women differ much as animals and plants do. Men and animals correspond, as do women and plants, for women develop more placidly and always retain the formless indeterminent unity of feeling and sentiment. (Georg Wilhelm Friedrich Hegel, Philosophy of Right, 1821)

> We will place nature on the rack and torture her secrets out of her. (Francis Bacon, 16th c)

> Naturalistic flowers are for children and the feminine spirit. Flowers best express the outward, the female. (Piet Mondrian, 1919-20)

"Stark data on women, 100 million are missing," The New York Times headlines informed us November 5, 1991. For the first time rumors of infanticide are translated into chilling numbers. There is statistical evidence that girl babies, by the "tens of millions" are being neglected or killed world-wide simply because of the social preference for boys.[1]

What does female infanticide tell us about society? The preference for boy infants is clearly not random but is fueled by society's choice or selection of an entire set of characteristics which it most values.

Within a patriarchal hierarchy, it is the male characteristics which are the most heralded; dominance, aggression, winning and control hold sway over society. Infant boys, those most capable of delivering these characteristics to society, like "sacred cows" have a priori respect and protection from harm (see Chart 1). Girl infants, however, represent another set of values which in our present patriarchal hierarchy are of lesser status (also see Chart 1).

Serving mostly as nurturers and primary-care givers, women's roles are very much devalued and are closely associated with biological functions and nature. Ecofeminists state that women and nature are inextricably linked and mutually devalued.[2] My research shows this societal devaluing of nature and women also includes plants.[3] This paper is an exploration of the linkage between plants, women and nature in our patriarchal society.

Whether or not there is intrinsic association between nature, feminine characteristics and women, they nevertheless have been assigned by culture a feminine connotation in the Western patriarchal hierarchy. My paper is a discussion of how these relative values have influenced human thought and art. For example, in Western thought the dominant school is one that gave rise to Newton and others who devalued nature with their

CHART 1. Gender Associated Characteristics in a Patriarchal Hierarchy

MASCULINE	FEMININE	MASCULINE	FEMININE
Large	Small	Power	Weakness
Superior	Inferior	Reductionist View	Holistic View
Culture	Nature	Cartesian View	Monist View
Mind	Body	Control	Submission
Intellect	Emotions	Ambition	Contentment
Ideas	Senses	Aggressive	Passive
Order	Chaos	Conquer	Adapt
Historical	Personal	Geometric	Organic
Pragmatic	Romantic	Linear	Cyclical
Traditional	Sentimental	Rectilinear	Curvilinear
Heroic	Everyday	Public Sphere	Private Sphere
Objective	Subjective	Hi-Technology Achievement	Domestic Activities
Rational	Irrational	Work	Home
Reason	Instinct	Competition	Nurturing
Analytic	Intuitive	Leader	Supporter
Spiritual	Material	Animals	Plants
Abstract	Concrete	Architectonic/Structural	Ornamental/Decorative
Epistemology	Ontology	Monuments, History Paintings	Still Life, Landscape Painting
Knowing	Being	Art	Craft
Transcendence	Immanence	Norm	Other
Independent	Dependent	Subject	Object
Separation	Connection	Science/Mathematics	Humanities
Domination	Cooperation	Professional	Amateur
Hierarchal	Interdependent	Left Brain	Right Brain
Scientific	Anecdotal		

mechanistic thought. In art, the fact that nature and landscape are not as highly valued as heroic monuments or history paintings in the West is directly linked to this dominant Cartesian view. I use trends in art history as illustrative guideposts of Western society's values.

Western art history reflects society's attitudes towards plants. Mostly considered foodstuff or decoration, plants were relegated in the hierarchy of artistic subjects as the least desirable. Still lifes in particular were viewed as a "woman painter's subject."

Bryson quotes from an 1860 journal, "Gazette des Beaux-Arts,"

> Male genius has nothing to do with female taste. Let men of genius conceive of great architectural projects, monumental sculptures and elevated forms of painting. In a word let men busy themselves with all that has to do with great art. Let women occupy themselves with those kind of art they have also preferred . . . the paintings of flowers, those prodigies of grace and freshness which alone can compete with the grace and freshness of women themselves.[4]

The traditional hierarchy of artistic subject matter up until the twentieth century was, in order of prestige–"History painting at the top–followed by portraiture, genre, still life and landscape."[5] It wasn't until the twentieth century that abstraction replaced history painting in the hierarchal order. Even today, "flower painting" is marginalized because of its association with the "feminine." Romantic, decorative, sentimental flowers, as art subjects, did not carry the "intellectual weight" of history painting or abstraction in the value system of the patriarchal hierarchy.

Thought incapable of "genius," women themselves were considered ill-equipped for the "intellectual rigour" necessary for the creation of important history paintings or abstractions.[6] Not only have women and plants been jointly marginalized and considered intellectually inferior, but "historical women" and "plants" are nearly completely excluded as monumental sculpture subjects. The devaluation of women, nature and plants by artists and society is confirmed by their absence in public sculptures.

MONUMENTS AS REFLECTIONS OF SOCIAL VALUES

Monuments have always manipulated and reinforced what society's members should admire or value. Placing a person or idea in heroic scale creates a powerful psychological effect. A historical figure, at say 20 feet tall, is an object of cultural status. Cast in bronze or carved in stone, a monument is permanent and immortal, forever separated and above mere humans. We as society's members are dwarfed in size by comparison, becoming literally submissive, like children looking up to an all powerful parent. We have all experienced this unique feeling of looking up to a monument. It is impressive.

Interestingly, the power of monuments is even more permanent than the politics they can symbolize.[7] Political leaders come and go, but their cultural effect continues through the continued presence of their monuments.

Who or what have we monumentalized throughout history? One can experience in the City of New York's Central Park the "memories" of what and who patriarchal society values. Of the hundreds of monuments of poets, generals, explorers, inventors and political leaders (417 to be exact), only three are women.[8] It is quite telling that these three are Mother Goose, Alice in Wonderland and Joan of Arc. Plants, in addition to women, are absent as "monumental" subject matter except as settings or props for human figures, like a tree stump for a general's foot.

What does this tell us about society? The plethora of male monuments clearly communicates patriarchal values–power, aggression, winning, intellectual achievement and adventure.

The fact that men far outnumber women as subject matter in public sculpture underscores society's bias for "masculine" ideals. Women, when they do occur in monuments, are depicted as insignificant or unreal creatures such as nude mythological sex objects (sometime chained or raped) or whimsical figures (Mother Goose or Alice in Wonderland).

Prominent in the 17th century were sculptures of women as allegories, titled "chastity," "beauty" and "grace."[9] These provocative, scantily clad images of women more clearly demonstrate women's sexual object role than their virtuous titles describe. Other disempowered or negative female images continued to occur into the twentieth century. Alberto Giacometti's abstract sculptures: "Woman with her Throat Cut," and the more realistic nude female titled, "The Invisible Object," frankly communicate patriarchal attitudes toward women.[10] Feminists have long complained about Gaston Lachaise's sculptures of what appears to be aggressively dismembered and beheaded women.[11]

Yet even worse, women are most frequently included in the promotion of patriarchal values such as war and "righteous violence" as a virtue. Joan of Arc leading the troops to battle and nude "Winged Victory" Goddesses are examples of how women, either historical or mythological, have unfortunately been forced to take on traditional male roles, characteristics or values in order to become suitable subject matter for monuments.

The lack of plants in monumental sculpture reflects plants' role in society as background or settings for humans. It would seem "pagan" or an ancient echo of nature worshippers to monumentalize plants. Western

Orthodox religions defeated animism long ago by creating a new order which placed man as the apex of creation. Women in this patriarchal hierarchy were below men, with plants below both women and animals. Once you start thinking in hierarchal terms–high/low, top/down–it is hard to stop. Our thoughts continue into inevitable, "logical" conclusions of good, bad, better, best humans.

This patriarchal hierarchy made it rational to conclude that man, being the most perfect and complete form (a belief clearly stated by Aristotle) was the best possible subject matter to hold and express monumental ideals.[12] These, of course, were white males. White male images hover 20′, 30′ tall over cities and towns, on horseback or with weapons; they are erect, powerful, some with fists held over their heads, others with fingers pointing in defiance or benevolence.

African-American males were depicted as unerect, powerless as in "Freedom" sculpture by Thomas Ball. Lincoln holds a book in his right hand with his left hand held low over the head of a crouched African male slave. Therefore, like Lincoln, those "historical" white men who distinguished themselves through battle or power were displayed in society as public heroes or role models for other men, women and children of all races to follow. The permanence of these monuments was part of the message–it was as if to say, "these white males and their ideals are lofty, right and true and they will live forever."

Even today with the competition from many communication mediums the power of monuments continues. This is why the photographic images of people toppling the monuments of Lenin in Russia seem so potent, and the ad hoc erection of the "Freedom Sculpture" in Tiananmen Square is so moving. "Mere mortals" are challenging and defying the power of giants and the a priori values of society–simply through their actions towards monuments.

Just as the erection or destruction of monuments are political acts, like the "Freedom Sculpture" in Tiananmen Square, to monumentalize women, feminine characteristics or roles, nature or plants, would be an affront to the established social order. As de facto appropriate as it has been to monumentalize men in our patriarchal society, it has been just as *inappropriate* to create monuments of women and plants.

In our patriarchal hierarchy "historical" women, in their traditional feminine societal roles, and plants are not important enough. They are too "natural," too "ordinary" and too "everyday" to have "heroic" social status. "Woman's work," after all, is a pejorative; to be able to do a "man's job" is always complimentary.

In keeping with patriarchal values, the Virgin Mary and Eve are probably the most prolific female images in Western sculpture. Eve serves as an appropriate subject matter for monuments because she is a reminder of the evil of woman's association with nature. St. Bernard of Clairvaux reinforced Christian doctrine with his belief that Eve was "the original cause of all evil, whose disgrace has come down to all other women."[13] The bottom line was that the team of women and nature (Eve and her apple) corrupted Adam and caused the fall of all men.

The Virgin Mary, in complete contrast, is monumentalized in history for her "immaculate conception." Mary, as the "Mother of God," did not recapitulate Eve's original sin for achieving pregnancy. She was revered by Christian fathers for acting *above* nature and the senses–a feat real women are powerless to recreate. The perfectly natural conception and birth process demonstrates, in Western theological terms, women's fated connection to nature and "lack" of spirituality.[14]

Certainly plants in a patriarchal hierarchy don't fulfill the requirements for "monumental" status. Plants don't "think" and can't move and, therefore, are considered insignificant when compared to humans. The expression "don't stand there like a potted palm," is like saying "don't just stand there stupid." We are so indoctrinated into this patriarchal hierarchal view that it seems rational. In fact, it is difficult to think outside of this dominant construction. When we do, we quickly discover that what seems "rational" is in actuality *irrational* and destructive, contrary to the interests of basic human survival in our society.

A case in point is the environment. Our patriarchal hierarchical attitudes and, thus, our actions have viewed nature as something to be conquered and exploited. Our arrogance is so great in this regard that we treat nature as if there is no end in the availability of clean air, water and plants. Since we see ourselves as so superior to nature it is difficult for us to imagine or admit the extent of our dependency. This is precisely the danger of hierarchal thinking.

Within the patriarchal hierarchal construction, anything or anyone who is dependent has reduced social status. People, as well as nature, have been casualties of this system. According to Murray Bookchin, in his book *The Ecology of Freedom, The Emergence and Dissolution of Hierarchy*, "the very notion of the domination of nature by man stems from the very real domination of human by human."[15]

DEPENDENCY THREAT TO PATRIARCHAL CONTROL

Western civilization has been based on the notion that "primitive peoples" and women are physically and mentally inferior and dependent on nature. Nineteenth century scientists offered scientific "proof" of the need for white European males to care for "morphologically inferior" humans such as women and people of color.

White males felt a "moral responsibility" to elevate women's and primitive people's dependency upon nature to a dependency instead on "higher" male culture. Among other spurious conclusions larger brain size and stature were seen as conclusive evidence and justification of the evolutionary "superiority" of Victorian white males.[16]

The message has been clear. In a patriarchal hierarchy having someone dependent upon you puts you in the "superior" position and gives you power and control over those in your charge. We are trained in our patriarchal society that dependency is "bad." This training begins in childhood, with sharp parental commands and social clues: "don't be a baby!" "don't hang on to your mother's skirts!" "be a man!" "be independent!" The list goes on. We quickly learn that dependency in a patriarchal hierarchy is equated with weakness and inferiority.

A great deal of the resentment against women can be attributed to the fears of dependency and the power of "mother" in our childhoods. The patriarchal hierarchy gives license to the rejection of dependency as something one outgrows (children) or is forever condemned to by evolutionary fate (women and "primitive people"). This echoes our primal fear of dependency on nature and "Mother Earth." The actual reality of dependency on women and nature is not recognized for it runs counter and is an affront to the control of the patriarchal system. To be dependent on women and nature in patriarchal terms is to give them power and control. When a social system as violent as the patriarchy is threatened by loss of control, it attacks, discredits and disregards. This litany describes what has happened to women and nature throughout history in a Western patriarchal hierarchy.

Unfortunately, our inability to reconcile ourselves to the realities of dependency has created a violent result. We have deceived ourselves about the lack of importance of women and nature; which is to say, we devalue in our lives the very factors in which we are *completely* supported and sustained.

Contrary to patriarchal thinking it is irrational, not rational, to not maintain and appreciate those life-giving factors for which there is no escape or alternative. Like it or not, we are completely dependent on plants, women and nature. It is important that we recognize this fact in order to change the negative behavior and consequences which have resulted from their devaluation. Hierarchal thinking is itself the culprit and must be replaced. Thinking in the context of the patriarchal hierarchy distorts judgements with the irrational priority of masculine values instead of an order which values all life-sustaining aspects.

In reality, plants' proper status cannot be understood or valued in the old-fashioned hierarchal order. "Human hierarchical superiority" masks the fact that there is no human or non-human life without plants.

In sculptures from pre-history the ancients expressed their concern for the important connection between humans and plants. According to Pamela Berger, early goddesses associated with nature and plants proliferated. Women's "birthing powers" were considered by early farmers as analogous to the fertility of the soil.[17] In sharp contrast with today, judging these pre-historical cultures by their artifacts, their beliefs seemed totally focussed on the "feminine"–not the masculine–as their "highest" cultural expression.

She cites goddess statuettes portrayed as giving birth found in ancient grain bins; or others holding or decorated with plants as examples of her thesis. Seeds were seen, in the context of woman and nature, to have the power of rebirth–a mystery similar to the female body which was considered not just magical but divine. As evidence, Donna Wilshire points to the archaic Greeks calling "the 'Divinity-within-Seeds' 'Kore,' daughter of Mother Earth."[18]

The "web of life" concept is a much more authentic construction of plants' essential role. Instead of the hierarchal emphasis on seeing only differences, the web of life model underscores connections, shared characteristics and destinies. Plants can now be seen as an important *part* of the complex interaction of the organic and inorganic; not as vertically ranked according to status as in a hierarchy, but a "holistic" vision which views all life and non-life as *interdependent*.

In addition, we now see that the human need and dependence on plants goes beyond what has been determined as their biologically essential role as food and oxygen sources. Recent research indicates that the plants considered decorative in our environment may be necessary for human health and sense of psychological well-being.[19]

Our need for and dependency on women is a correlative issue. It is women who bear us and *choose* to help us survive as infants.[20] Women, not men, take primary responsibility for cooking, cleaning, caring for the sick, young and old–the essential "dirty" work which supports human existence. Although these actions didn't seem important enough for monuments or history, it is these unwritten, unheralded roles of women which has allowed our human species to continue and survive.

DEVALUATION OF WOMEN AND NATURE RESULTS IN VIOLENCE

The degradation of nature, women and feminine values has reduced the quality of life for all humans. Just as the devaluing of nature created actions of violence against it, the devaluation of women and feminine values not only has created violence against women, but has resulted in an atmosphere of general violence in society itself.

When domination, aggression, winning and control are so highly prized, they begin to rule society by influencing individuals' behavior. According to Myriam Miedzian,[21] when you have popular societal role models such as "Rambo," violence is a continuing natural result, especially for males. Professor Miedzian relates frightening statistics in corroboration of her thesis of the linkage between masculine values and violence.

Approximately 1.8 million women a year are assaulted by men in domestic violence and 89% of all violent crimes are committed by males. Moreover, she states "that wars have always been and continue to be initiated and fought almost exclusively by men." From her research, Professor Miedzian concludes the masculine values, "toughness, dominance, repression of empathy, extreme competitiveness" are the key factors in criminal, domestic and global violence. Despite this violent masculine reality, nurturing, caring and other "niceties" which give us quality to life are considered "whimpy" and are relegated to the shadow world of the feminine out of public view.

More and more women are identifying with masculine values to be able to participate and compete in our patriarchal society. By extension, the statistics are beginning to reflect women's identification with the masculine by their increasing participation in violence. Although infrequent in comparison to male arrests, during 1985-1989 there was a 42% increase of women arrested for violent crimes.[22]

Like Joan of Arc leading the troops, women can be just as duped by the patriarchal values as men. Surprisingly, it is *women* who are primarily responsible for the murder or neglect of 100 million missing female infants because of their and society's preference for boys. Who would predict that the patriarchal bias for males would go so far and be so deliberate that 7½% of the expected global population of women would be missing and presumed dead simply because they weren't males?[23]

Even more shocking, however, is the apathetic public response to this revelation. Silence instead of outrage is a de facto condonment of female infanticide on the part of society. It can be argued that if the "feminine" were of social value the mass murder of female infants would not exist.

The danger of the patriarchal hierarchy is that when someone or something is devalued the one doing the devaluing becomes "desensitized." The person or thing devalued is psychologically remote from the devaluer; this separation impairs and can completely eliminate empathy or sympathetic response, traits necessary for "moral" behavior. This technique was used at the death camps during World War II. Prison guards were "trained" to accept the "valueless" and "sub-human" status of the Jewish prisoners and, thus, were psychologically able to carry out their assigned tasks cooperatively without thinking about the object reality of human suffering and mass murder.

The devaluation of nature and native peoples in the patriarchal hierarchy has created a similar deadening of feelings and a dangerously distorted morality. Looking at the history of the U.S. alone, one confronts our callous destruction of nature with Western technology, which also included the conquering and elimination of "primitive people" who were simultaneously devalued and murdered.

Throughout the world, organic societies recognized and respected their dependency on nature; their ethics and religion organized and perpetuated a cooperative, life sustaining relationship. Their culture was not based on human superiority of nature, as in the Western patriarchal hierarchy, but on the identification of nature as part of themselves. This important distinction of *connection* (feminine) instead of hierarchal *separation* (masculine) created harmony and non-violent behavior towards nature which, left untampered by Western technology and "civilization," would still be sustaining life for organic societies today.[24] Instead, the violence against nature has brought all humans and non-humans to the brink of global catastrophe.

In the final analysis, revaluing all of "the feminine," not just nature, is important for changing the patriarchy. This new direction is not to suggest the reversal of the roles of masculine and feminine characteristics but for interdependence and "holistic thinking" instead of masculine domination. Interdependence of masculine and feminine values would not only improve the quality of life for women, plants and nature but help the future survival of all humans and non-humans.

THE WORLDS OF WOMEN AND NATURE ARE ALIEN TO THE PATRIARCHY

Within a hierarchy we are conceptually separated from each other by notions of "lesser" or "greater" "value" and "status." The feelings of alienation induced from hierarchal separateness between humans is similar to the conceptual and, for most of us, literal alienation we have from nature.

Early patriarchal societies justified the control and domination of nature as the need for human "progress" and "survival." Nature today is mostly seen as a repository of raw materials for future manufactured goods or a resource for recreational activities. In summarizing Western patriarchical attitudes: nature is to be feared and fought, used and enjoyed–whether friendly or adversarial, wild or tame, nature is outside of ourselves like another world than our own.

Analogous to this alien world of nature is the separate world of women. Like inside a dark harem room, the domestic realm is closed, alien and devalued in comparison to the "outside" world of business and commerce, politics and war.

What occurs in the home is close to the cycles of nature itself: birth, nurturing the young, caring for the old, and death. The daily comings and goings within the domestic interior, for the most part, are unknown to history except in the context of the "outer world" declarations of birth, marriage and death notices published in newspapers. Both nature and the internal world of the "feminine" remind us of how close we are to the rhythms of life and our mortality.

This is in complete contrast to the patriarchal focus on man as intellectual, conqueror and transcendent being, identified not with nature or earth creatures but an all powerful, unchangeable "sky God." Most Western Orthodox religions throughout the world promise that humans will find "immortality" by joining their "sky God" in the after life; a reward for our human rejection and transcendence over earthly cycles and mortal flesh.

This patriarchal view has universal appeal. Thoughts of being connected to a power above nature gives us, psychologically, an illusionary sense of power and control over very real and unpredictable events in nature. These feelings of safe detachment from the mundane (obtained from identification with a transcendent "sky God") counter identification with nature which seems alien but eerily familiar, like going back to the darkness of the womb. Feeling identified with nature makes us feel mortal, flesh-bound and out of control. There is no heroic death, no glory in a cyclical view of life as birth, life, decay and rotting like plants which absorb back into nature. The linear view of birth, life and transcendence of the flesh seems much safer.

In addition to Western Orthodox religions, further psychological separation from nature was induced by science and philosophy. After Descarte taught us that man is the only creature who thinks, Newton's mechanistic view reduced nature to a lifeless machine based on mathematical laws. Science, mathematics and philosophy became transcendent "abstractions," demonstrations of mind over body which were hoarded and only shared between certain elite members of society, mostly white males.[25]

Scientists trained the mass public and themselves to think in terms of hierarchies. These hierarchies reflected the white male patriarchy and by intellectual justification helped maintain its power: food chains (big predator fish eat small weak fish), evolution theories (man as the superior species) and the world divided into scales ranging from the big to small (quantum, relativity theories).

This was culture. The way out of the uncertainty and transience of nature more than technology was education. For, in almost every discipline, it taught us that nature was part of the "everyday" available to everyone; plants, animals, women and "primitive people" . . . but culture was something of special status, a product of the "superior" mind, abstract and for humankind alone. The fact that only humans had culture and "better" humans had more "advanced" culture was de facto "proof" of Western civilization's supremacy. Since advancing culture was the gauge for "the ascent of man" the further we "progressed" from nature with technology and knowledge, the "better" and more "advanced" humans we became.[26] So we thought from our training in a patriarchal hierarchy's value system.

The yardstick from the patriarchal hierarchy was clearly linear; progress is man growing *away* from nature and primitive peoples. This linear bias of "progress" devalued women and people of color because they were seen as totally absent or minor players in the advancement of culture over nature. Plants were valued in culture only in terms of their subordinate relationship or usefulness to humans.

Women, in particular, have been exploited by this view both in modern and "primitive" cultures.[27] Daily maintenance activities done by women have continually created throughout time the platform of support for males in the home which enable males to go "out" into the world. This has evolved into the creation of "abstract" labors in culture, leaving "lower caste" males and women for material labors.

Women's work, however, is less valued than men's manual labors in our patriarchal hierarchy. Manual labor, such as ditch digging, has more status than women's work, though both are considered unskilled. Ditch diggers are compensated with a wage, pay taxes and are therefore considered as more contributing participants in society and culture. Since women's work is unpaid and occurs within the private realm (home) outside the mainstream of public life, it cannot compete in the patriarchal hierarchy which values monetary reward and power. It is within this context of public life, monetary reward and power that women's work and women themselves are negatively judged. Association with "women's work," whether by men or women, results in the lack of respect and status in the patriarchal hierarchy.

This underscores the point of this paper; it is not just women, nature and plants themselves which are devalued, but *feminine characteristics*. If women, for example, take on male characteristics, or men take on feminine characteristics, they are valued or devalued respectively. This has been one of the complaints of cultural feminists. The reality of the patriarchal hierarchy is that women will not make significant change in the patriarchal system just by their participation alone. As Ynestra King points out the "underlying assumption" of society is that "male is better."[28] By joining the system, women are expected to conform to male norms. The status quo of male culture must be challenged, not simply joined, in order for feminine characteristics to become in balance with the masculine and by extension begin to create their positive behavior and effects.

THE MASCULINE MODERNIST AESTHETIC VS. "FEMININE SPACE"

The masculine bias clearly affects aesthetics in our culture. Not only has it created the preference for history, paintings, and monuments which best perpetuate male characteristics; the masculine bias has also been communicated through the visual reality of modernism itself. Look at the masculine characteristics in Chart 1–objective, rational, intellectual, technological, spiritual; these are the characteristics assigned to early modernist art. Painter Piet Mondrian, considered the father of geometric abstraction, wrote of the "irrational," "random" and "capricious" qualities of nature (associating it with the feminine).[29] Mondrian believed that nature must be "subdued by man for greater technology."

Not unique to Mondrian or other modern artists, this aggressive need to control was a reflection of Mondrian's time. Technology looked like a utopian force for man's glory in the conquering of nature at the beginning of this century. Nature was a hindrance to man's "progress" or, as Mondrian believed, "veiled" the ultimate universal reality. To Mondrian geometry was perfect, spiritual; nature offered an imperfect reflection of these Platonic ideals.

He specifically writes of tree structures as examples of nature's "imperfect" geometry. The nearly straight "lines" in tree branches were interpreted as nature's failed attempt to achieve Euclidean perfection.

Nature as the "feminine" was considered inferior to the spiritual, masculine intellectual order of Euclidean geometry.

Culture continues to be visually dominated by this early modernist vision.[30] Grids, machined lines and edges map our technological world and our action upon nature. Think of our cities, our fields; grids or Euclidean geometries are the key way Western humans dominate space, control nature and develop technology.[31] Since nature's order is diametrically opposed to this man-created Euclidean geometric order, the effects upon nature are devastating. Grids and rows rule, creating the destruction of nature's natural rhythms. Growth, wholeness, connectedness and complicated patterns unique to nature's geometry are considered aesthetically inferior (romantic, sentimental, non-intellectual) in comparison to modernism's sleek line, reduced simplicity of form and hi-tech look. We are so conditioned to this masculine modernist aesthetic in our present culture that nothing else looks as "new."

I believe this dominating social preference for masculine characteristics can be likened to being stuck in one mode of a gestalt figure. Instead of seeing both the vase and the two profiles as in the classic gestalt figure we are stuck in the masculine mode, unable to shift our vision.[32]

The Euclidean geometric approach to nature in patriarchal terms was a bias connected to the feminine bias. Plants and the complicated phenomenon of nature seemed irrational compared to the sleek rectilinear lines which created masculine control and order. In our patriarchal hierarchy what wasn't rational and controllable by man was relegated to the feminine, sentimental and romantic. Euclidean geometry in art literally "acted out" this aggressive process, which the viewer would re-experience. Consider Mondrian's paintings of leafless trees in series–curvilinear branches are carved by lines which, over time, are reduced and subdued into "pure" line and right angle. Mondrian declared that the new plastic art "is hostile to natural man" . . . "we must *convert the curved to straightness*" . . . "in order to annihilate this naturalism."[33] We in the patriarchy have preferred to see the world as masculine; Euclidean geometry helped make it a reality.

In addition to its use in controlling nature with technology, it is important to note that Euclidean geometry was not only a dominant construction for modernist abstraction but was, since the beginning of Western art history, a major consideration in the making of paintings and sculpture. Since the ancient Greeks, art has been constructed with the Euclidean geometric system firmly conditioned in artists' minds. Whether it be the choice to make an art work in two or three dimensions or the selection of proportions in sculptural figures, or perspective in paintings, all of these decisions are made with Euclidean concepts, which results in Euclidean biased art constructions.

The paradigm shift, which is occurring in science and mathematics, is now challenging this dominating Euclidean world view.[34] With the advent of fractal geometry, there is now a new geometric alternative organized on very different principles (see Chart 2).[35] When artists use fractal geometry in their approach to art-making, nature and the feminine are echoed in their constructions, not just the man-made and technological. Unlike straight lines controlling space or the circles, lines and squares of man-made shapes, fractal geometry is representative of all of nature's forms.

Euclidean geometry is useful for the construction of man-made objects. When it comes to nature's complexity only fractal geometry can closely

CHART 2. Comparison of Euclidean and Fractal Geometries

EUCLIDEAN (and other integer dimensional geometries)	FRACTAL
• traditional (> 2500 yr) ancient through modern	• post-modern (~17 yr)
• based on characteristic size or scale	• no specific size or scaling
• suits man-made objects	• appropriate for natural shapes
• described by formula	• (recursive) algorithm
• has masculine character; used for controlling space and dominating nature with technology (masculine)	• displays both feminine and masculine characteristics; technological (masculine) it is also sympathetic with nature (feminine)
• contrasts with nature's forms and monitors space	• reflects nature's forms and growth dynamics
• on our macrocosmic scale represents only culture	• represents a combination of nature and culture
• dualistic, is part of the hierarchal values and Cartesian thought	• not dualistic, is free of hierarchal values; participates in both Cartesian and Monist thought

simulate nature's dynamics and irregularity to create images of clouds, mountains, flowers, trees. Nature, under this new paradigm, is not only feminine (nature, organic), but masculine (culture, geometric). "The feminine," excluded from history and public life, can now be geometrically visualized and experienced in art due to fractal geometry.

This new fractal geometric visualization of space and form breaks the fixed focus on the masculine (Euclidean) geometric expression in society as well as in art. Instead *both modes*, feminine (fractal) and masculine (Euclidean), can be switched into and experienced as an entire Gestalt.[36] Referring to Chart 1 again, look down the list of masculine vs. feminine characteristics and notice that "the masculine" serves as precise descriptions of the Euclidean geometric approach to visualizing space and form. Since the geometric models available have been so biased to the "masculine" in their organization and structure, existing geometries did not or could not express the "feminine." With the qualities and characteristics of Euclidean geometries being decidedly masculine, the "feminine" remained invisible.

With fractal geometry the "feminine" now has its own geometry which reflects *feminine* characteristics and values. Just as one's traditional concept of space is conditioned by Euclidean (therefore masculine association), so space which is defined by fractal geometry corresponds with feminine characteristics. My recent sculptures, "Women's Work," are experiments in making this "feminine space" visible.

As a challenge to the status quo devaluation of feminine characteristics, I've recently completed 8 monumental sculptures in a series called "Women's Work." Each made of a single bronze outline, like a drawing in space, these historical "women," with titles such as "Nina Vacuuming" or "Kiki Ironing," (see Photos 1-3), have alternating empty space or foliage as internal substance.[37] Their monumental 9-foot size is in shocking contrast to the subject matter.

As described earlier, plants, women, nature or the domestic have not had monumental status in the patriarchy and therefore have not been depicted as monumental "ideals" in sculpture. Whereas most female sculptures are titled in the abstract such as "Woman Bathing," my "Women's Work" sculptures state the actual woman's name in the title. These are real women doing (domestic work) and being (plants, one with nature, decorative) what has been assigned to them as "feminine" but now asserting their own space, their own reality in public view.

Creating power with their size, women's lack of public identity is made conscious and visible by these sculptures as witnessed by our own surprise, amusement or embarrassment at seeing ironing and other domestic

PHOTO 1. "Yves's Wife with Baby," 1991
103″ × 100″ × 30″

PHOTO 2. "Kiki Ironing," 1991
103" × 72" × 23"

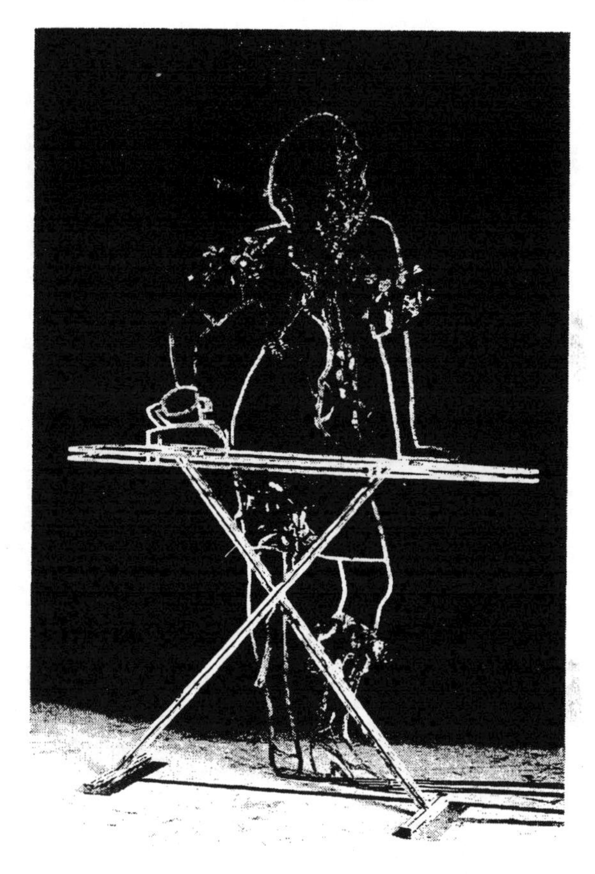

acts depicted on a heroic scale. "Feminine space" in my work is the invisible feminine characteristics and private lives of women made visible through the construction of fractal geometry. Fractal dimensions–those dimensions in-between 1 and 2 and 2 and 3 integer dimensions–unknown and indescribable through Euclidean geometry, are now visible through fractal geometry. Making fractal dimensions perceptible in my work is an analogy for making plants and the private lives of women perceivable in public sculpture.

Fractal scaling, another aspect of fractal geometry and "feminine space" is illustrated by plants within my sculptures. "Self-similar" plants

PHOTO 3. "Nina Vacuuming," 1991
106" × 53" ×59"

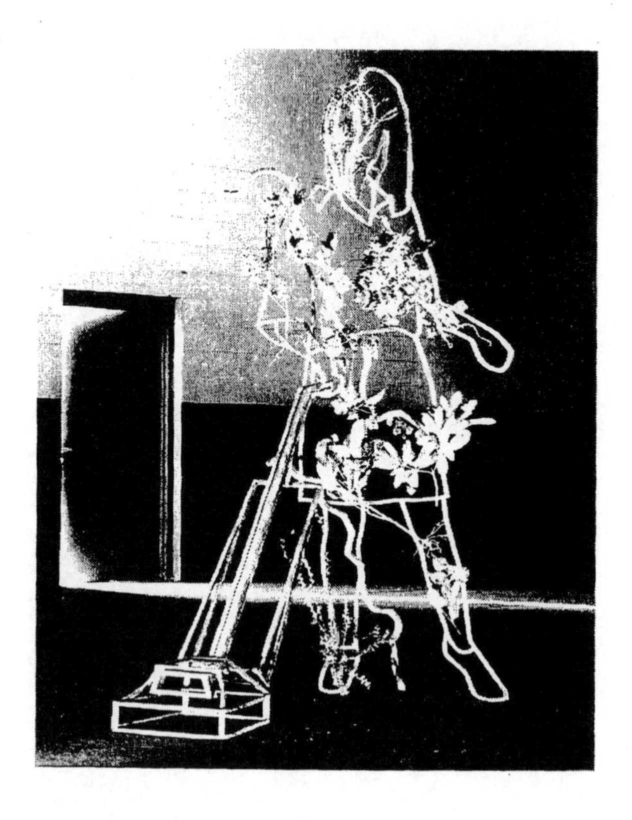

are a demonstration of nature's continuous repetition of limited forms expressed at various scales in seemingly unlimited ways–within this simplicity at the same time occurs complexity. Each plant, individualized and anecdotal, is also universal and geometric. Fractal geometry is a resolution of these polarities and feminine space is a visual expression of this dialectical "whole" (see Chart 1, holism is feminine).

Instead of being afraid of the feminine association with nature, I've embraced it. Plants, reflecting feminine curves of women's form and fractal characteristics in their structures, are important elements in the sculptures. Plants and space become the "stuff" of which women are made. The complexity of plants in my work illustrates the change from a hi-tech modernist aesthetic (Euclidean) to that of fractals. In the old paradigm, plants' complexity was anathetical to modernist principles. Considered irrational because nature's forms were irreducible into Euclidean shapes, plants were devalued and relegated in our patriarchy as "feminine."[38] This was an arrogant assumption based on the limits of our knowledge of mathematics. Just because nature appeared non-mathematical within the limited model of Euclidean geometry, we assumed it was inferior and thus feminine. Therefore, plants' complexity in art was seen as inappropriate subject matter for "high art," and most suitable for women artists as plants were decorative and sentimental like women themselves.

This "logical" deduction is changed due to fractal geometry. Instead of polarized as feminine within the masculine/feminine dualism, plants as well as other aspects of the feminine can be seen holistically–as an entire Gestalt. Fractal geometry expresses feminine characteristics *in addition* to the masculine. Therefore, the distinct boundary between the masculine and feminine is now unclear (see Chart 3).

Look at the feminine characteristics on Chart 1–decorative, romantic, nature, senses, emotions, instincts–all of these characteristics are traditionally evoked by the images of plants and women. Because of fractal geometry these patriarchally assigned feminine characteristics are now brought in concert with their masculine counterparts. Plants, for example, are no longer confined to the feminine but participate in characteristics from the masculine as well (see Chart 1). To illustrate this deconstruction of boundaries between the masculine and feminine, I have combined the two aesthetic experiences in my work–that of Euclidean and fractal geometries–into a distinct yet dialectical relationship. In the "Women's Work" series the natural complexity and visual (fractal) dynamics of the plants and curved women are in sharp contrast to the smooth machined-bronze outlines of the domestic objects which evoke a hi-tech (Euclidean) aesthetic.

Because, like everyone else, I am visually conditioned by our culture, I find it difficult not to prefer the hi-tech objects in my own sculptures. The complex plants in my work, even to me, evoke negative associations with the primal, irrational and decorative. Beneath my anxiety of what plants mean in our culture, I have had confidence in a deep intuitive knowing that plants are of profound significance. Fractal geometry is a universal, objective articulation of this inner private voice. In essence, fractal geometry is the manifestation of the universals behind the many individual forms of nature, giving a voice to their meaning and a status within our culture.

As passive and non-political as plants have been, they can now be fearlessly used in art as subversive to the status quo of patriarchal values and modernist aesthetics. In my work, plants introduce a new era of fractal

CHART 3. Patriarchal and Fractal Values

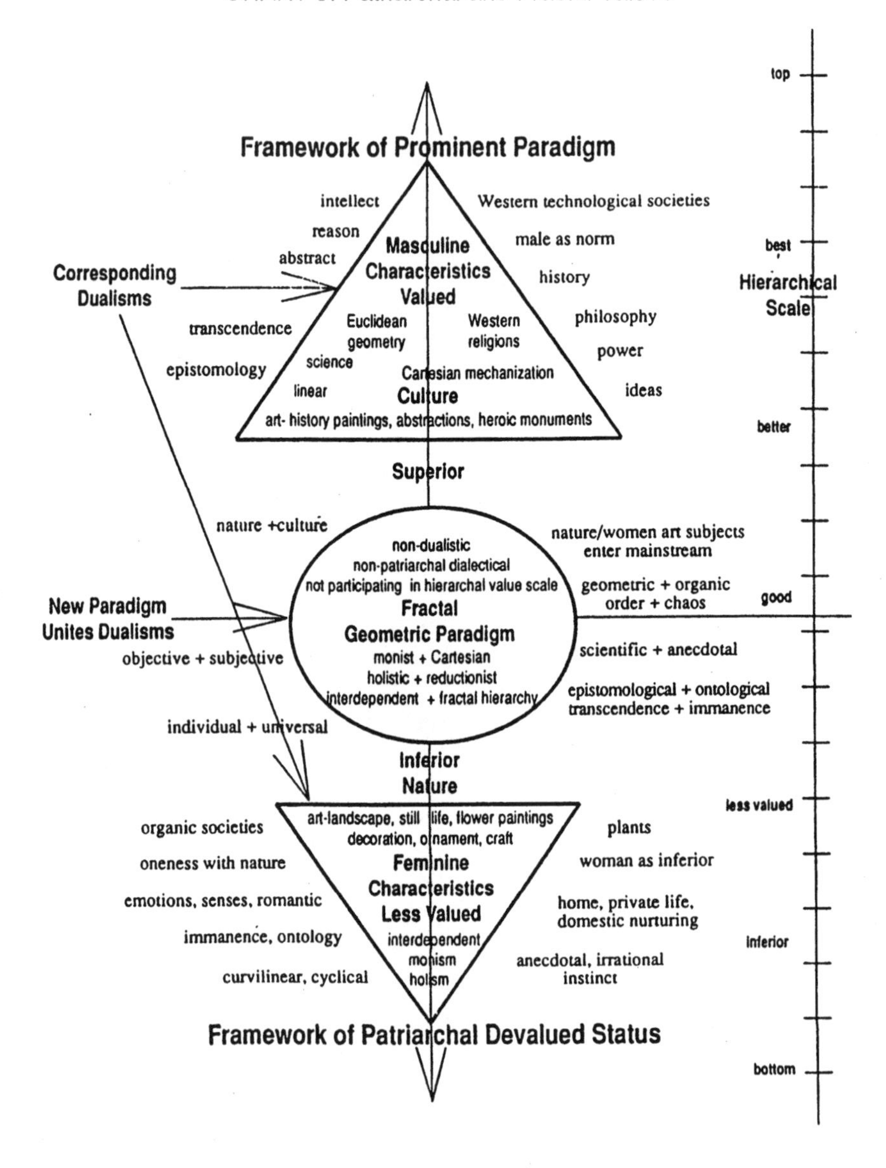

or post-modern aesthetics as a departure from the modernist and Euclidean structured bias in art.

CONCLUSION

Within the new paradigm, geometry is not just culture but nature. In addition to being rational, analytic and abstract, geometry is now intuitive, subjective and material. While it is Euclidean geometry that evokes Cartesian thought, when it's fractal, geometry is describable as a Cartesian *and* Monist view. This revolutionary elimination of dualistic thinking is a major resolution of one of the most basic philosophical conflicts (that of Cartesian vs. Monist thinking) and is in keeping with the post-modern attack on modernist thought, its hierarchies and dualisms.[39]

Large-small, top-bottom, up-down, are the concepts which have ruled in patriarchal hierarchal thinking, creating a distinct "geometry of values." What is valued is large, top and up; what is devalued is small, bottom and down. Before perspective in medieval painting, this hierarchal view was literally translated into use of space. God, Jesus, were depicted as large figures with angels, priests, wealthy patrons of the church, or nobles, all receding in size with peasants being the smallest at the bottom of the image.

Not ending here, perspective, a Renaissance invention, changed from this conceptual view of space based on a "geometry of values" to one of total geometric control. In perspective, all of nature–trees, mountains, humans–were dominated into bending and turning in order to conform to geometric projection lines. Straight lines, foreign and alien to nature, made space "rationalized" with every inch of the painted image completely congruent according to a single rigidly artificial view.

What I am arguing here is that existing geometric models perpetuate and co-exist with social values and are visualized in art. The essential patriarchal hierarchy is a geometric construction where values are distributed according to a set pattern, so internalized by us it is often not seen as a construction but confused as reality.

Geometry can be neutral–a tree has a top and bottom and in this non-hierarchical context, values are irrelevant–it would be absurd to think about whether the top or bottom of the tree is "superior." Unfortunately, Euclidean geometry, through philosophical argument or logical justification, has not been so benign. Values in the hierarchy, dualistic and hierarchical, are externally and internally structured in humans by conventional geometry.

As the semiotic movement believes language creates reality and must be changed to create new values, I believe the geometric paradigm must be changed before we can construct new values which alter the patriarchal system. As non-hierarchial and non-dualistic, the new fractal geometry could be a key factor in the revaluing of plants, women and nature in Western culture.

This new "post-modern geometry" would not only alter society's values but modernist aesthetics as well. Art such as monuments could reflect the "value" of feminine characteristics in addition to the masculine. This new adaptation of values as "holistic" would create a plethora of new artistic possibilities because artists would be able to shed the confines of internalized status-quo geometry which had silently structured their choices. Formerly unexpressed as having status in culture, future use of plants, women and nature images in art can be signs of the dismantling of their devaluation in the patriarchy.

Recently many idea "models" have been offered as solutions to the devaluation of women and/or nature which include the "spiritualizing of nature" and suggestions of not matriarchal but "partnership" social constructions.[40] The popular anthropological revision of pre-historical culture as "feminine" as judged by artifacts is, for some, inspirational but has created no present cultural change. None of the aforementioned models have yet to have significant social impact beyond a "new age" or a feminist audience. It remains frustrating that when "the feminine" is introduced to "history," it continues to be marginalized, never quite integrating into mainstream masculine knowledge or values. Since society has polarized into the singular appreciation of masculine characteristics there is no context for which feminine characteristics can be recognized as valuable. Fractal geometry creates this context within the present paradigm.

I believe it is only a new geometry–something with sacred respect in the patriarchy–which could have the impact, the credibility, to change the structure of "logical" devaluation of plants, women and nature to their "logical" valuation. After all, it was geometry which held "the feminine" in its "place." Dualistically opposite of the "masculine," the "feminine" was geometrically "controlled" as less valued from its position at the "bottom" of the hierarchal construction.

Beyond social systems or religions, mathematics in our patriarchal culture from Pythagoras to Bertrand Russell has been perceived as synonymous with "objective, rational truth." "Fighting fire with fire," fractal geometry holds the potential of revaluing feminine characteristics in the patriarchy because it is a paradigm shift of the geometry fundamental to our basic patriarchal beliefs. As argued in this essay, revaluing the feminine would be of paramount importance not only for plants, women and nature, but the world.

REFERENCES

1. Nicholas D. Kristoff, "Stark Data on Women, 100 Million are Missing," *New York Times*, p. C1 (November 5, 1991).

2. Carolyn Merchant, *The Death of Nature, Women, Ecology and the Scientific Revolution* (San Francisco, CA: Harper & Row, 1980). Judith Plant, ed., *Healing the Wounds: The Promise of Ecofeminism* (Philadelphia, PA., Santa Cruz, CA: New Society Publishers, 1989). Irene Diamond and Gloria Feman Orenstein, *Reweaving the World, The Emergence of Ecofeminism* (San Francisco, CA: Sierra Club Books, 1990).

3. Jennifer Bennett, *Lilies of the Hearth: The Historical Relationship Between Women and Plants* (Camden East, Ontario: Camden House, 1991). I wanted to include this reference even though it differs in focus from this paper. Bennett's excellent discussion is on women, plants, as associated with botany, gardening, horticulture, herbalogy.

4. Norman Bryson, *Looking at the Overlooked, Four Essays on Still Life Painting* (Cambridge, MA: Harvard University Press, 1990). Also Whitney Chadwick, *Women, Art and Society* (London: Thames & Hudson, 1990).

5. Chadwick, n. 5, p. 128.

6. Norma Broude and Mary D. Garrard, eds., *Feminism and Art History, Questioning the Litany* (New York: Harper & Row, 1982).

7. Samir al-Khalil, *The Monument, Art, Vulgarity and Responsibility in Iraq* (Berkeley, Los Angeles, London: University of Chicago Press, 1991).

8. "A Monumental Oversight," staff piece in *Glamour*, p. 110 (May 1991).

9. Bernard Ceysson, Genevieve Bresc-Bautier, Maurizio Fagiolo dell'Arco and Francois Souchal, *Sculpture: The Great Tradition of Sculpture from the Fifteenth to the Eighteenth Century* (New York: Rizzoli International Publications, Inc., 1987).

10. Antoinette Le Normand-Romain, Anne Pingeot, Rheinhold Hohl, Barbara Rose and Jean-Luc Daval, *Sculpture: The Adventure of Modern Sculpture in the Nineteenth and Twentieth Centuries* (New York: Rizzoli International Publications, Inc., 1986).

11. Louise Bourgeois, "Obsession," *Art Forum*, 85-87 (April 1991).

12. Donna Wilshire, "The Uses of Myth, Image and the Female Body in Re-visioning Knowledge." In *Gender/Body/Knowledge Feminist Reconstructions of Being and Knowing*, edited by Alison M. Jaggar and Susan R. Bordo (New Brunswick, NJ: Rutgers University Press, 1989) p. 93. Women to Plato were "recycled souls of inferior cowardly men." Ruth Berman in *Gender/Body/Knowledge*, p. 233. Aristotle calls women "mutilated mates," "emotional," "passive," "captives of their 'body functions' . . . therefore a *lower* species more like animals than men."

13. Henry Kraus, "Eve and Mary: Conflicting Images of Medieval Woman." In *Feminism and Art History, Questioning the Litany*, edited by Norma Broude and Mary D. Garrard (New York: Harper & Row, 1982) p. 80.

14. Adele Getty, *Goddess, Mother of Living Nature* (London: Thames & Hudson, 1990).

15. Murray Bookchin, *The Ecology of Freedom, The Emergence and Dissolution of Hierarchy* (Montreal, New York: Revised edition, Black Rose Books, 1991).
16. Cynthia Eagle Russet, *Sexual Science, The Victorian Construction of Womanhood* (Cambridge, MA: Harvard University Press, 1989).
17. Pamela Berger, *The Goddess Obscured: Transformation of the Grain Protectress from Goddess to Saint* (Boston: Beacon Press, 1985).
18. Donna Wilshire. In Jaggar, n. 12.
19. Roger F. Aldrich and Russ Parsons, "Influences of Passive Experiences with Plants on Individual Well-Being and Health." In *The Role of Horticulture in Human Well-Being and Social Development*, edited by Diane Relf (Portland, OR: Timber Press, 1992).
20. Perhaps this is why the "abortion" issue has raised such powerful social turmoil. In the patriarchal hierarchy the reality of women having complete "choice" of whether or not to procreate is a threat to the patriarchal system. In a patriarchal hierarchy it is a de facto woman's role, *not choice,* to have children. "Giving" women power over their bodies and roles makes them no longer dependent on the patriarchal system, thus a threat. Women choosing to work instead of having children is something the patriarchy could ignore–the termination of pregnancy as a woman's complete prerogative strikes primal fear of women out of patriarchal control and is a frightening reminder of the power of "mother" over us all.
21. Myriam Miedzian, *Boys Will Be Boys, Breaking the Link Between Masculinity and Violence* (New York: Doubleday, 1991).
22. Ibid.
23. Kristoff, n. 1.
24. Bookchin, n. 15. Bookchin discusses the fallacy of idealizing organic societies. The point I am suggesting here is that Western civilization can learn from organic societies, not that they should imitate them. Native American elders have often said that Westerners need to find their own truth, not adopt theirs. Also refer to Steve Wall and Harvey Arden, *Wisdom Keepers: Meetings with Native American Spiritual Elders* (Hillsboro, OR: Beyond Words Publishing, 1990).
25. Women, for the most part, did not have access to college and university education until the mid-nineteenth century. In fact, the trend of society was to discourage it. Russet (note #15) points to the political atmosphere, "the 'Declaration of Sentiments' in 1848 represented a full scale of assault on the status of women, including the legal death (murder, clarification mine) of women in marriage, their exclusion from higher education and the professions. . . ."
26. J. Bronowski, *The Ascent of Man* (Boston: Little, Brown & Company, 1973).
27. Jane Perlez, "Woman's Work is Never Done (Not by Masai Men)," *New York Times*, p. A4 (December 2, 1991).
28. Ynestra King, "Healing the Wounds: Feminism, Ecology and Nature/Culture Dualism." In Jaggar, n. 11.

29. Harry Holtzman and Martin S. James, eds., *The New Art–The New Life: The Collected Writings of Piet Mondrian* (Boston: G.K. Hall, 1986) pp. 8, 32, 54, 56, 73, etc. Mondrian liberally uses the word "capricious" to describe nature throughout his writing.

30. As demonstration of the cultural influence of a specific geometry, see Richard L. Gregory, *Eye and Brain: The Psychology of Seeing* (Princeton, NJ: Princeton University Press, 1990. 4th edition). As opposite of our Western rectilinear culture Gregory writes of the Zulu tribe in Africa as being a " 'circular culture,' their huts are round, they do not plough their land in straight furrows but in curves and few of their possessions have corners or straight lines." Gregory discusses how cultural geometries affect human perception. For example, due to the lack of visual experience of converging parallel lines in their landscapes, Zulus do not perceive the illusion of perspective as we do in the West. In addition, "primitive people make little or nothing of photographs."

31. When I refer to Euclidean geometry in this paper I mean to include all Euclidean and non-Euclidean geometries which are based on integer dimensions.

32. Thomas S. Kuhn, *The Structure of Scientific Revolutions* (Chicago: The Univ. of Chicago Press, 1972). Kuhn writes "led by a new paradigm . . . scientists see new and different things when looking with familiar instruments in places they have looked before." Kuhn likened this transformation–one mode of perception changed to another mode–as a Gestalt-like switching: "what were ducks in the scientists' world before the revolution are rabbits afterwards."

33. Holtzman and James, n. 29, p. 49 and 91. Refer also to p. 77.

34. Benoit B. Mandelbrot, *The Fractal Geometry of Nature* (New York: W.H. Freeman, 1983). John Briggs and F. David Peat, *The Turbulent Mirror* (New York: Harper & Row, 1989).

35. Chart is derived from one by Richard Voss, see Heinz-Otto Peitgen and Dietmar Saupe, *The Science of Fractal Images* (New York: Springer-Verlag, 1988), p. 26.

36. Kuhn, n. 32.

37. All women in this series are "portraits" of actual women (my friends) and are titled after them: Helen Alexander, Virginia Dean (her children Caroline and Andre), Nina Duran, Beth Durif and daughter Olivia, Kiki Kosinski.

38. Peitgen, n. 35. See essay by Richard Voss.

39. In philosophy the dualism between Cartesian thought and Monism has been up till now an unresolvable conflict, for example, the philosophy of Descarte vs. Bruno respectively. The post-modern attack on modernist dualism is discussed in Andreas Huyssen's essay, "Mapping the Post-Modern." In *Feminism/Post-Modernism*, edited by Linda J. Nicholson (London & New York: Routledge, 1990) p. 267.

40. For "the spiritualizing of nature," see Jim Nollman, *Spiritual Ecology: A Guide to Reconnecting with Nature* (New York: Bantam Books, 1990), one of many books available on the subject–in the context of art see Suzi Gablik, *The Re-Enchantment of Art* (London: Thames & Hudson, 1991). For "partnership model" and the anthropological revision of pre-history as "feminine" see Riane Eisler, *The Chalice and The Blade: Our History, Our Future*, (San Francisco: Harper & Row, 1987).

Chapter 4

Consuming a Therapeutic Landscape: A Multicontextual Framework for Assessing the Health Significance of Human-Plant Interactions

Nina L. Etkin

SUMMARY. Since earliest times, humans have used botanicals to affect health, and have brought into cultivation a great diversity of plants. For the large number that have been managed explicitly as medicines, pharmacognosists have uncovered substantial pharmacologic potential. Increasingly, we come to understand that these selfsame plants are used in other contexts as well–as foods and cosmetics, and to meet horticultural objectives. That this extends the range of circumstances through which people are exposed to active plant constituents helps us to comprehend better the complex paradigms through which people interpret their biotic environments. This paper offers a multicontextual framework for assessing the physio-

The author would like to thank Paul Ross for collaboration in all phases of research, and acknowledge the contributions of Ibrahim Muazzamu and Hurumi village. This work was supported by grants from: the NSF (BNS-8703734), Social Science Research Council, Fulbright Senior Research Scholars Program, the Bush Foundation, the University of Minnesota, and the Social Science Research Institute of the University of Hawaii; and by a faculty research appointment in the Department of Pharmacognosy and Drug Development at Ahamdu Bello University, Zaria. She also thanks the Kano State Department of Health and Wudil Local Government.

logic import of plant utilization through attention to the interdependent uses of plants by real populations in specific cultural contexts.

INTRODUCTION

Therapeutic landscapes are settings in which variously construed healthful circumstances are obtained to promote emotional, social, and physiologic health. These environments are natural (gardens, mountain summits) or have been naturalized through interior design of constructed places, and plants figure prominently to emblemize health. At present, most attention is paid to the positive effects of people-plant proxemics via passive/visual and participatory/cultivating interactions (e.g., Fisher 1990; Lewis 1990; Hagedorn 1991; Kaplan and Kaplan 1990; Honey 1991; Thoday and Stoneham 1989). I argue that the substance of place signifies as much for therapy as does the symbol, text, and gestalt of place: what is physically constituted in plants signifies at least as much as what plants mean.

Since earliest times, people have used botanicals to affect health, and have brought into cultivation or otherwise fostered the protection of a great diversity of plants. Many of these have been managed explicitly as medicinals, for which botanists and pharmacognosists have revealed a rich pharmacologic potential. Increasingly, we come to understand that those selfsame medicinals are used in other contexts as well–as foods, cosmetics, and herbicides, and to meet ornamental and other horticultural objectives. I argue that these other uses have health implications as well, over and above whatever may be obtained as an intrinsic benefit (Ulrich 1986, 1991) of the people-plant interaction itself.

The point of this paper specifically is to illustrate the merits of a multicontextual framework that demonstrates how overlapping uses of plants extend the range of circumstances through which people are exposed to active plant constituents. This is best informed by anthropological and other social science inquiry that is responsive to how indigenous peoples comprehend their universe and the influence that it has on their use of plants. My objective is to show how a culturally sensitive perspective allows us to more accurately assess the physiologic import of plant utilization by drawing attention to the interdependent uses of plants by real populations, in specific cultural contexts. Using the example of my own extensive research on plant use among Hausa in Nigeria, I illustrate how we can move ethnographic study beyond a preoccupation with metaphors, at the same time that we raise pharmacologic inquiry above the mere contriving of abstracted inventories of constituents and activities.

I turn first to a brief overview of anthropological studies of human-plant interactions.

BACKGROUND/HISTORY OF ANTHROPOLOGICAL INQUIRY ON PLANTS

Early anthropological inquiry on people-plant interactions tended to focus on core and periphery subsistence plants, with much effort devoted to domesticates. The evolution of agriculture, especially, captured the attention of social scientists who were keen to understand the ways in which past human populations met the challenges of food-getting and the complex transformations that were effected in the transition from gathering and hunting to sedentary agriculture (Farrington and Urry 1985; Harris 1989; Rosenberg 1990; Nabhan and Rea 1987; Stahl 1989). Less concern was accorded other "economic" plants, such as those used in the manufacture of tools and crafts, clothing, fuel, fiber, and residential and other construction (Balandrin et al., 1985; Fleuret 1980; Ford 1978). Attention to medicinals lagged further behind, eclipsed as these species were by the purported superiority of pharmaceuticals and other biomedical treatments. This had the effect, through at least the first half of this century, to foster disinterest in plants and other nonwestern therapeutic media. Still, on balance, the growing number of studies deepened the base of knowledge about plant use generally. Anthropologists, like their colleagues in other disciplines, came to appreciate the extent to which indigenous peoples are astute observers of their natural environments, and that they have intimate knowledge of both local botanic resource potential and the means to ensure its preservation (Oldfield and Alcorn 1991; Posey and Balee 1989; Rhoades 1986; Richards 1985; Warren, Slikkerveer, and Titilola 1989).

From those early beginnings developed more systematic efforts to identify useful plants and to describe in culturally relativistic terms their diverse applications by human populations. Fueled by the concern to move the study of indigenous medicine beyond exercises in the obscure, anthropologists engaged more rigorous ethnographic inquiry. This affirmed the more mundane aspects of day-to-day existence of indigenous peoples, at the same time that it elaborated the intellectual basis for scrutiny of the exotic, which had become the trademark of cultural anthropological study. From these efforts, we witness the development of a sizeable literature in ethnobiology and ethnomedicine (e.g., Balee and Daly 1990; Berlin 1990; Nabhan and Rea 1987; Stoffle, Evans, and Olmsted 1990). Embellished by anthropological linguistics, those studies have fostered a comprehension of how local populations structure their knowledge of the biological universe. They relate lexical and taxonomic categories to the cultural significance of plants, and provide a basis from which to analyze the structural principles of indigenous biological classifications cross-culturally. And the rich ethnography developed through long-term residential, and longitudinal, field study of the contexts in which indigenous medicines are engaged has helped us to understand how local knowledge of disease causation shapes preventive and therapeutic modalities.

Improved methodology notwithstanding, these studies remained flawed by the tendency in anthropology to treat plants only as cultural objects, in such a way that preoccupation with symbolic import excluded from serious consideration the biological activities of those plants: what medicines mean took precedence over how they affect the physical body. We have come to appreciate through more recent studies that a problem with such cultural models is that they obscure that the tangible actions of plants may be more directly related to the semiotics of medicines than are mere metonymic or other, arbitrary devices. In other words, great symbolic import is invested in particular plants precisely because they reduce fever or alleviate pain. That those plants are linked also to powerful spirits or other metaphors of healing may derive only secondarily from their physiologic action. Thus, encouraging a less culturally loaded view, a leading scholar in this field contends that the relationships between humans and plants are at base interactions between people and phytochemicals (Johns 1990).

Some ethnographers went further to accommodate the pragmatics of medicinal plant use and tried to link local vernaculars to botanical taxonomies. Whereas that had the potential to facilitate cross-cultural interpretations through the lingua franca of botanical systematics, in fact few of those researchers troubled to collect voucher specimens and relied instead on published taxonomies (e.g., Biggs 1985; Chambers and Chambers 1985; Weiner 1978). This ignores the dynamic nature of evolving landscapes and peoples' interpretations of them. Although the lenses through which people filter information about their environment may in some sense be "traditional," they are never immutable. Because names and plants change over time and space, reliance on previous identifications leads to inaccuracies, and makes it impossible to engage cross-cultural comparisons or to make reference to biologic activity which is recorded and indexed in the literature according to taxonomic designates (genus and species).

Regrettably, much of what began as more exacting inquiry into plant use deteriorated into disembodied inventories that lacked such behavioral details as selection, preparation, and therapeutic or nutritive strategies. These catalogues do more to decontextualize and homogenize people's experience with botanicals than they help to explicate preventive and therapeutic objectives, or to provide means for exploring them further through pharmacologic and other biomedical paradigms. Similarly, in the hands of phytochemists, generic constituent or activity screens that are conducted without regard for the therapeutic strategies employed *in situ* (e.g., Sukumaran and Kuttan 1991; Alice et al. 1991), do not, despite their own intrinsic merits, link bioscientific research to indigenous empirical knowledge.

Above all, regardless of the quality of data or their interpretation, inquiry into plant utilization has remained largely a *single-context* issue: study of medicinal, dietary, or ornamental plants is shaped primarily by the disciplinary affiliation of the researcher. For example, studies of indigenous medicine are preoccupied with matching therapy to symptom and typically fail to elicit additional contexts of use; dietary surveys measure plants consumed during "official" meal events and not otherwise; and plants used in adornment, leisure, and manufacture are not problematized beyond those applications to consider, for example, potential physiologic effects.

Contemporary Anthropological Inquiry. At present, anthropological interest in human-plant interactions has been refined through more elaborate theoretical formulations and methodologic initiatives. Specifically, effort is made to link the biological and behavioral antecedents of plant utilization with physiologic and social outcomes. That more sophisticated anthropological dialogue about people and plants is one voice in the coalescence of social science and pharmacologic concerns in an interdisciplinary study labeled *ethno*pharmacology (Etkin 1988, 1990; Rivier and Bruhn 1979; Holmstedt 1991; Holmstedt and Bruhn 1983). This prefixed study of drug chemistry is distinguished by comparative, cross-cultural analysis of indigenous pharmacopoeias developed by peoples whose understandings of health and illness are shaped by worldviews that are not altogether consonant with the paradigms of western science.[1] To this, anthropology brings a unique biobehavioral perspective that accords primacy to indigenous paradigms of health and therapeutics, at the same time that it underscores the intertexture of biology and culture by linking therapeutic behaviors to such external paradigms as botanical systematics, phytochemistry, and pharmacology. But here too, technical rigor and "thick" ethnography notwithstanding, the legacy of *single-context* approaches endures. Such atomized concerns as medicine-only, diet-only, etc., configure restrictive methodologies that cannot elicit the range of applications for a single plant. We are, thus, offered only a fragmentary–and thus inaccurate–depiction of people's experience with a particular species, one that is inadequate to the task of assessing the pharmacologic implications of that plant.

To redress those shortcomings, I underscore once more the significance of *multiple contexts* of plant use. If we are concerned with how constituents of a plant affect physiologic function, we need to be mindful of each circumstance of ingestion, application, and other intimate contact with that plant. This is illustrated by the following discussion which takes us from abstract principles to the specific example of my research in northern Nigeria.

MULTICONTEXTUAL PLANT USE BY HAUSA IN NIGERIA

Research Setting and Methodology

Since 1975 I have sustained, with the collaboration of Paul Ross, a comprehensive investigation of medicine, diet, and health among rural Hausa in northern Nigeria (e.g., Etkin 1979, 1980, 1981; Etkin and Ross 1982, 1991a). The general ecology of the region is a Sudanian undifferentiated woodland with savanna type vegetation of scattered shrubs and small trees, once interspersed by grass cover. At present, the natural vegetation is highly disturbed, following generations of intensive small-plot cultivation and livestock grazing, and landscape transformations of more recent origin–resulting from road construction and other extensions of infrastructure–that are related to the integration of this region into a broader economy. Yet, despite these pressures toward biotic simplification, Hausa have opportunity to select from among a large number of local species for their various needs.

Whereas our research originally centered on indigenous medicines, comprehensive interviews and observations of plant utilization drew our attention immediately to other contexts of use that increase the likelihood of human exposure to pharmacologically active constituents. In this way, we discerned the other-than-medicinal role of plants as foods, cosmetics, items of personal or environmental hygiene, objects of manufacture and construction, and the mediators of spirits, sorcery, and witchcraft. Our interest in what bearing these plants have on physiologic function refined the focus to plant uses that include ingestion, inhalation, application to skin or mucosa, and other occasions in which body tissues come into more than casual contact with plants.[2]

Because food is the use category for which there is greatest overlap with medicines, we systematized that inquiry during each field study through year-long dietary surveys conducted simultaneously with our investigations of local therapeutics (for details of methodology consult Etkin, Ross and Muazzamu 1990; Etkin and Ross 1991a; Ross, Etkin, and Muazzamu 1991).

The other categories–cosmetics and hygiene–typically are ignored by researchers who are more concerned with so-called "primary" function. But since these other uses constitute regular and repeated exposure to plant constituents, pharmacologic assessments must argue in favor of their inclusion as well. The point is that, collectively, these different contexts of use account for considerably more contact with pharmacologically active plants than any single application.

Let us examine each of these use categories in turn.

Plant Medicines

Preventive and therapeutic strategies in this Hausa locality are dominated by an herbal pharmacopoeia–this despite the dramatic incursion of biomedicine over the last decade and the marked increase in availability and widespread use of pharmaceuticals (Etkin, Ross, and Muazzamu 1990). Extensive interviews combined with observations of therapeutic and preventive events elicited the details of plant selection, preparation, application, and desired outcome. A representative, village-wide interview schedule supplemented practitioner- and specialist-focused studies to provide a broader (exoteric rather than esoteric) base for compiling the knowledge that more accurately represents preventive and therapeutic modalities that affect the majority of villagers on a day-to-day basis.

The findings reveal 3165 distinct (nonreplicate) medicines. Table 1 lists the more common of the 211 categories for which these preparations are administered. Of those medicines, 89% (2801) include a total of 374 plants. Three hundred forty-five of all plants are used for diseases that I have termed "overtly physiologic" (Table 2). These are illnesses whose etiology Hausa attribute to naturalistic phenomena such as dirt or excessive cold, and whose prevention and treatment range among dislodgement of disease substance, neutralization of cold or corollary characteristics, palliation of pain or discomfort, and finally symptom resolution.

Heeding the very inclusive Hausa definition of medicine–*magani:* literally anything that affects, influences, or repairs something–necessitates attention also to plants used in the management of disorders attributed to witchcraft or the intervention of extrahuman agency. Two hundred sixty-six of all plants are so used for the mediation of sorcery, spirits, and

TABLE 1. Most Common Categories for Hausa Medicinal Preparations

HAUSA NOSOLOGY	GENERAL ANALOGUE	n medicinal preparations
ISKA	Spirit caused illness	104
SAMMU	Sorcery	103
CIWON CIKI	General GI disorders	94
CIWON CIKIN KABA	Acute GI disorders	90
SHAWARA	Hepatitis, malaria	72
KYANDA	Measles	70
MAYE	Witch caused illness	66
ZAZZABI	Febrile disorders	64
DANSHI	Pediatric malaria	59
MAYANKWANIYA	"Wasting," anemias	59

witches, which is accomplished largely through preventive measures but can be treated after the fact as well. To the extent that these intercessions include ingestion, topical application, fumigation, inhalation, and other close contact with plant constituents, these contexts of use also have physiologic implications. I do not propose that the management of spirits, witches, and sorcery entails a unique constellation of plants, for it does not. I problematize the issue here precisely because most researchers have not concerned themselves with plant applications whose rationale and objectives fall outside the context of disorders of naturalistic origin. Further, Hausa exposure to constituents of these plants increases dramatically because they actively engage these imperceptible threats to well-being on at least a daily basis, and not just when the fact of illness is established.[3]

With medicine thus writ large, my first consideration of multicontextual use directs attention to plant medicines used also in diet.

Plants in Medicine and Diet

Food plants overlap most prominently with medicines (Etkin 1986; Etkin and Ross 1991b).[4] Indeed, all but 5 of the 119 plants (96%) that this

TABLE 2. Plant Medicines (1987-88)

INTENDED USE	"SEMI-WILD" PLANTS		ALL PLANTS	
	n plants	n remedies	n plants	n remedies
Overtly physiologic	254	1854	345	2275
Other	215	452	266	526
TOTALS	272[a]	2306	374[a]	2801[b]

[a] totals are less than the sums due to multiple plant uses.
[b] excluded are 5 remedies that specify plant location or situation rather than species.

population identify as foods are included among the master list of 374 medicinals. This conflated category of "medicinal-foods" should not be construed to mean that for Hausa the therapeutic properties of plants guide dietary decisions. To the contrary, medicines and foods are conceptually distinct: in a preventive or therapeutic context, a plant is explicitly "non-food" and is ingested alone or with other plants as part of a composite medicine, or added to the food of the ailing individual. The same plant in a dietary context is explicitly "nonmedicine"; it is regarded instead for its nutritive value and, as part of a meal, is consumed by all household members who "eat from same pot." Parts used and preparation also covary with culinary and medicinal objectives.

To distinguish further, Hausa knowledge of the health potential of medicinal plants is highly elaborated relative to interpretations of the healthful qualities of foods. These tend to be ambiguous and limited to such assessments as "this food strengthens the body," or "fortifies the blood." This is very different from the embellished discourse that informs the medicinal uses of these same plants. In therapeutic context, a plant is understood to dislodge phlegm, promote disease egress, change the locus of complaint, and so on. For foods, criteria applied in their evaluation are palatability, satiety, texture, and the like–measures that are not central to the selection of medicinals.

The principal dietary cultigens in this area are sorghum, millet, groundnut, and cowpea. While these are of course not chemically inert, they are less likely than noncultigens to manifest significant pharmacologic activity. That Hausa recognize the relative neutrality of these plants is reflected in their being regarded as generically nourishing, and their use only as vehicles for the administration of medicines rather than as explicitly therapeutic in their own regard. In view of this, my assessment of the pharmacologic potential of food-also plants use was suitably narrowed to "semi-wild"[5] plants: plants which are not deliberately cultivated but which are nonetheless influenced by human actions–as, for example, when they are intentionally overlooked during weeding. Hausa use 254 "semi-wild" plants for "overtly physiologic" disorders. Whereas these "semi-wild" food-plants contribute relatively little of the standard dietary measures such as calories and protein, they are important sources of micronutrients and embody a range of pharmacologic activities,[6] a point to which I will return below.

The use categories "cosmetics" and "personal hygiene" are discussed next.

Cosmetic Plants

Among cosmetic applications, the adornment of eyes and coloring of skin are routine. Red is the color of choice for hands, feet, forearm, and lower leg; and for the lips and teeth. Blackening agents are applied to accentuate the eyes, and to highlight designs incised or drawn on the face and cicatrized onto the chest, upper arms, and back. The only noncoloring agent among plants in Table 3 is *Arachis* (groundnut), the oil of which is used several times each day to impart a sheen to the skin. Applications to the eyes, face, and mouth typically are reproduced daily; the others are renewed as they fade, after a week or so; especially elaborate adornment is the convention for ceremonies. Whereas women typically engage cosmetics and include their daughters at an early age, men and boys seldom use such items, and then usually in contexts that are difficult to separate from medicinal efforts. Table 3 illustrates the extent of overlap between the cosmetics category and others. Twenty species comprise this group, all of which have medicinal uses; 5 are food plants, too; and 3 are used also in hygiene.

Plants Used in Personal Hygiene

Plants that Hausa employ for personal hygiene are divided among the daily routines that center on bathing the body, attending to teeth and mouth, and hair treatments. Locally manufactured soaps and cleansing oils are confected with plants. Men and women typically bathe twice per day. Soaps and oils cleanse the skin; rinses are used for the mouth and teeth; and chewing sticks represent especially prolonged contact with the oral

TABLE 3. Hausa Plants Used for Cosmetics: Overlapping Contexts of Use

Genus species	Hausa	Med	Food	Cos	Hyg
Acacia nilotica (L) Willd ex Del	GABARUWA	x		x	
Acacia nilotica (L) Willd ex Del var tomentosa (Benth) AF Hill	GABARUWA	x		x	
Anacardium occidentale L	KANJU	x	x	x	
Arachis hypogaea L	MAN GYADA	x	x	x	x
Argemone mexicanca L	KWARKO	x		x	
Bombax buonopozense P Beauv	GURJIYA	x		x	
Cola nitida (Vent) Schott & Endl	GORO	x		x	
Cola acuminata (Pal) Schott & Endl	GORO	x		x	
Commiphora africana (A Rich) Engl	DASHI	x	x	x	x
Datura innoxia Mill	ZAKAMI	x		x	
Datura metel	ZAKAMI	x		x	
Diospyros mespiliformis Hochst	KANYA	x	x	x	
Feretia canthioides Hiern	KURUKURU	x		x	
Feretia apodanthera Del	KURUKURU	x		x	
Indigofera arrecta Hochst ex AR	BABA	x	x	x	x
Lawsonia inermis L	LALLE	x		x	
Nicotiana tabacum L	TABA	x		x	
Portulaca oleracea L	DABURIN SANIYA	x		x	
Thelepogon elegans Roth	LADANBALI	x		x	
Trianthema portulacastrum L	DABURIN SANIYA	x		x	

mucosa as they are retained in the mouth long after purposeful, active tooth brushing has ceased. Hair treatments apply only to women and girls since the conventions of personal hygiene for men and boys dictate shaving the head, usually weekly. Items used for the hair include soap, oils, and various dressings that assist plaiting. Of the 16 plants in Table 4, all are used as medicines, 6 are used in diet, and (as noted above) 3 have cosmetic applications.

Pharmacologic Implications of Plants Used in One or More Contexts

This discussion of only four categories of use illustrates the extent to which plants are used multicontextually. Increasing the number of use classifications would expand the various inter-category permutations exponentially. While this gives us some appreciation for the extent to which people come into contact with plants, still we cannot fully comprehend the health implications of such overlap without attention to how exposure to those plants affects physiologic function.

From a strict pharmacologic perspective, whether or not people consider that they are engaged in therapy does not signify so much as does the fact of contact with active botanicals, for whatever reason. This is not to

TABLE 4. Hausa Plants Used for Personal Hygiene: Overlapping Contexts of Use

Genus species	Hausa	Med	Food	Cos	Hyg
Anogeissus leiocarpus (DC) Guill & Perr	MARKE	x			x
Arachis hypogaea L	MAN GYADA	x	x	x	x
Azadirachta indica A Juss	DARBEJIYA	x			x
Boerhavia diffusa L	GADON MACIJI	x			x
Boerhavia repens L	GADON MACIJI	x			x
Commiphora africana (A Rich) Engl	DASHI	x	x	x	x
Euphorbia lateriflora Schum & Thonn	BI DA SARTSE	x			x
Euphorbia balsamifera Aiton	AIYARA	x			x
Glossonema nubicum Decne	TATARIDA	x			x
Glossonema boveanum Decne	TATARIDA	x			x
Indigofera arrecta Hochet ex AR	BABA	x	x	x	x
Khaya senegalensis (Desr) A Juss	MADACI	x			x
Salvadora persica L	SHIWAKA	x	x		x
Vernonia colorata Drake	SHIWAKA	x	x		x
Vernonia amygdalina Del	SHIWAKA	x	x		x
(Unidentified)	SABULIN SALO	x			x

devalue local constructs of context, but instead to augment that ethnography with what we can learn by applying basic physiologic and pharmacologic principles. In this light, the significance of any pharmacologic action is elevated at least on the strength of the multicontextual use of plants. Beyond that, how plants are prepared, which parts are used, and how they are combined further confound the interpretation of pharmacologic potential. To address that adequately, we would need to embrace the vagaries of variable activity per plant part, growing location, and season; the influence of dilution, heating, and other preparatory modes on activity; and the potential antagonism or potentiation that might obtain between constituents of the same plant or of different plants. How should we make sense of all this short of full pharmacologic appraisal, which is beyond the scope of this paper? One way to deconstruct that complexity is to focus attention on a single class of activity, or on a specific disorder.

To illustrate inquiry through a *generic activity category*, I cast the plants of the present discussion in the light of antimicrobial action. If the pharmacologic and botanic literatures are an indication, broad spectrum screens for antimicrobial action are a popular vehicle for assessing medicinal plants (Rios, Recio, and Villar 1988). Implicit in such studies is that among populations who rely most on plant medicines, infectious diseases figure prominently in morbidity and mortality. Still, antimicrobial assessments typically are conducted without regard for the specific objectives of people on whose advice the plants were collected. Catalogues of plants tested against arbitrarily selected pathogens go only some distance toward helping us comprehend the outcomes of plant therapy. It signifies, for example, whether an infection is integumentary (e.g., leprosy, scabies, measles, streptococcal skin infection); upper respiratory (e.g., diphtheria, streptococcosis, sinusitis); parenchymal respiratory (e.g., tuberculosis, bacterial pneumonias, aspergillosis); renal (e.g., pyelonephritis); digestive tract (e.g., cholera, amebiasis, typhoid); digestive gland (e.g., hepatitis, enteritis, yellow fever); or skeletal (e.g., osteomyelitis). And the problem is compounded by disregarding contexts of use other than therapeutics; this results in researchers projecting, in error, only a small likelihood of therapeutic success in view of low titers of antimicrobial constituents. As limitations of time (space) preclude full treatment of the plants under consideration, I will restrict further comments to plants that overlap the contexts of use considered here. Not only is this more manageable, but also it more aptly illustrates my point that one cannot properly assess the implication of plant use unless all circumstances of exposure are explored.

My focus now narrows to the 8 plants used in 3 or more of the contexts medicine, diet, cosmetics, and hygiene (Table 5). Of those, 7 contain antimicrobial principles.[7] The one that does not is groundnut (*Arachis hypogaea*) which, Hausa argue, is more a vehicle than a primary actor in therapeutics. Four of these plants are used in oral hygiene as chewing sticks and medicinally to treat gum disorders: *Commiphora africana* (A Rich) Engl, *Salvadora persica* L., *Vernonia amygdalina* Del., *V. colorata* Drake. For the first three of these, pathogens against which they are effective are localized to the oral tissues (mixed saliva microorganisms, streptococci, lactobacilli). The fourth is effective against pathogens of the lower gastrointestinal tract (amoebae and helminths), suggesting that its efficacy in maintaining dental health may be related more to mechanical abrasion

TABLE 5. Hausa Plants Used in at Least 3 Contexts

Genus species	Hausa	Med	Food	Cos	Hyg
Anacardium occidentale L	KANJU	x	x	x	
Arachis hypogaea L	MAN GYADA	x	x	x	x
Commiphora africana (A Rich) Engl	DASHI	x	x	x	x
Diospyros mespiliformis Hochst	KANYA	x	x	x	
Indigofera arrecta Hochst ex AR	BABA	x	x	x	x
Salvadora persica L	SHIWAKA	x	x		x
Vernonia colorata Drake	SHIWAKA	x	x		x
Vernonia amygdalina Del	SHIWAKA	x	x		x

which physically dislodges cariogenic microorganisms. Of the five plants used cosmetically, all but groundnut are bactericidal to staphylococci and streptococci that colonize the skin and that would otherwise be a more important risk in cosmetic cicatrization. Similarly, conjunctivitis of streptococcal and other origin may be prevented or improved with the cosmetic application of these plants.

While I have not presented a full valuation of these plants across their ranges of activity and use, this documentation better links pharmacologic action to indigenous medical strategies and outcome than do the more extensive catalogues that pay scant attention to local therapeutic objectives and that ignore multiple contexts of use.

In addition to working through generic activity categories as I have just illustrated, another strategy is to focus on a *single disease*. As an example, I report my analysis of Hausa fever medicines (Etkin 1979, 1981; Etkin and Eaton 1990; Etkin and Ross 1991a). I refined a larger sample of 126 by applying the criterion of multicontextual use, and selected 82 species that appear in diet as well, thereby increasing people's exposure to any intrinsic benefit. Twenty-three of those plants (29%) demonstrate antimalarial activity against cultured *Plasmodium falciparum* (Etkin and Eaton 1990; Iwalewa et al. 1990) (Table 6). Similarly, for a symptom constellation labelled here "gastrointestinal disorders" (Table 7), we found that of 53 plants used both as food and medicine, a significant number demonstrate properties that relieve a variety of GI complaints: 66% have astringent action, 43% contain bitter principles, 64% contain gum resins and mucilages, 45% and 32% respectively contain fixed and volatile oils, and 64% of the plants exhibit specific action against both micro- (bacteria, fungi, amoebae) and macroparasites (helminths) (Etkin and Ross 1982).

In both the fever- and GI-centered examples, the percentages expressed reflect the overlapping categories medicine and food. If I project these plants against the same multiple framework I constructed earlier in this paper, the overarching medicine-food category again subsumes other contexts of use: notably, the fever-food plants significantly overlap the coloring agents for plaited mats and leather crafts; and the GI-food plants include a subset of species commonly manipulated (including ingestion) to assure farming success. Logically, the inclusion of other categories of use will document even more extensive exposure to therapeutically beneficial (and other) plant constituents.

CONCLUSION

This multicontextual framework tells us much more about the potential physiologic import of plants than any single-category study. Whereas

TABLE 6. Activity of Plant Extracts Against Cultured *Plasmodium falciparum*

Genus species	Hausa	(Part)	I	II NaCl	II Chlor	III
Acacia nilotica (L) Willd ex Del	Gabaruwa	(root)	+	+	−	+
Agelanthus dodoneifolius (DC)	Kauci	(whole)	+	−	−	+
Artemisia maciverae Hutch & Dalz	Tazargade	(whole)	+	−	+	−
Cassia occidentalis L	Majamfari	(root)	+	−	+	+
Cassia occidentalis L	Majamfari	(leaves)	+	+	−	+
Centaurea perrottetii DC	Dayi	(whole)	+	−	+	−
Chrozophora senegalensis (Lam)	Damagi	(whole)	+	+	−	+
Cyperus articulatus L.	Kajiji	(root)	+	−	+	+
Diospyros mespiliformis Hochst	Kanya	(leaves)	+	−	−	+
Erythrina senegalensis DC	Minjirya	(root)	+	−	+	−
Feretia apodanthera Del	Kurukuru	(leaves)	+	−	−	+
Ficus ingens (Miq) Miq	Kawari	(bark)	+	−	−	−
Ficus polita Vahl	Durumi	(leaves)	+	+	−	+
Ficus ovata Vahl	Cediya	(bark)	+	−	+	+
Momordica balsamina L	Garahunu	(whole)	+	−	−	−
Piper guineense Schum & Thonn	Masoro	(fruit)	+	+	−	−
Psidium guajava L.	Gwaiba	(leaves)	+	+	−	+
Securinega virosa Baill	Tsa	(leaves)	+	+	−	+
Sorghum spp.	Dawa	(root)	+	−	−	−
Syzgium aromaticum (L) Merr&Perry	Kanumfari	(clove)	+	+	+	+
Terminalia avicennioides G&P	Baushe	(leaves)	+	+	−	+
Thonningea sanguinea Vahl	Kulla	(root)	+	+	+	+
Xylopia aethiopica (Dunal) A Rich	Kimba	(fruit)	+	−	+	−
Zingiber officinale Roscoe	CittarAho	(rhizome)	+	−	+	+

TABLE 7. Fifty-Three Plants Used to Treat Gastrointestinal Disorders and in Diet

Action/Constituent	%
Astringent	66%
Bitter Principles	43%
Gum Resins & Mucilage	64%
Fixed Oils	45%
Volatile Oils	32%
Antimicrobials	64%

disciplinary boundaries typically prevail and confine most studies to one purportedly "primary" application, I have demonstrated here that appreciation for overlapping plant uses helps us to comprehend better the complex paradigms through which humans interpret, and in turn shape, their biotic environments, and what impact that has on their health. The therapeutic landscape can be assessed along complex longitudinal and horizontal continua.

Whereas I have inflected this discussion toward an adaptive mien by focusing on beneficial plant constituents, I note too that one can apply this multicontextual perspective to examine toxicity and other untoward effects of plants, since these too are fundamental concerns for assessment of health benefit/risk (e.g., Abebe 1992; Airaksinen et al., 1986; Aresculeratne, Gunatilaka, and Panabokke 1985; Bye and Dutton 1991; Shah, Qureshi, and Ageel 1991; Nyazema 1986; Spencer et al. 1987).

In the Hausa case, the pharmacologic implications of overlapping categories of use that entail medicine, food, hygiene, and to a lesser extent cosmetics are salient for the population generally. One can refine analysis even further by isolating these and other categories of use that might covary along such axes as gender, age, social status, and occupation. For example, since farming, leathercraft, and weaving are largely the domain of men, intimate and regular contact with plants used in the application of insecticides and dyes would more likely affect men than women. Conversely, more salient for women are plants used in the manufacture and decoration of plaited mats, for cleaning cooking utensils, and for some veterinary objectives.

As a final statement regarding the importance of multicontextual frameworks for the assessment of human-plant interactions, I wish to tie discussion to more global issues that center on environmental protection. Concern to preserve *biodiversity* has entered the discourse of conservation efforts largely at the macro level, outside of specific social contexts and without full appreciation for what such diversity affords in human cultural terms. Conservation strategies center on the purportedly "interesting" and "important" species which have been identified through economic and political agendas and through western scientific paradigms, rather than through local cognitive categories. Where species preservation has been promoted for potential sources of natural products, such as plant-based medicines, these are regarded virtually exclusively for their contribution to the pharmaceutical industry, disregarding their significance in indigenous therapeutics or otherwise. As I have demonstrated here, attention to multiple contexts of use increases the significance of individual plants exponentially. We should embellish our analytic framework with specific data in order to appeal to "development planners" for more cogent and culturally germane programs that seek not only to understand the importance of plants from local perspectives but also to comprehend that importance across multiple contexts of use.

NOTES

1. This does not imply that biomedicine shares no common features with other medical systems (regarding etiology, prevention, etc.) nor does it construe biomedicine as a monolith always and everywhere the same.

2. Like most peoples who practice intensive cultivation and whose use of land is governed by the restraints of limited land availability, Hausa do not engage in ornamental horticulture. But there exists an explicit *aesthetic* about the appearance of cultivated plots with regard to such features as precision of line, balance and texture, and how fastidiously weeding and other tasks of maintenance are conducted. Those sensitivities apply especially to complex, intercropped farms which include alternating furrows and ridges of groundnuts, cowpea, and sorghum or millet; bordered by guinea grasses, semi-cultigens, and cultivars such as hibiscus and henna; and interspersed with opportunistic clutches of plants that are deliberately not removed during weeding in view of medicinal, minor culinary, and other uses. One could argue that some nonspecific health benefit derives from this aesthetic aspect of people-plant interactions as well (Fisher 1990; Kaplan and Kaplan 1990; Thoday and Stoneham 1989; Ulrich 1986, 1991).

3. This is the weft of a fabric of social control that pervades all interrelationships. There is a constant, at times palpable, concern that malevolence born of spurned affection or envy will invite the disaffected to engage in sorcery (a learned magic) against one's person, family, or property. So too, one desires protection against witches (people who have inherited supernatural powers, typically to cause harm to others), whose potency is especially insidious since one is never sure of the identity of witches. And the malfeasance wrought by spirits is legion in a world where there is persistent tension to appease one's personal spirits and to deflect harm from others. This oversimplifies, of course, but the point is that these imperceptible entities are actively engaged as a matter of course in the daily goings on of Hausaland.

4. We have argued elsewhere (Etkin and Ross 1991c) that for this population the medicinal uses of plants likely preceded, and later informed, the selection of these species for food.

5. We have discussed at length elsewhere the medicinal and nutritional significance of purportedly "wild" plants for Hausa and other populations (Etkin and Ross 1991c).

6. The relative lack of human manipulation of these plants, compared to full domesticates, affords them higher titers of allelochemicals which serve the plant as protectants against herbivory and as allelopaths which delimit growth of other plants near them. These properties do not go unnoticed by humans who have ap-

propriated those chemical actions for medicinal and other purposes. These observations are further elaborated by the details of preparation whereby certain patterns of cutting or cooking may diminish only partially what are considered to be toxic allelochemicals.

7. Data contained herein regarding antimicrobial and other plant activities were compiled through NAPRALERT (Natural Products Alert Data Base), a computerized natural products data base that contains extensive information on the pharmacology, ethnobotany, and chemical constituents of plant parts and extracts. This is maintained in collaboration with the World Health Organization by Dr. Norman Farnsworth, University of Illinois College of Pharmacy.

REFERENCES

Abebe, W. 1992. Adverse effects of traditional drug preparations. *Jour. Ethnopharm.* 36:93-94.

Airaksinen, M., M. Peura, P. Ala-Fossi-Salokangas, S. Antere, J. Lukkarinen, M. Saikkonen, and F. Stenback. 1986. Toxicity of plant material used as emergency food during famines in Finland. *Jour. Ethnopharm.* 18:273-96.

Alice, C.B., V.M.F. Vargas, G.A.A.B. Silva, N.C.S. de Siqueira, E.E. Schapoval, J. Gleye, J.A.P. Henriques, and A.T. Henriques. 1991. Screening of plants used in south Brazilian folk medicine. *Jour. Ethnopharm.* 35:165-71.

Aresculeratne, S.N., A.A. L. Gunatilaka, and R.G. Panabokke. 1985. Studies on medicinal plants of Sri Lanka. Part 14: Toxicity of some traditional medicinal herbs. *Jour. Ethnopharm.* 13:323-35.

Balandrin, M.F., J.A. Klocke, E.S. Wurtele, and W.H. Bollinger. 1985. Natural plant chemicals: sources of industrial and medicinal materials. *Science* 228:1154-60.

Balee, W., and D.C. Daly. 1990. Resin classification by the Ka'apor Indians. *Adv. Econ. Bot.* 8:24-34.

Berlin, B. 1990. The chicken and the egg-head revisited: further evidence for the intellectualist bases of ethnobiological classification. In *Ethnobiology: Implications and Applications.* ed. D.A. Posey and W.L. Overal, pp. 19-33. Belem, Brazil: Museu Paraense Emilio Goeldi.

Biggs, B. 1985. Contemporary healing practices in East Futuna. In *Healing Practices in the South Pacific.* ed. C. Parsons, pp. 108-28. Honolulu: Institute for Polynesian Studies, Brigham Young University-Hawaii.

Bye, S.N., and M.F. Dutton. 1991. The inappropriate use of traditional medicines in South America. *Jour. Ethnopharm.* 43:253-59.

Chambers, A., and K.S. Chambers. 1985. Illness and healing in Nanumea, Tuvalu. In *Healing Practices in the South Pacific.* ed. C. Parsons, pp. 16-50. Honolulu: Institute for Polynesian Studies, Brigham Young University-Hawaii.

Etkin, N.L. 1979. Indigenous Medicine among the Hausa of Northern Nigeria: Laboratory Evaluation for Potential Therapeutic Efficacy of Antimalarial Plant Medicinals. *Med. Anthropol.* 3(4):401-429.

______ 1980. Indigenous Medicine in Northern Nigeria. I. Oral Hygiene and Medical Treatment. *Jour. Prev. Dentistry* 6:143-149.

______ 1981. A Hausa Herbal Pharmacopoeia: Biomedical Evaluation of Commonly Used Plant Medicines. *Jour. Ethnopharm.* 4:75-98.

______ 1986. Multidisciplinary Perspectives in the Evaluation of Plants Used in Indigenous Medicines and Diet. In *Plants in Indigenous Medicine and Diet: Biobehavioral Approaches.* ed. N.L. Etkin, pp. 2-29. Gordon and Breach Science Publishers (Redgrave). New York.

______ 1988. Ethnopharmacology. *Ann. Rev. Anthropol.* 17:23-42.

______ 1990. Biological and Behavioral Perspectives in the Study of Indigenous Medicines. In *Medical Anthropology: Contemporary Theory and Method.* ed. T.M. Johnson and C. Sargent, pp. 149-58. New York: Praeger.

Etkin, N.L., and J.W. Eaton. 1990. Antimalarial Activity of Plant Medicines against Cultured *Plasmodium falciparum.* Unpublished Research Results.

Etkin, N.L., and P.J. Ross. 1982. Food as Medicine and Medicine as Food: An Adaptive Framework for the Interpretation of Plant Utilization among the Hausa of Northern Nigeria. *Soc. Sci. Med.* 16:1559-1573.

______ and ______ 1991a. Recasting Malaria, Medicine, and Meals: A Perspective on Disease Adaptation. In *The Anthropology of Medicine.* 2nd edition. L. Romanucci-Ross, D.E. Moerman, and L.R. Tancredi, eds. pp. 230-258. New York: Praeger.

______ and ______ 1991b. Should we set a place for diet in ethnopharmacology? *Jour. Ethnopharm.* 32:25-36.

______ and ______ 1991c. Pharmacologic implications of "wild" plants in Hausa (Nigeria) diet and therapeutics. Paper presented as part of the symposium: "Eating on the Wild Side: The Pharmacologic, Ecologic, and Social Implications of Using Noncultigens" at the Annual Meeting of the American Anthropological Association. Chicago. 20-24 November.

Etkin, N.L., P.J. Ross, and I. Muazzamu. 1990. The Indigenization of Pharmaceuticals: Therapeutic Transitions in Rural Hausaland. *Soc. Sci. Med.* 30:919-928.

Farrington, I.S., and J. Urry. 1985. Food and the early history of cultivation. *Jour. Ethnobiol.* 5:143-157.

Fisher, K. 1990. People love plants and plants heal people. *Amer. Hort.* 69:11-15.

Fleuret, A. 1980. Nonfood uses of plants in Usambara. *Econ. Bot.* 34:320-33.

Ford, R., 1978. Ethnobotany: historical diversity and synthesis. In *The Nature and Status of Ethnobotany.* ed. R. Ford, pp. 33-49. Ann Arbor, Michigan: University of Michigan Press.

Hagedorn, R. 1991. Horticulture as a prescriptive tool for behavioural change. *Prof. Hort.* 5:129-133.

Harris, D.R. 1989. An evolutionary continuum of people-plant interaction. In *Foraging and Farming: The Evolution of Plant Exploitation.* ed. D.R. Harris and G.C. Hilman, pp. 11-26. Boston: Unwin Hyman.

Holmstedt, B. 1991. Historical perspective and future of ethnopharmacology. *Jour. Ethnopharm.* 32:7-24.

Holmstedt, B., and J.G. Bruhn. 1983. Ethnopharmacology–a challenge. *Jour. Ethnopharm.* 8:251-56.

Honey, T.E. 1991. The many faces of horticultural therapy. Part I. *Amer. Hort.* 70:37-43.

Iwalewa, E.O., L. Lege-Oguntoye, P.P. Rai, T.T. Iyaniwura, and N.L. Etkin. 1990. *In vitro* Antimalarial Activity of Leaf Extracts of *Cassia occidentalis* and *Guiera senegalensis* in *Plasmodium yoelli nigeriensis*. *W. Af. Jour. Pharmaceut. Drug Res.* 9:19-21.

Johns, T. 1990. *With Bitter Herbs They Shall Eat It: Chemical Ecology and the Origins of Human Diet and Medicine.* Tucson: University of Arizona Press.

Kaplan, R., and S. Kaplan. 1990. Restorative experience: the healing power of nearby nature. In *The Meaning of Gardens.* eds. M. Francis and R.T. Hester. pp. 238-243. Cambridge: MIT Press.

Lewis, C.A. 1990. Gardening as healing process. In *The Meaning of Gardens.* eds. M. Francis and R.T. Hester. pp. 244-251. Cambridge: MIT Press.

Nabhan, G.P., and A. Rea. 1987. Plant domestication and folk-biological change. *Amer. Anthropol.* 89:57-73.

Nyazema, N.Z. 1986. Herbal toxicity in Zimbabwe. *Roy. Soc. Trop. Med. Hyg.* 80:448-50.

Oldfield, M.L., and J.B. Alcorn, eds. 1991. *Biodiversity: Culture, Conservation, and Ecodevelopment.* Boulder, Colorado: Westview.

Posey, D.A., and W. Balee, eds. 1989. *Resource Managements in Amazonia: Indigenous and Folk Strategies.* Bronx: New York Botanical Garden.

Rhoades, R.E. 1986. Using anthropology in improving food production: problems and prospects. *Agric. Admin.* 22:57-78.

Richards, P. 1985. *Indigenous Agricultural Revolution.* London: Hutchinson.

Rios, J.L., M.C. Recio, and A. Villar. 1988. Screening methods for natural products with antimicrobial activity. *Jour. Ethnopharm.* 23:127-49.

Rivier, L. and J.C. Bruhn. 1979. Editorial. *Jour. Ethnopharm.* 1:1.

Rosenberg, M. 1990. The mother of invention: evolutionary theory, territoriality, and the origins of agriculture. *Amer. Anthropol.* 92:399-415.

Ross, P.J., N.L. Etkin, and I. Muazzamu. 1991. The greater risk of fewer deaths: an ethnodemographic approach to child mortality in Hausaland. *Africa* 61(4): 502-512.

Shah, A.H., S. Qureshi, and A.M. Agell. 1991. Toxicity studies in mice of ethanol extracts of *Foeniculum vulgare* fruit and *Ruta chalepensis* aerial parts. *Jour. Ethnopharm.* 34:167-72.

Spencer, P.S., P.B. Nunn, J. Hugon, A.C. Ludolph, S.M. Ross, D.N. Roy, and R.C. Robertson. 1987. Guam amyotrophic lateral sclerosis-Parkinsonism-Dementia linked to a plant excitant neurotoxin. *Science* 237:517-22.

Stahl, A.B. 1989. Plant-food processing: implications for dietary quality. In *Foraging and Farming: The Evolution of Plant Exploitation.* ed. D.R. Harris and G.C. Hilman, pp. 171-194. Boston: Unwin Hyman.

Stoffle, R.W., M.J. Evans, and J.E. Olmsted. 1990. Calculating the cultural significance of American Indian plants. *Amer. Anthropol.* 92:416-32.

Sukumaran, K., and R. Kuttan. 1991. Screening of 11 ferns for cytotoxic and antitumor potential with special reference to *Pityrogramma calomelanos*. *Jour. Ethnopharm*. 34:93-96.

Thoday, P.R., and J.A. Stoneham. 1989. Amenity horticulture and its contribution to the quality of life. *Prof. Hort*. 3:5-7.

Ulrich, R.S. 1986. Human responses to vegetation and landscapes. *Landsc. Urb. Plan*. 13:29-44.

Ulrich, R.S. 1991. Effects of interior design on wellness: theory and recent scientific research. *Jour. Hlth. Care Int. Design* 3:97-109.

Warren, D.M., L.J. Slikkerveer, and S.O. Titilola. 1989. *Indigenous Knowledge Systems: Implications for Agriculture and International Development*. Ames: Iowa State University Press.

Weiner, M.A. 1978. Nutritional Ethnomedicine in Fiji. Unpublished PhD Dissertation. Department of Anthropology, University of California, Berkeley.

Chapter 5

Adoption and Abandonment of Southeast Asian Food Plants

Lyndon L. Wester
Dina Chuensanguansat

SUMMARY. The people of Southeast Asia have, since prehistory, used a great variety of plants and were the first to domesticate some species now of global importance. In addition to the main staples and cash crops, a large number of plants are still grown in kitchen gardens, collected and transplanted from the wild or casually tended where they volunteer. A survey of plants sold in markets shows the natural floristic richness but also suggests considerable variation in the array of plants offered for sale from place to place. Cultivation of some of their traditional herbs and vegetables by Southeast Asian immigrants in the US has provided a source of income, as well as a supply of familiar foods and flavorings to them as they adapt to their new environment. In the process, it has provided the opportunity to enrich the cuisine of the population at large. However, rapid economic development in Thailand has been coupled with social and environmental change and concern is raised for the preservation of both the genetic diversity of these plants and knowledge of their use.

Mainland Southeast Asia is generally viewed as the land between the two great civilizations of China and India whence it received cultural innovation. However, DeCandolle (1882) and Vavilov (1926) recognized the region was an important center of diversity for cultivated plants. Sauer (1952) deduced that, with its abundance of wild plants and strongly seasonal climate, it was an ideal location for early development of the first fishing-farming communities based on vegetatively reproduced plants, which, he hypothesized, had predated cereal monoculture of wet rice. Archaeological evidence from Thailand now shows inhabitants were lively experimenters with indigenous plant resources by 10,000 BC (Gorman 1969; Glover 1977; Solheim 1970, 1972). Furthermore, technological innovations like metallurgy may have arisen in this region before they did in China and much of India (Bayard, 1980).

An indication of the first agricultural revolution is perhaps observable in the swidden agriculture persisting in the region today and in the kitchen gardens immediately around dwellings. Here the diverse assemblage of crops and animals raised together in a garden complex are structurally and functionally similar to a natural ecosystem (Fernandes and Nair, 1986; Soemarwoto et al., 1985). In rural Thailand, an extraordinary number of food plants are cultivated, casually tended or collected directly from the wild and this diversity is reflected in the wealth and variety of food plants used in Thai cuisine. The delicate blending of hot, sweet, sour, salty and bitter tastes, the use of many fresh herbs, vegetables and spices accounts, in part, for the popularity Thai food enjoys today in the West (admittedly in a tamed and modified form).

The Thai society itself, however, is experiencing extremely rapid economic development which affects all aspects of day to day life. As wet rice and cash crop monoculture expands, forested areas diminish, population grows, and migration increases within the region and beyond, there should be some serious concerns about the effects these changes are having on the use and variety of foods which, throughout history, have served as the basis for stability of a subsistence agricultural system.

In a survey of plants in markets throughout Thailand, Jacquat (1990) recorded 381 species. In the northeastern province of Kalasin alone, Moreno-Black (1991) compiled a preliminary list of 165 species and a survey of several villages (Yongvanit et al. 1990) found 152. In nearby Khon Kaen Province, Pei (1987) recorded 115 (Table 1). What is remarkable is not just the total numbers, but the dissimilarity of the sets of species they found, implying great variation in the use of plants over relatively small distances.

Northeastern Thailand was selected as the location for this research because of the unusual conditions there and because it is culturally similar to neighboring Laos, the place or origin of many recent Southeast Asian immigrants to the US. The region of sixteen provinces that make up the Northeast extends over a plateau underlain by sandstone and, traditionally, has been more impoverished than other areas of Thailand. This results, in

TABLE 1. Numbers of species of secondary food plants recorded in surveys

Author	Location	Number of species
Jacquat	Market plants Thailand	381
Moreno-Black	Markets Kalasin Province	165
Yongvanit et al.	Dong Mun Home Gardens Kalasin Province	115
Pei	Phu Wiang Watershed Khon Kaen Province	152

part, from the generally poor quality of the soil derived from sandstone. The area is also ringed by mountains that extract moisture from rain-bearing monsoon airstreams, thus making the plateau more susceptible to drought than any other part of mainland Southeast Asia. The region is referred to as Isaan and has its own distinctive language and culture within Thailand akin to neighboring Laos. Historically, the area has been under the control of Khmer, Mon, Thai and Lao kingdoms at various times. Today, the majority of the population is ethnically Lao. The rural character of Isaan culture is sometimes viewed with a certain amount of disdain by those in the more prosperous and urbanized areas of central Thailand. Yet, in the Isaan region, one readily sees the traditional practices and attitudes reflecting the ancient agrarian cultures of Cambodia, Thailand, Laos and Vietnam.

The staple food of the region is glutinous rice. Protein comes mainly from fish, insects, crustaceans, amphibians, reptiles and mammals complemented by a wide variety of plants that serve as the main sources of vitamins and minerals (Somnasang et al., 1986). The multitude of wild plants collected from the forests (now much diminished) are held in high regard. It is common practice to transplant useful species to more convenient locations such as home gardens and waysides. Many plants are identified as either a domesticated (baan) or a wild or forest (paa) variety. Other public areas, waysides, drainage ditches and ponds also yield useful plants. Fruits, spices and vegetables seem to be the most favored plants for home gardens, but numerous other species used for floral arrangements, medicine or construction are also common. Plants and seeds are traded actively within and between the villages and with contacts far afield. Produce not consumed by a householder may be traded or given as gifts, and also frequently sold to provide a small cash income of perhaps $10 per year for a household (Yongvanit et al., 1990).

A survey of nine markets in Northeastern Thailand, as well as from the capital cities of Bangkok and Kuala Lumpur in Malaysia, was conducted in July 1991. Species of all fresh produce was surveyed in the entire market, and when possible at several times during the day and on different days. Problems with identification of fresh produce presented some difficulty. For example, the most palatable vegetables are often juvenile leaves of trees and herbs, and these fragments are not easily identified to the species level with confidence. Common names may give clues to identity but in Thailand, as elsewhere, they frequently differ from place to place.

A total of 178 species were encountered in the survey, of which only five were found in every market, onion (*Allium cepa*), asparagus bean (*Vigna unguiculata* var. *sesquipedalis*), loofa gourd (*Luffa* spp.), and two New World contributions, the pumpkin (*Curcurbita moschata*) and chile pepper (*Capsicum annum*), were found in every market. The number of plants per market usually ranged from 50 to 80 and a maximum of 91 were observed in one large Bangkok market (Table 2). This compares to about 70 species typically found in a large US supermarket. The difference, however, is that the variation from market to market in Thailand is very high. Of all the species observed, 66 appeared in only a single market

TABLE 2. Comparison on numbers of species in markets of various urban places

City/town	Number of species	Population
Bangkok	91	6,000,000
Khon Kaen	85	127,000
Korat	78	250,000
Sakon Nakon	73	25,000
Kuala Lumpur	73	919,000
Udon Thani	69	83,000
Tha Bo	67	25,000
Tha Bo	66	25,000
Narathiwat	57	38,000
Nong Kong	54	c10,000
Kham Perm	35	---

(Table 3). The Bangkok market, mentioned above, had the highest number of these restricted species (17) although many temperate fruits and vegetables (turnip, spinach, beetroot, cherry, plumb, pears) were found in this market, which caters to foreign nationals and Thais with more cosmopolitan taste. The same is true for Kuala Lumpur, another large capital city, with the distinction that this city lies in a region with higher rainfall and less seasonality than Thailand, and some different species should be expected. Every market, even the smallest found in a tiny village, had some unique species, many of which are still undetermined.

Although the New World chile pepper (*Capsicum annum*), tomato (*Lycopersicon esculentum*), pumpkin (*Cucurbita moschata*), papaya (*Carica papaya*), manioc (*Manihot esculenta*), and coriander (*Coriandrum sativum*) have contributed in a spectacular way to Thai cuisine, the great majority (80%) of the food plants found in the markets today are Old World in origin and 40% are tropical Asian. The proportion of plants from various provenances varies little from market to market. Fruits make up more than half the species found in the markets, vegetables (leaves and stems) make up about a third, and roots and tubers the rest (Table 4).

In Thai society, the preparation and consumption of food is a prominent part of social interchange. Meals, typically eaten in small groups seated in a circle, punctuate daily life. Eating is considered one of the joys of living. Often several cooks contribute to a meal and great pride is taken in using as many available ingredients as possible at any given time. People take particular pleasure in seasonal foods and the discovery of rare or unusual items stimulates great excitement and is a source of gratification for the collector.

TABLE 3. Numbers of plants observed in a single market

City/town	Number of species found in a single market
Bangkok	16
Kuala Lumpur	12
Tha Bo	7
Korat	6
Sakon Nakon	5
Udon Thani	5
Tha Bo	4
Khon Kaen	2
Nong Kong	2
Kham Perm	2
Narathiwat	1

TABLE 4. Parts of plant used

City/town	Fruit	Leaf vegetable	Root or tuber
Bangkok	59	26	12
Kuala Lumpur	58	11	30
Tha Bo	58	35	8
Korat	46	38	15
Sakon Nakon	49	36	15
Udon Thani	48	39	13
Tha Bo	43	42	15
Khon Kaen	56	33	11
Nong Kong	59	33	7
Kham Perm	65	23	11
Narathiwat	50	35	14

As with previous immigrant groups who came before them to the United States, Lao and Vietnamese have brought along a dozen or more culinary herbs and salad plants. Kuebel and Tucker (1988) noted nine formerly rare or unknown species being used by Vietnamese immigrants in Texas and all of these, as well as several others, are now being grown commercially in Hawaii, where a lucrative trade in these herb plants has developed. They are primarily used to supply Southeast Asian communities in the northern states and Canada for both home and restaurant use, but are being offered with increasing frequency in local markets serving other populations. Asian basils, mints and other pot herbs probably represent the next wave of foods to be adopted by the general public around the country, similar to the acceptance of some traditional Chinese vegetables now found regularly in most supermarkets.

Not only have Lao and Vietnamese introduced some of the major herbs used in their daily cooking, but growing these on a commercial basis has become a source of survival for some of the urbanized families who have settled in areas of the US with warmer or subtropical climates. This is apparent in Hawaii where a number of middle class or professional immigrants from Laos have turned to farming when they were unable to find employment similar to the work they did before leaving.

An intensive survey of ten families conducted to investigate social and economic adaption of refugees in Hawaii, showed that language problems and lack of transferable credentials prevented them from finding jobs matching their training and experience. Although 80% had urban salaried occupations in Laos, in the US, most were forced to take various low-prestige and low-paying jobs and found it difficult to support their families. Wives, who may have had small children, less training, and less knowledge of English, were not able to provide supplemental income through outside employment.

Turning to horticulture served many needs. It enabled families to work together as a group, provided employment for those who would not otherwise be able to contribute economically, and brought about an increase in economic independence for each family as a whole. Children and elders could help out and those who wished to continue holding outside jobs did so. It also enabled some of the families to produce familiar plant foods not otherwise available in their adopted home.

The advantage of horticulture, over other small businesses such as restaurants and small shops, was the relatively low overhead and capital input required to get started. Certain fast growing crops, such as basil and green onions, provided a quick return on capital invested, and machinery such as tractors was available through borrowing from helpful local neighbors or inexpensive rental. The farmers interviewed also indicated that Laotians have a cultural attachment to land and nature, are very independent people, and working in extended family groups is traditional. Farming for approximately 50 Lao families in Hawaii, where climate and landscape have some similarity to Laos, has been a realistic alternative despite the fact their previous experience had been mostly urban.

Agricultural extension agents, and several social service organizations, assisted the Lao farmers in establishing their markets and coping with problems of production, pests and disease. Gaining access to land was not difficult, initially, but it was generally rented on a month to month basis. The lack of security of tenure on leased land, and the need for rapid cash flow, led producers to concentrate on a few short term crops and little was invested in capital improvements which could not be moved. Particularly notable is the observation that no significant kitchen gardens were established around these temporary quarters. In two cases where individuals have been able to obtain long term leases on land, they have invested in permanent improvements, planted a number of perennial crops, including trees, and have created kitchen gardens around their houses rich in species, including many new to Hawaii (Table 5). They expressed eagerness to obtain additional familiar plants from Laos, especially the wild (paa) varieties, clearly held in high regard.

The spread of a dozen or so important herbs, vegetables and fruits common in Southeast Asia seems to be taking place wherever large populations of Laotians, Thais, Vietnamese, Cambodians and Filipinos have settled in suitable climates. Although this is desirable for protection of

TABLE 5. Southeast Asia food plants or herbs recently introduced to Hawaii. None of these plants were listed in Neal (1965) or St. John (1973) the most recently published comprehensive surveys of cultivated plants in Hawaii.

Acacia insuavis Nielsen	cha oom
Coccinia grandis Voigt.	ivy gourd
Citrus hystrix DC.	kaffir lime
Houttuynia cordata Thunb.	heart leaf
Limnophila aromatica Merr	phak khayaeng
Pandanus amaryllifolius Roxb.	pandanus
Piper lolot DC.	lolot
P. sarmentosum Roxb.	chaa phluu
Polygonum odoratum Lour.	knotweed
Solanum stramonifolium Jacq.	ma euk

genetic diversity, it is a tiny representation of the great array of food plants in the source region, which number in the hundreds. Replacement or degradation of forests has severely limited the possibilities for collecting from the wild and many of these plants are known to be difficult or impossible to cultivate. Development has made economic and social change possible and stimulated wholesale migration from rural areas.

Although typically relocated rural people retain ties with their villages, the number of young and middle-aged adults spending much of their working life away is increasing rapidly. As the groups traditionally most responsible for passing on cultural information to the young disappear from village life, much of the knowledge of local plant resources could be lost with the turn over of one or two generations. Urban life also influences the culture by homogenizing tastes and changing attitudes to rural patterns of living. The spread and insuing influence of television on attitudes and preferences is likely to be profound, especially in rural areas. Even in cases where people return to their village of origin in their retirement years, as more are now doing, they are no longer the same as when they left, and in fact, may be highly influential in bringing urban values and cultural practices to the village culture. Refugees, who have settled in the US and turned to agriculture for employment, make an important donation to the crop diversity in their adopted lands. Having had to leave Laos has made many more conscious of the rich plant world they left behind. But, they are no longer in a good position to preserve the agricultural heritage of their former homeland.

More purposeful investigation is needed to determine how the existing knowledge of the nature and uses of local indigenous food plants is being retained, transferred or lost as the circumstances of the lives of people undergo vast transformation brought about through urban and foreign migration, rapid social change and environmental destruction. Of major concern is whether, or to what degree, secondary food plant species themselves are being lost by extinction or becoming so rare that they are now ignored or forgotten as a resource. Andrade-Aguilar and Hernandez-Xolocotzi (1991) have shown that socioeconomic change has been accompanied by loss of diversity for the common bean and the same process is likely to be operating in many parts of the developing world. It is crucial to determine what measures can, and should, be taken to slow or reverse this process.

Terms such as "natural food," "environmental food" or "secondary food plants" have been used for the foods that were gathered from the areas surrounding habitations. But the fact that there is no generally accepted terminology for these supplementary foods indicates the lack of consideration accorded this important source of nutrition and the role these foods play in enhancing the lives of rural people. It is hoped that this research will stimulate interest in collecting and cataloging these vital foods and the human knowledge of their uses.

BIBLIOGRAPHY

Andrade-Aguilar, J. A. and E. Hernandez-Xolocotzi. 1991, Diversity of common beans (*Phaseolus vulgaris*, Fabaceae) and conditions of production in Aguascalientes, Mexico, *Economic Botany*, 45:339-344.

Bayard, Donn. 1980, East Asia in the Bronze Age, In: *The Cambridge Encyclopedia of Archaeology*, pp. 168-173, A. Sherratt (Ed.) Cambridge University Press, New York.

DeCandolle, A. 1883, *Origine des plantes cultivees*, Biblioteque Scientifique Internationale No. 44, G. Bailliere et Cie, Paris.

Glover, Ian C. 1977, The Hoahabinian: Hunter-gatherers or early agriculturalists in South East Asia? In: *Hunters, gatherers and first farmers beyond Europe*, pp. 145-163, J.V.S. Megaw ed. Leicester University Press.

Gorman, Chester F. 1969, Hoabinhian: A pebble-tool complex with early plant associations in Southeast Asia, *Science*, 163:671-73.

Jacquat, Christine. 1990, *Plants from the markets of Thailand*, Editions Duang Kamol, Bangkok.

Kuebel, K.R. and Arthur O. Tucker. 1988, Vietnamese culinary herbs in the United States, *Economic Botany* 42:413-419.

Moreno-Black, Geraldine. 1991, Traditional foods in Northeast Thailand: Environmental, cultural and economic factors, *Annual meeting, American Anthropological Association*, November, 1991.

Neal, Marie C. 1965. *In gardens of Hawaii*, Bernice P. Bishop Museum Press, Honolulu.

Pei, Sheng-ji. 1987, Human interactions with natural ecosystems: The flow of minor forest and other ecosystem products of Phu Wiang watershed, Northeast Thailand, pp. 129-148, *Ecosystems interactions of a rural landscape: The case of Phu Wiang watershed, Northeast Thailand*, Southeast Asian Universities Agroecosystem Network.
Sauer, Carl O. 1969, *Agricultural origins and dispersals: The domestication of animals and foodstuffs*, M.I.T. Press, Cambridge, Mass.
Smitinand, Tem. 1980, *Thai plant names*, Royal Forest Department, Bangkok.
Solheim, William G. 1970, Northern Thailand, Southeast Asia, and world prehistory, *Asian Perspectives*, 13:145-161.
Solheim, William G. 1972, An earlier agricultural revolution, *Scientific American*, 226:24-41.
Somnasang, P., P. Rathakette and S. Ratanapanya. 1986, *Natural food resources of Northeast Thailand*, Khon Kaen University, Khon Kaen.
St. John, H. 1973, List and summary of the flowering plants in the Hawaiian Islands, Pacific Tropical Botanical Gardens, Memoir 1, Lawai, Kauai, Hawaii.
Trisonthi, Chusie. 1992, Edible wild fruits in Northern Thailand, *Annual Conference, Society of Ethnobotany*, Washington D.C. 25-27 March.
Vavilov, N.I. 1926, Studies on the origin of cultivated plants, *Bull. Applied Botany* 16:139-248.
Yongvanit, S., T. Hom-ngert and K. Kamonpan. 1990, Homegardens in Dong Mun National Forest Reserve: A case study from Ban Na Kam Noi, Kalasin Province, pp. 53-76, In: *Voices from the field*, C. Carpenter and J. Fox (Eds.) East West Center, Honolulu, Hawaii.

Chapter 6

Gathered Food and Culture Change: Traditions in Transition in Northeastern Thailand

Geraldine Moreno-Black

SUMMARY. The traditional diet in Northeastern Thailand is widely known for its non-domesticated, gathered food component. In this region, gathered food, especially plants, can act as a link among individuals, families and community members. Women's gathering activities create an intensive interaction between members of the community and the plant resources. These interactions are being affected by the recent increase in the marketing of gathered food resources. Changes are occurring in the selection and management of certain plants, in gathering practices, in dietary patterns, and in social relationships involving reciprocity and exchange of gathered food. These findings are discussed in the context of increasing participation in the cash economy and regional environmental changes.

The intellectual domain, which has been developed to cultivate an understanding of the significance of the role of plants in human culture, has been predominately shaped by the work of ethnobotanists. The focus

---

This study was part of a larger project focusing on the marketing and use of non-domesticated, indigenous plants and animals in the Northeast. The project was supported by grants from The Wener Gren Foundation for Anthropological Research, The National Geographic Society and a Fulbright Hayes Southeast Asian Regional Research Award.

of this body of information has been in four primary areas: (1) attending to the concerns of human-environment manipulation and its affect on the domestication process; (2) giving voice to indigenous peoples so that "indigenous knowledge systems" can be preserved; (3) recording the importance and utilization of "wild/nondomesticated" species, especially those with biomedical and pharmacological concerns in mind; and (4) studying the relationship of human cultural and biological evolution and adaptation to environmental factors (Etkin, 1988; Posey and Overal, 1990a; 1990b; Sarukhan, 1985).

Social scientists and biologists studying people-plant relationships have increasingly located their investigations in local markets (Bye and Linares, 1983; 1990; Jacquat, 1990; Pei, 1988; Wester and Chuensanguant, 1992; Whitaker and Cutler, 1966). Markets represent a place of intense interaction between people and their food resources (Pei, 1988). People require these items to meet biological, cultural, and economic needs and often rely on organized exchange structures to obtain them (Bye and Linares, 1983). In addition, as Pei (1988) aptly notes, it is the nature of the market to be a place where the selection pressure on specific items can be seen. This selection on specific items–i.e., selection by people because of specific culturally defined qualities of the items–can result in more intense interactions between the plant or animal population and the human populations involved. Certain plants are continually tested, evaluated, and demanded because of their recognized values, properties, and effectiveness (Bye and Linares 1983). The market allows for increased selection of plants, more intensified interactions within the environment, the modification of habitats, selection and maintenance of certain plants, and changes in how individuals exert control over each other within their social context for the use of these resource areas.

In this paper, then, I discuss the changing relationship among people and the plants they utilize in their environment by focusing on gathering and management strategies utilized by vendors of nondomesticated food plants. I specifically compare collection and management practices that are utilized when plants are used for home consumption and for sale in the market. Since the market vendors in Thailand are primarily females, I also build on gender sensitive assumptions and focus on women as gatekeepers in the environment. Specifically I focus on women's gathering practices as behaviors that create an intensive interaction with the physical and social environment. These activities result in the selection, consumption, and exchange of plants, as well as of other non-domesticated, gathered items. The work is based on my interpretations and analysis of data collected in Northeastern Thailand.

THE SETTING

The Northeastern part of Thailand (Isaan) is the largest of the country's four major geographic regions. It contains one third of the nation's population and is usually considered to be the poorest and least economically developed. A gently sloping plateau of undulating mini-watersheds and flood plains characterize the region, which also includes a zone of hills and upland areas most pronounced in the western and the southern part of the region (Hafner, 1990). Low soil fertility and erratic rainfall explain the relatively low agricultural productivity of the region; floods and drought periods are common in the area, and only about one fourth of the farms are situated near reservoirs or other permanent water sources (Hafner, 1990; Grandstaff et al., 1986). The semi-arid environment greatly influenced the traditional subsistence system and other adaptations to the habitat.

Traditionally the people in the Northeast appear to have adjusted to variability in these habitat factors through the development of a combined subsistence system in which the reliance on the staple rice crop was complemented by a large input from wild food (Pradipasen, 1986; Somnasang et al., 1987; Tontisirin et al., 1986). My recent research on the dietary practices in this region indicates that the people still rely heavily on non-domesticated resources (Moreno-Black, 1991). Their diet is characterized by a staple core of glutinous rice and fish or fish products. They embellish this core diet with a wide assortment of local plants and animals (Moreno-Black, 1991; Moreno-Black and Leimar, 1988). These important items contribute valuable nutrients and, coupled with a wide number of cooking methods, add variety to a potentially monotonous diet. These items are collected from a variety of resource areas such as forests, upland fields, rice paddies, gardens, and house areas, as well as various water sources such as canals, ponds, swamps, rivers, and dam areas (Moreno-Black, 1991; Somnasang et al., 1987).

Deforestation in Thailand is occurring at a rapid rate; the proportion of forested area to total area has changed from 62% in 1940 to 30% in 1982 (Ramitanondh, 1989) and to 18% in 1990 (Ganjanapan, 1992). In the Northeast, however, it was estimated that in 1982 only 15% of the land area was still under forest cover (Hafner, 1990). In the Northeast, upland cash cropping has increasingly become the foremost source of cash for the population. Three crops have dominated the areas being planted with cash crops: corn, kenaf, and cassava. Although the amount of rice paddy land in the region where I was working has not been drastically affected by the incorporation of these cash crops into the production system, the amount of forested land has been drastically affected. The spread of these crops, especially cassava since the 1970s, has significantly contributed to the decline of forested land and forest resources because these crops are well suited to the marginal drought conditions and poor soils of the upland and dry dipterocarp forests which predominate in the region (Hafner and Apichatvullop, 1990; Hafner, 1990).

Consequently, for the people of Isaan, social change and economic development are complex problems. Three main factors currently impinge on these processes: (1) development programs do not appear to be taking the importance of gathered, nondomesticated plants and animals for rural village economy into consideration when attempting to implement change; (2) land availability and accessibility are increasingly being challenged as rural village lifestyle competes with agricultural and urbanization schemes; and (3) forests and other utilized biomes are being depleted as national needs are expanding.

Certainly the people of Northeastern Thailand are currently experiencing all of these problems; indeed the Thai government has made its 1987 "Greening of Isaan" program one of the top priority development programs. This development plan includes increased agricultural projects, crop diversification, small businesses, and increased access to the Lam Pao Dam, which was built in the 1970s. Despite the good intentions behind this and similar programs, many people are critical of the attempts to rehabilitate the environment and stimulate or sustain economic growth. Their concerns and the gloomy prospects which have been forecasted are echoed in Hafner's concluding remarks when he says: "Without a broad-based effort to overcome the varied social, institutional, and economic barriers to a stable and productive forest resource system, the once rich forest resources of the Northeast will survive only in the memories and folklore of the Isaan people" (1990, p. 90).

With this picture of the current environmental situation as a backdrop, this paper focuses on the effect some of these factors are having on the Isaan people. My discussion centers on women's marketing of non-domesticated plants. In particular I will focus on how the use of nondomesticated plants as a source of income is related to small scale changes in reported behavior. These changes appear to be occurring as a result of the increased use of these plants as a source of cash rather than just for home consumption. These changes can be seen to be indicative of potential modifications in the procurement practices and uses of these local resources. The information comes primarily from data collected from women vendors in 13 markets in central Isaan as well as information obtained from women in one focus village.

METHODS

The information in this paper comes from 214 vendors at 13 of the 15 markets which were studied in Northeastern Thailand. The data were collected in the cool, hot, and rainy seasons of 1990-1991. All of the interviews were conducted at the market during market hours while the vendors were selling their items. Interviewing was halted when necessary in order to allow the women to conduct a transaction and every effort was made to sit with the women, behind their items so as not to block the narrow, often busy isles or to block particular items from view. The sample of vendors who were interviewed was not a random sample since my primary concern was to obtain information about the different items themselves.

The data discussed in this paper were collected as part of a larger study on the prevalence and use of nondomesticated items in the central Isaan region. The markets were observed on several different occasions for different types of information, and data was also collected in one specific village. The information presented here came from one set of market observations. On the days when the item survey was conducted, each market was first scanned for vendors who were selling items which had not been recorded in other markets or items which had no associated interview data. Once the market was scanned and the first interviews were conducted an attempt was made to select vendors randomly.

Each vendor was questioned by using an interview which contained primarily open-ended questions. The information requested included: (a) brief demographic information on the vendor; (b) plant name; (c) price per unit of purchase; (d) collection site and resource area status (i.e., public, private, etc.); (e) collection methods, including whether permission was needed to collect the item and whether this differed when gathering for home consumption or for selling; (f) management and propagation practices; (g) uses; (h) methods of cooking or preparation. Specimens, usually in the units they were sold, were purchased. Color slide photographs were taken of the plant specimens. Color, laminated photographs were also made from the slides and used in the focus village to confirm or expand the survey information. The specimens that were appropriate for preservation or propagation were deposited with the Herbarium at Khon Kaen University.

RESULTS

Two hundred and fourteen women vendors of plant resources at the 13 markets surveyed participated in this study. The age of the vendors ranged from 7 to 76 years of age with a mean of 45 years (sd = 17) (Table 1). One hundred and four different items, consisting of plants, algae and mushrooms, were being sold by these vendors.

Ninety-five percent of the women indicated they had collected the items they were selling by themselves and 5% stated they obtained their items from relatives or friends. The items being sold by the 214 vendors were obtained from a variety of different locations. I have grouped these into several categories and indicated the percent of the items obtained from each locale type (Table 2). Fifty-three percent of the women indicated that they owned the resource areas, 32% collected their items from public areas and 17% utilized areas which they identified as belonging to relatives or other people (Table 3).

I attempted to gain insight into the influence of marketing behavior on gathering practices by eliciting information focused on comparisons between collection for home consumption and collection for marketing. The consumption of non-domesticated plants has recently been documented (Moreno-Black, 1991; Moreno-Black and Leimar 1988; Pradipasen, 1985; Somnasang et al., 1987; Tontisirin, 1986). However, the practice of selling nondomesticated items has not received much attention. Consequently a comparison of practices used when gathering for home consumption and for selling was made. The data indicate that several themes can be used to distinguish practices surrounding collection and management of plants for

TABLE 1. Age Distribution for the Sample of Women Vendors (N = 209). Mean Age = 45 (S.D. = 17) years

AGE	PERCENT OF SAMPLE
7-17	9%
18-29	9%
30-40	25%
41-50	15%
51-60	24%
61-76	18%
	100%

TABLE 2. Habitat of Procurement and Percent of Items Obtained in Each Locale.*

LOCALE	PERCENT OF ITEMS
Forest	28%
Paddy	23%
Garden	23%
Water Sources	22%
Multiple Sources	2%
Upland Cultivation Area	1%
House Area	1%
TOTAL	100%

* One vendor indicated that she purchased the items that she was selling and was omitted from this table.

home consumption and for cash value. Two features which appeared to be distinguishing are: (1) whether the items are being manipulated, i.e., transplanted and/or managed such that the "human-item" interaction is beginning to change or intensify and (2) whether differences surround the collection of items when they are being sold, as compared to being used for home consumption.

The most marked difference occurred in reference to collection locales (Table 4). Eighty-nine percent of the women indicated they felt it was acceptable to collect food from resource areas other than their own if they were collecting for eating purposes, while only 50% of the women vendors felt it was acceptable to collect items for sale from resource areas that belonged to other individuals. Of these vendors, forty percent indicated that this practice was not appropriate and 8% said it was acceptable under certain conditions.

When comparisons were done on the responses of women concerning their willingness or ability to collect items from other individuals' resource areas for eating versus selling, a number of different factors emerged. First, it became clear that the location of the items was an important factor. The women considered forest and other public resource areas places where gathering for eating or selling could occur. Some women indicated that even if the forest was once public but now owned, it was still acceptable to collect there. The paddy field and garden areas,

TABLE 3. Ownership of Resource Areas for Collection of Plants for Market Sale (N = 209).

Resource Area Vendor Owned	55%
Public Resource Area	34%
Resource Area Owned by Relative	5%
Resource Area Owned by Friend	2%
Resource Area Owned by Nonclarified Other	2%
Miscellaneous (e.g. = rented)	2%
TOTAL	100%

TABLE 4. Acceptability of Collecting from Resource Areas Owned by Others (N = 214).

	FOR EATING	FOR SELLING
YES	89%	50%
NO	8%	40%
SOMETIMES	3%	8%
NOT SURE		1%
MISSING		1%
TOTAL	100%	100%

however, were more problematic. There was a trend toward the paddy field being less acceptable for collecting for selling, especially if the items were transplanted there and/or if the paddy field owner was also a vendor. This may be related to the transformation of these items into the category of "*baan*"[1] in the mind of the villagers. The women indicated that sometimes wild items were planted in their resource areas by using seeds. They further indicated that if a person had to buy the seeds from someone, then the item was not available to be collected for selling. Sometimes the general statement that anything that a person grows (in this sense more general than saying cultivated plants) was not appropriate to collect for selling.

Second, it appeared that factors related to the timing of collection became influential in determining if it was appropriate to collect items from resource areas that belonged to other people. There was some feeling that when items were first transplanted, it was not appropriate to collect them, but once they got established and were abundant, it was permissible. Also, it was not deemed appropriate to collect in another person's paddy when the rice crop was young because the rice plants could be destroyed.

Third, it was apparent that the quantity being collected was an important factor. Quantities collected for home consumption were usually quite small while the women needed to obtain much larger amounts when marketing these items. This was especially true since the prices for most of the green plant material (leaves, stems, buds, etc.) were quite low (1-2 Baht per handful). They frequently voiced a critical consideration that the amount needed for selling exceeded what was appropriate for taking from others.

The type of item collected also emerged as an important factor. Fruit, bamboo, and tubers were more likely to be considered inappropriate to collect from others' resource areas while green plants, mushrooms, and flowers did not consistently get mentioned. This may be related to the fact that they are favored items, the inherent nutritional value of these items, or to the fact that they are often transplanted.

The concept of management of non-cultivated items, then, is particularly interesting given the growing trend of utilizing these items for their market value and reluctance to gather them from areas that belonged to other individuals. Transplanting items into areas that are intensively utilized or occupied by humans marks one part of a process of management and manipulation which may ultimately have long term effects on the human plant relationship.

Fifty-five percent (n = 63) of the vendors indicated the item they were selling was to their knowledge transplantable. They indicated that transplanting was done for selling purposes, to decrease the distance for obtaining the item, especially in light of the decreasing amounts of forest and the increasing amount of land used for agriculture, and because people like to eat the item (i.e., it is delicious or popular). Fifty-three percent (n = 62) of these women indicated the items were transplanted from their own resource areas to other areas that were their own; 33% (n = 39) indicated the items were transplanted from public places; 8% (n = 9) obtained the items from friends or relatives (Table 5). As indicated earlier, their ability to identify plants as being transplanted was closely associated with their current locale and was also related to patterns of usage or availability by the population at large.

There did not seem to be any differences concerning transplanting among the different age categories except in the group over 60 (X^2_1 = 10.08, p = .002). In this group a significantly greater number of the respondents indicated their items had been transplanted than those who were not selling transplanted items. It seems likely this would be the case since the elderly women may have acquired more land on which to transplant, or they may have had more years in which to accumulate more transplanted items. It is also possible that they relied more on transplanted items since it required less energy in terms of collection distance and time.

The women used or recognized a wide variety of management and manipulation techniques for the nurturance and maintenance of these items. Twenty-seven percent (n = 58) of the vendors indicated that some form of care was given to the plants; 72% of these women indicated that some form of watering and/or fertilizing was being done. In some cases the care was more benign; however, some indicated they would attempt to

TABLE 5. Pathway for Transplanting Non-Domesticated Food Plants.

PATHWAY FOR TRANSPLANTING	
From Self-Owned Area to Other Self-Owned Area	54%
From Public Area to Self-Owned Area	33%
From Relative or Friend Owned Area to Self-Owned Area	9%
From Unknown Owned to Self-Owned	4%

build fences around the plants for protection against buffalo, or they described specific practices involved in the preparation of the soil prior to transplanting and care afterward. They reported methods of propagation including collecting and planting seeds, planting cuttings or tubers, and transplanting small plants.

The women in the focus village who are also vendors of non-domesticated food items gave me a number of insights concerning the transplanting of these items. These women spoke of their concern over the deforestation they have witnessed in their lifetime and are still seeing. They spoke of how many plants and animals there had been in the past and of how easy it had been to obtain food then. They also indicated there were some items which were difficult to procure now. They said they often transplanted items to their house gardens because they were afraid the plants were becoming extinct. They wanted to be able to have the items so they would transplant them before the areas were cut down. The village women and the market vendors indicated they utilized a variety of strategies for transplanting and relocating plants to various locales including paddy fields, upland gardens, and house gardens. The movement of these items to garden areas, however, marks the beginning of a change in the relationship between the people and the plant. Once they are located in the personal compounds of a family, the plant becomes part of the area that is associated with private ownership status. Availability of these items to other villagers continues but on a more restricted basis.

CONCLUSIONS AND DISCUSSION

The results of the market survey indicate the emergence of some interesting patterns that differentiate gathering for home consumption and cash value. Among these patterns are the influence of ownership of resource areas, the changes in attitudes concerning access to resource areas which were privately owned by other individuals, the types of items being collected, and the high proportion of items which were identified as being transplanted.

This research has begun to uncover some of the changing patterns of resource and habitat use as they are related to the consumption and marketing of local, indigenous food items. A focus on "nondomesticated" food resources provides a focus on not only a regular component of the diet but on those items which give Isaan diet its distinguishing character. By focusing on this component we can begin to see not only what is happening to Isaan tradition and culture but how the diet and procurement of food are changing due to environmental, cultural, and economic factors.

Some researchers propose that successful management of renewable common property resources would maintain ecological sustainability while reducing the traditional user pool as little as possible (Buck, 1989). This can be seen in the maintenance of the traditional practices of sharing resource areas when gathering for home consumption. Entrepreneurs, however, are more often ego focused and environmentally exploitative (Buck, 1989). The data in this project indicate that rather than completely exploiting the environment the women vendors are shifting into different personal relationships, decreasing their sharing of resources, claiming ownership, and increasing their management and manipulation of non-cultivated resources. Similarly Grandstaff et al. (1986) reported that public access to trees which provided food, medicine, lumber and animal fodder and were growing in privately owned paddies varied in different regions of the Northeast. They also found that planted trees were often considered and used differently than the naturally occurring trees. The impact of deforestation and the increased reliance on a cash based market economy appear to be important factors in the Grandstaff study and in the study I am reporting here.

The increased reliance on a cash based economy in the Northeast has changed the lifestyle considerably. It has changed the environment as forests have disappeared, and water sources have become degraded and it continues to affect in more subtle ways the use of the environment and social patterns. This can be seen, in the data reported here, in the process and increased reliance of moving resources into paddy and garden areas, for example, and managing them rather than creating and maintaining common resource areas.

Some researchers may view a focus on female marketing strategies and food procurement activities as stagnant by comparison with the radical nature of the social, cultural, and economic changes that are occurring in the region. However, I believe that by reinterpreting or viewing differently the impact of these patterns, we can begin to reinterpret the relationship of women's work, environmental and cultural change (Guyer 1991). A focus on such patterns highlights changes in food use, food procurement practices, and environmental relationships; it highlights the links between culture change and environmental problems and it allows us to perceive the power of gender in shaping–rather than simply submitting to–forces of change (Guyer 1991).

Today the task for social scientists and botanists is to shift the perspective of our research to include current issues which are relevant to the changing academic climate and current world problems. One approach to creating this shift is the development of a theory of praxis, similar to what Van Esterik has proposed for nutritionists and social scientists engaged in food research (1991). To this end our research should focus on an integration of plant systems within a particular social and historical context, provide both micro and macro perspectives, build on gender sensitive assumptions about women as gatekeepers of environmental systems, and should be broadly reflexive, encouraging critical reflection on how one system's use and ideas about botanical systems impinges and affects others.

Consequently, by focusing on the micro-processes involved in the use of these gathered foods and the practices surrounding their procurement, we can gain insight into the transformations which are occurring in this society. These behaviors not only can symbolize change, but can also create and sustain many of these changes. Food and food procurement practices can generate and transform many aspects of social life: they can create social relationships, affect social control, instill power or prestige, and they can alter human-environment relationships and the rules governing these relations (Fajans 1988). By focusing on these activities, we see that cultural practices concerning the management and gathering of these resources are beginning to change and so are the relationships among the people who use them.

AUTHOR NOTE

The author wishes to thank Ms. Prapimporn Somnasang, Ms. Watana Akanan, and Ms. Pisamai Homchampa for their assistance during this project. Their commitment and collegiality were an inspiration to her. She would also like to thank the staff of the National Research Council for their assistance and willingness to allow her to conduct this project over an extended period of time. The author is especially grateful to all of the market vendors and villagers for their patience, help, information and hospitality. She thanks her family, Ed, Tovah and Simca for being so understanding and willing to endure long periods of separation, and Karen Clausen, Guido Bodolioli, Lauren Spitz, and Debby Coulthard who coded with her and encouraged her throughout this project. Finally gratitude is expressed to Mrs. Doris Wibunsin and Mrs. Apiram of the U. S.-Thailand Educational Foundation for their help and encouragement.

NOTE

1. The word "*baan*" means village, however it is commonly used to refer to plants which are either domesticated or so commonly transplanted that they are considered to be grown by the local people.

REFERENCES CITED

Bye, R. A. and E. Linares 1990, Mexican market plants of 16th Century. I. Plants recorded in *Historia Natural De Nueva Espana. J. Ethnobiol.* 10: 151-168.

Bye, R. A. and E. Linares. 1983, The role of plants found in Mexican markets and their importance in ethnobotanical studies. *J. Ethnobiol* 3:1-13.

Fajans, J. 1988, The transformative value of food: A review essay. *Food and Foodways* 3: 143-166.

Ganjanapan, A. 1992, Community Forestry in Northern Thailand: Learning From Local practices. Paper presented at Workshop on Sustainable and Effective Management Systems for Community Forestry. Bangkok, January, 1992.

Grandstaff, S. W., T. B. Grandstaff, P. Rathakette, D. E. Thomas, and J. K. Thomas, 1986 Trees in the paddy fields of Thailand. In: *Traditional Agriculture In Southeast Asia.* Westview Press, Boulder.

Guyer, J. 1991, Female farming in anthropology and African History. In: *Gender At The Crossroads of Knowledge.* Feminist Anthropology in the Postmodern Era. M. di Leonardo (ed) Berkeley: University of California Press.

Hafner, J. A. 1990, Forces and Policy Issues Affecting Forest Use In Northeast Thailand 1900-1985. In: *Keepers of The Forest.* M. Poffenberger (ed). Kumarian Press, West Hartford, Conn. p. 69-94.

Hafner, J. A. and Y. Apichatvullop. 1990, Migrant Farmers and the Shrinking Forests of Northeast Thailand. In: *Keepers of The Forest.* M. Poffenberger (ed). Kumarian Press, West Hartford, Conn. p. 187-219.

Jacquat, C. 1990, *Plants From The Markets Of Thailand.* Duang Kamol: Bangkok.

Kimber, C. 1978, A folk context for plant domestication: Or the dooryard garden revisited. *Anth. J. Canada* 16:3-11.

Moreno-Black, G. 1991, Traditional Foods in Northeastern Thailand: Environmental, Cultural and Economic Factors. Paper presented at the American Anthropological Association Meetings, Chicago 1991.

Moreno-Black, G. and L. Leimar. 1988, Wild Food Gathering As An Economic And Nutritional Strategy In Northeastern Thailand. Paper Presented at the Conference of the Northwest Regional Consortium for Southeast Asian Studies. Eugene, Oregon Sept. 30-Oct 2, 1988.

Posey, D. and W. L. Overal. 1990a, Ethnobiology: Implications and Applications. Vol I. Proceedings of the First International Congress of Ethnobiology (Belem, 1988). Museu Goeldi, Belem, Para, Brasil.

Posey, D. and W. L. Overal. 1990b, Ethnobiology: Implications and Applications. Vol II. Proceedings of the First International Congress of Ethnobiology (Belem, 1988). Museu Goeldi, Belem, Para, Brasil.

Pradipasen, M. et al., 1985, Nangrong Dietary Survey. Institute for Population and Social Research. Mahidol University. ISPR Pub. No. 95.

Pei, Sheng-ji. 1987, Human interactions with natural ecosystems: The flow and use of minor forest and other ecosystem products of Phu Wiang, Northeast Thailand. Southeast Asian Universities Agroecosystem Network (SUAN). Khon Kaen University, Thailand.

Ramitanondh, Shalardchai. 1989, Forests and Deforestation In Thailand. In: *Culture and Environment In Thailand: A Pandisciplinary Approach.* A Symposium of The Siam Society. p. 23-50.

Sarukhan, J. 1985, Ecological and social overviews of ethnobotanical research. *Economic Botany* 39 (4):431-435.

Soemarwoto, O. et al., 1985, The Javenese home garden as an integrated agroecosystem. *Food and Nutr. Bull.* 7:44-47.

Somnasang, P. et al., 1987, The Role Of Natural Foods In Northeast Thailand. In: *Rapid Rural Appraisal In Northeast Thailand: Case Studies.* S. Subhadhira et al., eds. Khon Kaen University.

Tontisirin, K. et al., 1986, Annual Report 1985/1986: Thailand Food Habits Project. ASEAN Sub-Committee on Protein: Food Habits. ASEAN-AUSTRALIAN Economic Cooperation Program.

Van Esterik, P. 1991, Perspectives on food systems. *Reviews in Anthropology* 20:69-78.

Wester, L. and D. Chuensanguansat. 1992, Adoption and abandonment of Southeast Asian food plants. Paper presented at: *People-Plant Relationships: Setting Research Priorities.* A national symposium sponsored by: Rutgers University Cooperative Extension, Amer. Soc. Hort. Sci., Amer. Assoc. Bot. Gardens and Arboreta, Amer. Hort. Therapy Assoc. and People-Plant Council. East Rutherford, N. J. April 24-26, 1992.

Whitaker, T. W. and H. C. Cutler. 1966, Food plants in a Mexican market. *Econ. Botany* 20:6-16.

Chapter 7

People, Plants and *Proto-Paysage*: A Study of Ornamental Plants in Residential Front Yards in Honolulu, Hawai'i

Toshihiko Ikagawa

INTRODUCTION

This paper discusses how ethnicity and/or cultural background is reflected in front-yard gardens in Honolulu. The idea for this study grew from my own experience in Hawai'i. I often strolled through a residential neighborhood near the University when I first arrived in 1981. Along the way I saw many homes with stone lanterns, neatly trimmed shrubs and interesting rocks carefully arranged in the style of Japanese gardens (Figure 1). At first I thought, "They must be the homes of Japanese Americans, descendants of immigrants of many decades ago. These people are still maintaining their home culture in a foreign country." I soon realized, however, that although these gardens appeared Japanese, I had never seen others like them in Japan.

There are two points to this puzzle: (1) The style appears Japanese, but is not Japanese, and; (2) In Japan, in general, gardens are not seen from the street. I asked myself: What does this mean? How has this come to be?

FIGURE 1. Japanese-style corner of the front yard (Honolulu, 1992).

CONCEPTS

This study deals with three garden styles: American (European), Japanese and Chinese. In these cultures, the term "garden" does not share exactly the same meaning. In English, the term may mean a kitchen garden, an ornamental garden or an out-door living space. American-style residential gardens tend to serve all of these purposes. Thus, these gardens are largely an extension of the living area. Traditional Japanese-style (residential) gardens, on the other hand, are designed for appreciative viewing from the living space. It is an ornamental garden mainly composed of rocks and waterfalls with plain non-colorful plants. Traditional Chinese-style gardens are designed to walk through leaving everyday life behind. They are ornamental gardens of interesting architectural structures, flowering trees and unusually shaped rocks with ponds designed exclusively for elite intellectuals. For ordinary people in traditional Chinese houses, the outdoor living space is simply a courtyard called *tianjing* (skywell).

Gardens are part of "cultural landscape" which reflects people's tastes, values, aspirations, and even fears, in tangible, visible form (Lewis 1979). It is where all cultural elements including history, philosophy, art and

* Know Chinese vs. Japanese vs American.

architecture intersect plant life and the physical environment. We can read this landscape as we might read an autobiography. It will probably be more truthful because we are less self-conscious when we "write" it (Lewis 1979).

In Hawai'i's multicultural society, the conceptual framework has to provide (1) a mechanism to describe how the elements of foreign cultures have come to Hawai'i, and (2) the base for an interpretation of the relationship among different cultures. Thus, the framework revolved around two concepts: (1) the transported landscape (Anderson 1967), and (2) *proto-paysage* (Berque 1990).

The "transported landscape" concept suggests that when a group of people immigrate to a new place, they tend to bring their traditional ideas, in this case, garden style, with them. If the group maintains its "group identity" (Isaacs 1968) in a multicultural society, the transported landscape should distinguish the appearance of front-yard gardens of residents from different ethnic groups. *The Melting Pot* (Zangwill 1909) advocates cultural amalgamation in the society, while "cultural pluralism" (Kallen 1924) suggests that people live in both primary and secondary cultures (Sandberg 1974). Thus, a partial assimilation and acculturation may result.

The concept of "*proto-paysage*," or original landscape, proposes that humans have some universal common denominator, regardless of ethnicity or cultural background, in regard to landscape. That is, there is a shared reason why we like or dislike a particular landscape. The "habitat theory" (Appleton 1975) suggests that aesthetic satisfaction with a given landscape stems from the spontaneous perception of signs which indicate environmental conditions favorable to survival. The image of the original human habitat, a safe environment "to see without being seen" (Lorenz 1952, 193), probably at the boundary between forest and grassland with scattered trees, shrubs and open plains, is still with us.

The importance of the concept of original landscape is the perspective it provides. For example, a neatly secluded Japanese garden of Katsura and a vastly open French garden of Versailles appear very different. It seems to be reasonable to regard them as dichotomous. From the perspective of the human habitat, however, they are not opposite, but different expressions of the same common denominator. Possible interpretation is that "to see" is more vocal in Versailles, while "not be seen" is emphasized in Katsura. Thus, some common features may underlie inter-cultural varieties of garden style (Orians 1986). This perspective should be applicable between residential gardens in New England and those in Japan, or among those of different ethnic groups in Hawai'i today. This concept, with other concepts discussed above, suggests that the multicultural appearance of a garden in

Hawai'i may not be an exception but the norm because of the possible atavistic amalgamation of different cultural traditions.

QUESTIONS AND METHODS

Major questions addressed in this study are: How has the ethnic background of residents influenced the appearance of their front yards? What similarities exist? What differences prevail? What garden elements and which traditions have been transported to Hawai'i, and how have people dealt with such multicultural situations? If we can answer these questions, then we will better understand the relationship between plants, people and the "culture" of Hawai'i.

To find the answers, I compared the front yards of two dominant ethnic groups, Japanese and Chinese, and a third, mixed but mostly Caucasian group labelled "Other" in three neighborhoods on the island of O'ahu. The three neighborhoods, Pearl City, Palolo Valley and Aina Haina, are on mostly flat land and receive about the same amount of rainfall. They are all predominantly middle class neighborhoods. I randomly sampled the front yards of 150 houses, 50 each from the three neighborhoods. At each lot, I observed: (1) the presence/absence of plant species, (2) plant cover (m^2), (3) plant height (m) and (4) the use (function) of each plant species. The uses included hedge, lawn and foundation plantings. The Japanese species I was looking for included those commonly mentioned in books about Japanese gardens. The most common plants are pine, azalea, (maple), (cherry blossom), bamboo, camellia, (Japanese apricot, often called "plum"), juniper and (holly). (Plants in parentheses were not recorded in this study, as they are poorly suited to the tropical environment.) Of the 43 Japanese species checked for, 19 occurred within the study sites.

The ethnic background of residents was determined based on surname recognition (Greenbaum and Greenbaum 1981). Surnames were obtained from (1) questionnaire survey, (2) telephone directory and (3) tax key maps. Priority was set in the above order. Out of 150 questionnaires sent, 100 came back.

There is an advantage to pursuing this type of study in Hawai'i in that most ethnic groups live in similar types of houses in similar sized lots. Therefore, the space available for front yards is comparable. At the same time, however, many traditional garden styles are closely related with architectural styles. Thus, I did not expect an exact replica of any foreign garden styles in Hawai'i.

RESULTS

This paper is a progress report of my dissertation and most of my analyses and discussion are still limited to Japanese garden traditions. I began with Japanese elements because (1) I am familiar with the culture, and (2) the ethnically Japanese group has been the dominant or second dominant group in Honolulu for most of the past century. I will also briefly discuss Chinese elements. However, probably the most fundamental questions of the study will be left unanswered. These questions are: How much of the "American" garden traditions have been transported to Hawai'i? What is the current status of native Hawaiian garden traditions?

Let me return to the puzzle mentioned at the beginning. A traditional Japanese garden is secluded to achieve tranquility. It provides the resident an environment in which to contemplate and relax (Eliovson 1971). Following this tradition, most residential gardens in Japan are fenced in. In Honolulu, as in most of the United States, gardens are typically found open to the streets (see Figure 1).

It is also common to find red gravel used in garden settings because this lava material is readily available in Hawai'i. However, in Japanese stone gardens, which probably provided the motif, gravel represents the ocean as well as a sacred place where *kami* (a god) descends to. Only white gravel is used. To a person only familiar with traditional Japanese gardens, the use of red gravel as a substitute seems very strange.

The most significantly different variables among ethnic groups were (1) the presence/absence of certain Japanese species and (2) the number of Japanese species in each front yard ($p < .01$). More than twice as many Japanese species were noted in the front yards of ethnically Japanese (1.5 spp/lot) than in the front yards of others (Chinese: .7 spp/lot, Other: .6 spp/lot). However, only 70% of the front yards of ethnically Japanese included any of the 19 typical Japanese species.

Species with statistically significant occurrence in the front yards of ethnically Japanese were sacred bamboo (*Nandina domestica* Thunb.), cycad (*Cycus revoluta* Thunb.) and oriental hawthorn (*Rhaphiolepis umbellata* (Thunb.) Mak.). Sacred bamboo is usually planted on the side of the front entrance. Cycad is often combined with rocks and gravel. Oriental hawthorn grows in a neat spherical shape without pruning. It is important to note that although 80 to 90% of the total frequencies of these species occurred in the front yards of the Japanese group, their total frequencies were not high, only 10 to 20% of 150 lots. Thus, the presence of these species in a front yard strongly indicates that a resident is ethnically Japanese, but the absence does not always mean that a resident is not Japanese.

An interesting finding was the substitution of some common Japanese species by other species. For example, a pine tree is an indispensable component of a Japanese garden. In Hawai'i, however, it is rare and often substituted by ironwood (*Casuarina equisetifolia* L.). This is most likely because some Japanese plant species do not grow in Hawai'i. (Likewise, most Chinese plants, especially those from the north, do not grow in Hawai'i.) Such substitutions may be an interesting subject of future study.

An interesting species whose occurrence did not vary significantly among ethnic groups was Arabian jasmine or *Pikake* (*Jasminum sambac* (I.) Ait.). Although this plant is a favorite of Chinese people (Neal 1965), its frequency did not vary significantly among three ethnic groups. Surprisingly it was least frequent in yards of ethnically Chinese (total frequency = 37/150 lots; Japanese: occurred in 26.8% of Japanese lots, Chinese: 18.5%, Other: 25.0%). Probably it has become a favored "Hawaiian" plant as its name indicates. *Pikake* means "peacock" in Hawaiian. Both the flower and the bird were the favorites of the beloved Hawaiian Princess Kaiulani who died as a young woman (Neal 1965).

One surprise came from the presence/absence of hedges. Considering the secludedness of traditional Japanese gardens and "introverted" Asian traditions, I expected more hedges in the front yards of Japanese, and maybe in Chinese, residents. However, the frequency overall, as well as in each ethnic group, was near 50%. Even combined with the presence/absence of a fence, the result did not change much. The presence of a hedge may have more to do with the territorial marking behaviors of humans than of culture (but see Greenbaum and Greenbaum 1981).

A particularly American feature of a front yard is a grassy lawn. The presence/absence of lawns did not vary significantly among three ethnic groups, although the Other group had the highest frequency (total frequency: 105/150 lots; Japanese: 64.8%, Chinese: 74.1%, Other 82.7%). An interesting point is that the Chinese group had a higher frequency of lawns than did the Japanese group although traditional Chinese gardens have no grassy lawn (Morris 1983), while there are some in the Japanese tradition (Mori 1984).

Another variable which varied significantly among ethnic groups was the number of middle-sized plants (1 to 2 m tall). They occurred more often in the Japanese group (6.2 plants/lot) than in other groups (Chinese: 3.5 plants/lot, Other: 4.8 plants/lot).

CONCLUSIONS

My preliminary conclusion is that some Japanese garden traditions can be conspicuously found in the front yards of ethnically Japanese people in

Hawai'i. These include both plant species and garden ornaments. Sometimes the Japanese element was extensive. In other cases it was visible only in a small Japanese-style corner of the front yard. In such cases, the appearance of Japanese front yards is more similar than different to that of other ethnic groups. Chinese traditions are less visible. This might relate to the exclusively elite nature of traditional Chinese gardens–perhaps few members of this elite group are migrated to Hawai'i. It may reflect the fact that interracial marriage among Chinese is high in Hawai'i. These preliminary findings raise the question to the ubiquitous application of the concepts of transported landscape and original landscape to all ethnic groups. Each culture may have a different tendency to transport and assimilate traditions, contributing to the cultural variations that exist today.

REFERENCES

Anderson, E. 1967. *Plants, man and life*. rev. ed. Berkeley: University of California Press.

Appleton, J. 1975. *The Experience of Landscape*. London: John Wiley & Sons.

Berque, A. 1990. *Nihon no hukei, seio no keikan: soshite zokei no jidai*. Translated by K. Shinoda. Tokyo: Kodansha.

Eliovson, S. 1971. *Gardening the Japanese way*. London: George G. Harrap & Co. Ltd.

Greenbaum, P.E. and S. D. Greenbaum. 1981. Territorial personalization: group identity and social interpretation in a Slavic-American neighborhood. *Environment and behavior* 13:574-89.

Isaacs, H. R. 1968. Group identity and political change. *Survey* 69: 76-98.

Kallen, H. M. 1924. *Culture and democracy in the United States*. New York: Boni and Liveright.

Lewis, P. F. 1979. "Axioms for reading the landscape." In *The interpretation of ordinary landscapes: Geographical essays*, edited by D. W. Meinig. New York: Oxford University Press.

Lorenz, K. 1952. *The King Solomon's ring: new light on animal ways*. Translated by M. K. Wilson. New York: Mentor.

Mori, O. 1984. *Teien*. Tokyo: Kondo-shuppansha.

Morris, E. T. 1983. *The gardens of China: History, art, and meanings*. New York: Charles Scribner's Sons.

Neal, M. C. 1965. *In gardens of Hawaii*. new and rev. ed. Honolulu: Bishop Museum Press.

Orians, G. H. 1986. An ecological and evolutionary approach to landscape aesthetics. In *Landscape meanings and values*, edited by E. C. Penning-Rowsell and D. Lawenthal. London: Allen & Unwin.

Sandberg, N. C. 1974. *Ethnic identity and assimilation: The Polish-American community case study of metropolitan Los Angeles*. New York: Praeger Publisher.

Zangwill, I. 1909. *The melting pot*. New York: Macmillan.

Chapter 8

The Gardens of Hikone, Japan: Studying People-Plant Relationships in Another Culture

Lorisa M. W. Mock

SUMMARY. Looking at the role plants and horticulture play in other cultures can not only inform us and increase our understanding of other cultures, but can also introduce us to new ways of appreciating and interacting with plants and making the most of the benefits of horticulture in our own home culture.

This study, as an example, focuses on the small city of Hikone, Japan. It describes and examines the role of elite and vernacular gardens and public open spaces in community life and the various horticultural activities of the population. Issues such as social stratification, quality of life improvement, civic pride and the effects of modernization are stressed.

INTRODUCTION

Today's lightening-fast communications allow people around the world the ability to speak to each other instantaneously. Unfortunately, comprehension has not always increased at the same rate. It remains the task of those in the field of cross-cultural studies to produce constantly updated literature, providing information about other cultures to facilitate well-informed communication to help avoid the build-up of tensions and misunderstandings.

One of the ways to accomplish this is to focus on the elements of commonality between cultures, rather than always emphasizing the differences. Approaching a people by looking at their relationship with plants and the place gardens play in their culture is both an easy entree and very revealing about the culture.

This study begins with a discussion of the role of plants and gardens in Japanese society. This encompasses the broadly held attitudes and ideals and social activities that occur throughout the Japanese archipelago. The second portion of the study centers on the small city of Hikone, looking at how the residents of old and new neighborhoods garden, and other social activities centered around plants.

The information for this study on people-plant relationships in a small town in Japan was gathered over a two year residential stay in Japan. It is a collection of observations, interviews, and discussions. The aim was to highlight the role of plants and gardens in the everyday life of the Japanese. Japan's gardens embody a very large portion of the spirit of Japan and therefore occupy a very important position of significance in the society.

THE ROLE OF PLANTS AND GARDENS IN JAPANESE CULTURE

In order to understand the role plants and gardens play in the city of Hikone, we must first look at the position they take in Japanese society at large. Beyond the classic representations of Japan, the Pagoda, girls in kimonos, and the smooth-running corporations, Japan is famous for its aesthetic achievements, including its gardens. The gardens of Japan serve many purposes. In both the past and the present, they indicate power over people and over nature. They are places of pleasure–once the playgrounds of the aristocracy, now open to the general populace. They are representations of nature, glorifying the natural in often very unnatural ways. They are three-dimensional landscape paintings–living postcards of classical views of Japan, China, and mythological places. They are places of great spirituality, and they play an integral role in the religious life of Japan.

Japan's gardens are a living remnant of her historical heritage, and yet, the newest efforts in landscape design not only carry on the centuries-old art but are expressive of her newly acquired position of prominence and power in world affairs. In all manifestations of the art, Japan's gardens symbolize the relationship of the populace with nature and plants.

The Japanese relationship with plants ranges from the scientific to the spiritual. While they are at the forefront of plant breeding and seed production, many of the horticultural practices handed down verbally from gardener to gardener, in secrecy, from the beginnings of gardening in Japan are still in use, partly because of the strength of tradition in a very conservative field, partly because of the ritual associated with gardening, and partly because these methods work.

Plants and gardens feature heavily in the two main religions of Japan, Shinto and Buddhism. For example, the lotus is a ubiquitous symbol in Buddhist painting and sculpture, and a lotus plant is usually grown in the temple courtyard. Flowers are used to decorate altars. Gardens are featured in most temples and shrines; they are used for meditation, contemplation, and generally refreshing the spirit. Many temples and shrines are famous for particular plants, such as peonies, hydrangeas, or a venerable pine. Japanese people will travel a long way to see these in their seasons.

As an ideal, the Japanese would surround themselves with plants, and garden every iota of available space. In the overcrowded cities, this is increasingly the prerogative of the wealthy, but there are still many opportunities for people to commune with plants, in parks and gardens and urban scapes which increasingly include greenery.

Plants, flowers and natural themes feature heavily in the arts, particularly in painting with Chinese influence. Also from Chinese cultural routes, *Ikebana* (flower arranging) is a means of appreciating natural beauty and enjoying flowers close at hand. It is customary for the Japanese to place an arrangement in the entryway of the house where it can be seen by callers and welcome family members home. When a husband returns from his overly long, over-worked day, the sight of the calm repose of a most deceptively simple flower arrangement can soothe his nerves and marks that he can shed all the tensions of the day with his shoes and coat and relax in a beautiful and well-ordered place.

Ikebana also appears with a scroll painting in the honored place in a room, a niche called the *tokonoma*. In traditional homes this is the only source of decoration in the room, and in more awkward social gatherings, the only source of conversation. To be properly polite, a guest should initiate comments about the flower arrangement and any other works of art displayed, discussing their merits in some detail.

Other arts incorporate plants, such as in Noh theater, where a painting of a pine tree graces the back of the stage and serves as the only set. In poetry and literature the dominant imagery is associated with plants and flowers. One of the classics of Japanese literature is a collection of poems written by ordinary people. A botanical garden in the ancient capital of Nara features examples of all the plants mentioned in those poems.

In the traditional dress, the kimono, and in other textile art, flowers and plants are the most common motif. They are depicted with botanical accuracy and great attention is paid to the correct season of the plant. A summer kimono has summer flowers, like dahlias, the autumn kimono would feature chrysanthemums, and one would never see a rose combined with a cherry blossom.

Flowers and plants are not considered necessarily effeminate. The powerful warrior class, the Samurai, often had flowers such as the hollyhock, the cherry blossom, the camellia, as their family insignia. To this day, the crest of the Imperial family is the chrysanthemum, quite a far cry from the rampant lions and eagles of the European aristocracy. Historically, garden-making has been the pastime of the aristocracy, and later, the Samurai, and is, therefore, associated with power and wealth.

Flowers are present at all major functions, both as decoration and as presentations to participants. At an elementary school graduation ceremony, as an illustration, a bouquet of flowers is given to each graduate with the diploma. The children in the graduating class will each present the teacher with a flower, making a sizable bouquet. In exchange, the teacher will give a small bouquet to each child. Huge displays of flowers in big wreath arrangements on easel-like stands are used at weddings and, particularly, funerals. Fresh flowers are put on graves and are offered on the family altar to ones' ancestors. Gift-giving is an extremely complicated element of Japanese society, but flowers are always an appropriate gift.

The Japanese have special relationships with particular plants. Rice is by far the most revered. As the staple of their diet, that would make sense, but beyond that, rice is given a status approaching that of a deity. It is literally the food of the gods, in that it is presented as offerings to temples and shrines. Everyday a bowlful is presented to the family's ancestors at the family altar. There are separate words in Japanese for rice growing in the field, raw rice and cooked rice, and, in polite society, rice is always spoken of in the honorific tense. The origin of many of the festivals and holidays center around the rice-growing cycle. Even today, the holiday at the first week of May, called Golden Week, is when the rice is planted–at least in central Honshu. Rice is a strong symbol of wealth. Income used to be measured in *koku*–the amount of rice necessary to sustain a man for a year. It is associated with well-being, and being well-fed. One of the Japanese words for rice also means meal and, indeed, rice makes the meal.

"Returning to the rice" is one of the many terms for dying. Rice is considered by the Japanese as one of the main identifiers of being Japanese, hence they will resist importation of any other rice.

Some plants are nationally famous, for example, the 500-year old Japanese Black Pine in Hamarikyu-en, a public garden in Tokyo. Trips are made to gardens on the basis of what is in season: Plums, Cherries, Iris, Lotus, Wisteria, Hydrangeas, Peonies, Japanese Maple fall color, Chrysanthemums. Gardens will be known for featuring one or more of these. Each of these plants, besides being appreciated for their beauty, are held in a kind of melancholy poignancy by the Japanese. A relatively young woman will say she loves the peonies because they remind her of summers of her youth.

The most poignant plant in Japanese society is probably the Cherry tree. The cherry blossoms evoke a sense of melancholy in the Japanese soul. They symbolize the ephemeral nature of life. This makes for a feeling of sadness rather than of joy at this sign of renewed life. The Cherry blossom features heavily in Japanese poetry and literature and is a major symbol of national identity, even to the point that the song "*Sakura*" (Cherry Blossom) is practically an unofficial anthem. The approach of the *Sakura Zensen* (Cherry Blossom Front) is an event of national interest and is reported on the national evening news. The predictions for when the cherry trees will blossom is highly accurate and necessary to make arrangements for the cherry blossom appreciation parties held at this time.

Many other plants have symbolic value. In art, architecture and textile designs, the combination of plum, bamboo, maple, and pine is used. They represent spring, summer, fall and winter, respectively, as well as several positive attributes, such as longevity, perseverance and new life. New Year's decorations consist of pine, bamboo, and plum. In stories, bamboo groves are frequently the settings for intrigue, change and mystery; the childless old man and old woman find a child in a bamboo stalk. In one famous tale, the pine in a garden is used by the patient laborer as a ladder to heaven. Plants gain much of their symbolism from folklore.

Through these stories and by example, Japanese children are taught from a very early age to respect plants and nature, in general. Horticulture is included in the curriculum. And even when they are learning to read, many of the basic first words are plant-related. Indeed, every first grader can read the word *kiku*, while we would have a bit of difficulty teaching "C" is for chrysanthemum to our six-year olds.

Throughout their school years, Japanese students are taken regularly to gardens and other places of cultural interest. By the time they reach adulthood, they are well-versed in the symbolism and meaning inherent in the many different types of gardens. Visiting gardens and temples and other places of natural beauty is a national pastime. People travel in family groups or in large tour buses, hurriedly shuttled and herded through a usually quite ambitious itinerary.

In this way, the gardens and parks serve very important social functions besides being the repository of a cultural heritage. Many families go to relax in parks and gardens to escape their too-small living quarters and cramped cities. Groups go on tours to temples and their gardens partly for the interest, but largely because it is a time to relax with their friends and meet new people. The tours are regarded as extending ones' education as well. Courting couples often visit gardens, as they are respectable places to be seen in attendance. Grandparents, spending the time with their grandchildren that they did not always have for their own children, bring their little charges to enjoy the beauty and peacefulness of the gardens.

To say that the Japanese who visit gardens always behave in a reverent and restrained way would be quite wrong. They behave in parks much the way other people do, enjoying the weather and their friends–often quite boisterously. Most groups are a little more restrained in the temple gardens, particularly the meditation gardens. They are, on the whole, very careful of the plants in the gardens and stay to the paths, and, apart from the occasional case of name-carving on the trunk of a tree, only cause destruction by their sheer numbers.

An indication of how much the Japanese appreciate plants and gardens, and not just their own, would be the very favorable response to the 1990 Osaka Flower and Greenery Expo. Real enthusiasts visited the Expo more than once, in spite of the expense; it was a status symbol to have gone. Companies entertaining business associates treated them to a visit. Furthermore, there was a spill-over effect of related horticultural activities, like flower shows, competitions, and television programs, in addition to the many posters featuring flowers, plants and gardens and the live plant displays in public areas, such as train stations, to advertise the event.

The Expo featured garden displays from all over the world. The theme of the event was the idea that humans can live in harmony with nature. Unlike the bulk of traditional Japanese garden styles, the displays did not recreate nature and had plants in anything but natural situations. One section of the main display was a golf course placed adjacent to an artificial lake, dubbed the Sea of Life. In the wake of populist movements to have no more golf courses on Mt. Fuji because of the contamination of the water supply by herbicides and fertilizers, the organizers wanted to prove that "man could coexist with nature." One week after the Expo opened, however, all the fish in the Sea of Life were dead, killed by the chemicals used on the golf course.

The Japanese are very appreciative of, and consider themselves very close to nature, but demand a perfection which puts them at the top of the list for amounts of money spent on pesticides used in food production. In recent years several consumer organizations have evolved from food cooperatives formed by housewives seeking foodstuffs produced with less or no pesticides. This movement has been a catalyst for bringing many more women into the political arena.

Plants and gardens, then, feature heavily in the religious, social, and political life of Japan. Gardens are a highly respected aspect of the Japanese cultural heritage, places for important social interaction and sites of the basic and continuing education of the Japanese people. As a nation, the Japanese are very aware of plants, of their seasons and habits, and plants are both functionally and symbolically important in the culture.

THE GARDENS OF HIKONE

To further understand the role of plants and horticulture in Japanese society, let us concentrate on the town of Hikone and examine the part plants play on an everyday basis and through the different aspects of the life of the citizens.

Hikone is a town of 100,000 people on the shores of Lake Biwa, Japan's largest, and the world's third oldest lake. It is famous for its castle, which is one of the only three remaining non-reproduction castles in Japan. The area is one of great natural beauty which has been historically important in commerce and stands on the road between the ancient capitals of Kyoto and Tokyo. The lake is surrounded by mountains and the plains, mostly marshland reclaimed from the lake, are productive farmland.

This geography of very flat plains surrounding the lake, girded by mountains, makes for a very sharp delineation between the cultivated "safe" land and the "wildness" of the mountains. People rarely go wandering in these mountains; they tend to visit designated spots and temples to appreciate natural beauty rather than cut new paths. Equally, they are very familiar with cultivated plants and often label all others "mountain" plants.

The appreciation of nature among the Hikone citizens appears to be highly romanticized. When they go to visit a place of natural beauty, it is a very controlled experience–one does not wander very far from the car, paths are well maintained, with concrete handrails fashioned to look like rough logs. At many of these places, including the shores of Lake Biwa, litter and other evidence of human passage are a high profile feature. Part of this is caused by the very fast evolution to being a plastic throwaway society, whereas before bamboo and other biodegradable containers were used. Even more crucial, though, is that the Japanese do not tend to see themselves as being separate from nature and therefore do not see a conflict with the detritus of humanity being overlayed on the natural scene. Efforts, however, are being made to clean up Lake Biwa and its beaches and generally reduce litter in public areas.

Hikone is a rapidly modernizing city. The city center has recently been renovated. It now has tall, modern buildings, wider streets, broad sidewalks and generous plantings of trees and shrubs. The beautification scheme is gradually being carried out on all the major thoroughfares of the city. Most citizens welcome these changes, pleased with the clean and pretty image and the convenience of the new center. Others communicated a sadness at the loss of the sleepy, rural feeling of the town before the renovation. In any case, the changes the citizens have made indicate a much increased empowerment to influence their own environment and are a source of great civic pride.

Hikone castle, its associated garden, Genkyu-en, and the cherry trees surrounding the moat are also sources of pride for Hikone citizens. Most residents only visit the garden when they have visiting friends and relations. One group agreed that they all felt a tremendous feeling of wealth knowing that they could go to the garden if they wanted. The Ii family, long the lords of the fief of Hikone, and mayors of the city until two years ago, gave the castle and garden to the city in 1945. Prior to that time, the ordinary townspeople would have never been able to visit the garden as it was the inner sanctum of the Lord Ii. The garden represents to the present day citizen a feeling of holding a piece of the common wealth.

The Ii family still owns one of the original villas and gardens. There are other privately-owned elite gardens in the city, unattainable to the average citizen, as they are surrounded by high walls. These gardens usually contain a pond or some illusion of water, large rock works that represent mountains, venerable pines, an assortment of azaleas, and a variety of garden ornaments such as lanterns and stone water basins. The larger gardens will have some open area, either graveled or grassed. Most of the elite gardens of Hikone are rarely over a quarter acre. Having space for a large garden and the element of privacy are indications of wealth. Entrance to most of the temple gardens used to be restricted, making them elite gardens; however, almost all temples have now opened their doors and charge the admission that funds their operations.

Garden maintenance is very expensive. The climate is mild with high rainfall, so plants grow quickly and need radical pruning at least twice a year. One Hikone resident, who had been raised in Canada and then returned to Japan to inherit his father's large house and garden, worked steadily in his garden on weekends and was very proud of it. He often complained, however, of the high cost of having the gardeners in to do the heavy pruning. The style of gardening, a regimented natural, that uses potentially large trees in small areas, demands a great deal of maintenance work.

There is a tacit feeling that an owner of a good sized garden has an obligation to keep it tidy. Conformity is very strong in Japanese society, and to have the neighborhood older ladies shaking their heads and tisking over your garden wall is a very bad sign. So it is that gardening and other yard care has a certain zealous fervor to it. Trees and shrubs are cut back to an extreme, and every single unwanted plant is cut out one by one. After the garden crew has been through a group of company-owned houses, for example, the yard has an almost scoured look to it.

The gardening crews hired for the elite, middle class, and corporate gardens do most of their work by hand. Large, mechanized equipment is impractical in these small spaces, and is also considered undesirable. The trees are pruned or clipped by hand by the men of the crew, while the women do all the weeding and raking and removing of prunings. Weed-whackers are the one power tool used, and they are used a good deal, even doubling as lawn mowers as there are very few large expanses of grass in these gardens.

Gardening has been an elite hobby for centuries and to this day, designers and other elite gardeners are held in high esteem. The apprenticeship for gardening used to be very long, involve extensive travel to see the famous natural sights, and would include the learning of the garden techniques handed down orally from gardener to gardener. A certain amount of magic surrounded learning garden lore and even now there is a mystique that surrounds the art of gardening. After all, if a garden is considered a place of spiritual importance, it is only natural that a person would be a little nervous of getting it right. Interestingly enough, though, many gardeners in the garden crews are farmers, earning extra income in the off-season, and generally people who work outside are not held in general esteem.

It seems, then, that the average citizen has mixed attitudes towards gardeners, gardens and plants, and nature in general. The art of gardening is mystified, but gardeners are not always admired; plants are revered but at times pruned with something approaching viciousness, and nature is romanticized but not well preserved. However, when one looks more closely at the gardens of ordinary, middle-class citizens, unambivalent feelings of love, of nurturing and sheer joy in gardening can be observed.

The streets of the historic area of Hikone are very narrow, flanked by deep ditches to accommodate the rainfall and crowded in by a continuous line of house fronts; often the glass sliding doors of the entryway of the house open directly onto the road. The farther back the house is from the road, and the more dividing lines between public and private spaces, the more elite the house is. The average house has a path of about a meter and a half between the road and the actual entry to the house. Within that width and extending around the corner of the house there will be a gardened area. Small trees, such as the plum yew, camellias and azaleas are most often used in these gardens and are arranged with large rocks and gravel. These form what can be described as a three-dimensional representation of a landscape painting which is viewed from inside the house, with the garden wall as the backdrop. These gardens are considered part of the interior decoration of the house, in a way; they are not used for outdoor recreation.

This category of garden is best described as the middle class garden–most Japanese describe themselves as middle class. There is a great deal of tradition and conformity that determines the design of a garden of this type. One gardener in the study kept a portion of his garden separate from the traditional part so that he could spend his free time experimenting and being creative. It is very important to have the right kind of garden for one's status. A woman with a young family was waiting until they had enough saved up to do the garden "properly." In the meantime, the area around her house was practically bare.

In the older houses, where the door opens onto the street and allows entry into a very large entryway area where business is conducted, what garden there is will be around the back of the house, not visible from the street. One unassuming house in the neighborhood studied had a grove of twenty-five carefully pruned, very old pine trees behind the house. It is considered appropriately modest to keep one's garden hidden, but there is an historic reason for hiding the house's garden. The tax laws and sumptuary laws of the 1600's were very strict. It was wise to hide whatever assets one had; to flaunt it was to lose it. Today's changed values and increased wealth has meant that gardens have come out into the open a bit more.

A garden belonging to someone who has benefited from the recent boom economy will have all the elements of an elite garden: rocks representing mountains, some water, or illusion of water and some open space. Classic Japanese gardens often have the symbols of the crane and the tortoise represented in either rock or clipped shrubs. In the nouveau riche garden, these are so lacking the usual subtlety, that they seem almost caricatures. The gardens in this category frequently include some Western element. One notable garden had a very traditional mountain cascade rock arrangement with a rock bridge, but out on the grassy area were stone figurines of Snow White and the Seven Dwarfs. All in all, though, the improvement of the economy since the war has meant that more and more attention is being paid to outward display. The areas of peoples' houses visible to the public are being dressed up with decorative gates and plant displays.

In the newer areas of Hikone, more space is being allowed around the houses for gardens. Another way to look at this is that the houses are no longer being built to fill the whole lot. Green space around the house, as much as three meters deep, and construction that allows for light and breezes to enter the house has become preferred. The houses are still surrounded by walls for privacy, but often these walls are breached with fencing or hedging, allowing the passage of air and the admiring glances of neighbors. So it is that privacy has been sacrificed for a new comfort of living environment and outward show of that comfort.

The local town paper ran a series of short articles exhorting citizens to plant hedges instead of building walls, to plant vines to cover existing walls and to generally make more green areas in their neighborhoods to improve environmental and life quality. Resistance to having public green areas and too much plant material is the high cost of maintenance. As a general rule, the city takes care of the important public areas, such as the castle, moat area, downtown and sportscenter. The local neighborhood associations get together and take care of the small public areas in the neighborhood, like playgrounds, sitting areas, and neighborhood association building grounds. The neighbors meet on a Saturday morning, turnout is amazingly good, and they descend like locusts on the greenery. The work is done amidst much chatter, joking, and is done without much leadership, each person taking it upon himself to do what he thinks needs to be done. All greenery that is not shrub or tree is scrubbed out, the shrubs and trees are hacked back and everything is raked. This approach results in a rather uneven maintenance–from jungular to barren.

The popularization of the car has changed the gardens of Hikone neighborhoods. One cannot purchase a car without proof of having a place to park it. Available open spaces are given over to parking lots, then, and people find room in their front yards to park a car, sacrificing garden and, in some cases, living space. To accommodate the car, new construction has literally turned the houses around; the car space and garden are on the side of the house. The gateway permits the passerby a much better view of the garden, once the inner sanctum. In the new neighborhoods, roads are built with the car in mind. While still not allowing for on-street parking, the roads are wider, and are flanked with sidewalks rather than deep ditches. The life of the individual household does not spill out into the streets as it does in the older areas.

In the old neighborhoods, many of the streets are only just wide enough for a medium-sized car. This lack of space, along with the fact that the sunniest space in a crammed neighborhood is in the streets, means that many hobby gardens are in full public view. Some seem to be decorative, others purely recreative, but all demonstrate a great deal of determination on the part of the residents to grow something and to generally enhance the quality of their lives and living areas. Some gardens are almost Western style mass displays of flowers, others rotating displays of pot plants, changing with the seasons, and others flowers or bonsai being raised for competition or sale. Whereas just a few pots perched on the edge of the drainage ditch can stretch the definition of garden, there is no doubt that they represent the spirit and love of gardening and determination to garden in spite of the lack of space.

When walking in the early evening, one can find people tending these plant companions: watering, weeding, pinching back. Sometimes caring for the plants seems to be just an excuse to stand outside, to greet neighbors, to see who is coming home, who is doing what, and enjoying the end of the day, with the chores all done.

The display of plants outside a home often reflected their caretaker's character and lifestyle. A very pretty display, neat and expertly cared for, belonged to a very efficient, attractive housewife who was highly involved in neighborhood life. In comparison, a few weatherbeaten pots holding a few pathetic, beleaguered plants belonging to an aging woman whose ministrations to the plants involved the barest of care and a great deal of nagging and approbation as she tended them, carefully keeping an eye on what her neighbors were doing. In another case, an extensive collection of amaryllis in a gradually worsening state, told of an ill tender who got well, came back and a miraculous recovery of the plants ensued. Some plants lined up in regular intervals along a fence belonged to an older gentleman, who, each evening, wearing the traditional kimono and *geta* sandals, would march past his plants as if he was conducting a review of his troops. Of course, none of his plants had one petal or leaf out of place and they almost seemed to stand at attention.

Tucked in among the houses and fields, on land impossible to use for parking cars, are people's vegetable and cut flower plots. Vegetables are quite expensive, which would be reason enough to grow one's own, but most people said that they gardened for the pleasure of it, for the exercise, and as something the whole family could do together. Interestingly enough, the word in Japanese for the garden around your house or a display garden is different from the word for vegetable garden. They are seen as quite separate–vegetable gardening is much more approachable to the average citizen.

The different gardens of Hikone vary a great deal in their uses and meanings. The elite gardens demonstrate power and position. The public gardens indicate the empowerment of the citizenry. Middle class gardens in the old and new neighborhoods effectively improve the living environment of the residents. And the hobby and display gardens, as well as the vegetable plots, are the result of the love of plants and gardening and desire to improve life quality.

As in many Japanese towns and cities, temples and shrines feature in every Hikone neighborhood. They are havens of greenery. Even those temples and shrines in cramped neighborhoods afford one a feeling of spaciousness because of the magnificence of the architecture, the size of the accompanying gardens, and the size of the trees. Temples typically have gardens featuring many different types of plants; but most usually have pine, lotus, wisteria and azalea. Shrines are almost always set in a grove of Cryptomeria. Whereas most trees in other parts of the city are carefully pruned and their growth is checked, the trees in the temples and shrines are left to get quite tall, adding to the perception of the location as a place of power. The largest and oldest of these trees are girded with belts of straw to mark them as a place where the gods are likely to be.

This religious association with plants is further extended to the streets of Hikone through the small roadside shrines and altars which are all over the city. These are decorated with fresh flowers, usually bunches of long-lasting cut flowers sold in prearranged posies at the supermarket. People stop to pray at these shrines and altars on the way to the store, in a quiet moment of the day–or, in the case of the bank manager, quickly while on coffee break. The same type of flower bouquets are used to decorate family graves, and are sometimes placed at sites where someone died in a road accident. On an everyday basis, flowers are a symbol of reverence, care and remembrance.

Another major feature of the neighborhoods in Hikone are the schools. The school grounds themselves offer a large proportion of the neighborhood's open space, and they usually include shrubbery and flower displays. From the first years of elementary education, the children learn about plants through learning of botanical facts and processes, dissecting plants to identify different parts, and observing plants growing in the wild.

The curriculum also includes extensive sections on horticultural and agricultural industries.

The children also learn how to grow and take care of plants. Both the indoor and outdoor displays of flowers are raised and maintained by the school students. With a partner, a student will raise several plants from seed, such as salvia. These will be put in a planter and they form part of the display in the garden. In a typical elementary school grounds the garden also includes plants bedded out, a pond with lilies and water hyacinths, vegetable plots and some permanent shrubbery. All of these require ongoing care and produce a display that must inspire a certain amount of pride and feeling of accomplishment in the children.

In addition to raising flowers to decorate the interior of the school, the students are responsible for cleaning the school building. This seems to instill a sense of stewardship and respect for their surroundings. Twice a year, parents go into the schools on a Saturday morning to help with the cleaning of the building, and the maintenance of the outside grounds. This, too, is a social event and contributes a great deal to the solidarity of the parents as a group.

Horticultural activities, whether they are visiting a temple to see the peonies, or admiring a neighbor's potted plant display, or working together on the plants in the local playground, provides many opportunities for the different generations to relate in a positive way. These activities can also help other groups to develop cohesion. On the very local level there are the neighborhood work days that help neighbors to relate well, and on a city-wide level are such things as the cherry blossom appreciation parties. Inspired by the beauty of the blossoms and the warmth of the *sake*, rice wine, people renew both business and social contacts and take care of necessary politicking. On an individual basis, conversation about a garden or a prized plant or flower arrangement can ease relations, start relationships and allow a Japanese person to relate to anyone–even one from outside their culture.

CONCLUSION

In historical and modern Japanese society, plants play an important role in cultural expression through the gardens, the arts, and symbolism. The religious and spiritual life of the Japanese is centered around gardens and plants and inspires a reverence for nature. Meanwhile, though, gardening and usage of plants in Japan is one marker of social class and demonstrates a paradoxical relationship with nature which is made up of appreciation, control, and abuse. Japanese children are brought up to understand and appreciate plants which should prepare them to be caring and responsible for their environment and results in a populace that is horticulturally quite knowledgeable. Plants and gardens are the basis of a great deal of social interaction and contribute to a continuity of cultural meaning in Japanese society.

Chapter 9

From Open-Mindedness to Naturalism: Garden Design and Ideology in Germany During the Early 20th Century

Joachim Wolschke-Bulmahn
Gert Gröning

SUMMARY. The introduction of exotic plants from other continents during the past centuries affected German garden culture and many exotic plants found their way into German gardens. But the increasing nationalism in the late 19th century influenced this international orientation in garden culture. A 'horticultural' conservatism emerged, which introduced nationalistic and racist ideas into garden design. This paper will discuss this development with the example of Willy Lange, one of the most important garden architects in Germany in the early 20th century. He developed concepts of the so-called "nature garden" as expressing what he considered to be 'truly German' garden art.

Gardens and plants are often considered to be free of ideology and to be immune to political temptations. Many of those engaged in garden design placed and place emphasis on their political abstinence. This paper will discuss a specific aspect of the recent history of German garden design– the ideology and the political beliefs on which particular ideas about garden design were based. Gardens and plants served some German landscape architects in the late 19th and early 20th centuries for strengthening nationalistic and racist feelings.

But first of all the history of garden culture in Germany was marked by open-mindedness and an international orientation. The discovery of the so-called New World by Europeans led during the past centuries also to an increasing interest in the vegetation of other continents. The enthusiasm about the introduction of unknown exotic plants from continents like America and Asia since about 1600 affected European garden culture enormously. This is also to be understood in front of the fact that the vegetation of Central and Northern Europe during the past 20,000 years following the last glacial period is characterized by a limited number of plant species. The majority of plants used, for example, in German gardens are to be termed 'foreign.' Besides agricultural plants such as potatoes and corn, numerous ornamental plants, shrubs and trees like dahlias, chrysanthemums, tulips, lilacs, rhododendrons and chestnuts were introduced and shape the appearance of most of our gardens today (Figure 1).

The technical development of the nineteenth century, the displacement of the sailing ship by the steam boat and the construction of railroads shortened both distances and time horizons. The systematic exploitation of flora from other continents in European garden culture increased rapidly. For example, in the nineteenth century numerous expeditions were organized to search for unknown plants in other continents precisely in order to make them available to European garden culture. Many of the plants imported since the sixteenth century were introduced between 1800 and 1900.

Thus, in the nineteenth century the traditional "Grand Tour" to Italy and Greece was replaced by the "Grand Tour" of modern times, to Africa, America and Asia. This new "Grand Tour" influenced the various fields of science and also of art. An impressive example of the influence these new forms of the "Grand Tour" exerted on European art is the symphony "From the New World" by Antonin Dvorak. In September 1892, Dvorak left Europe for an extended stay in the USA. As a result of the enormous onrush of new impressions he composed early in 1893 his symphony "Aus der Neuen Welt," in which he also introduced elements of native music (1).

Also the field of garden design was influenced by such impressions. On their journeys numerous garden designers and garden writers became acquainted with the plants of these other countries as well as with landscape scenes of unknown beauty. How impressive such journeys have been may be shown in the case of the Irish garden writer William Robinson. At the end of the 1860s he visited the Alps, and in 1870 he published as a result of this journey the book "Alpine Flowers for Gardens," "to all intents and purposes the first British book on the subject. In that year

FIGURE 1. Title-page of Joannis Burmanni's *Rariorum Africanarum Plantarum* (Amsterdam, 1738).

JOANNIS BURMANNI,

Med. Doct. & in Horto Medico Amstelaedamensi Botanices Professoris,

RARIORUM AFRICANARUM

PLANTARUM,

Ad vivum delineatarum, Iconibus ac descriptionibus illustratarum

DECAS PRIMA.

AMSTELAEDAMI,

Apud HENRICUM BOUSSIERE.

MDCCXXXVIII.

Robinson visited, and delighted in, the United States of America, and also produced 'The Wild Garden'" (2).

In this book Robinson expressed his impressions, for example, as follows: "It is that which one sees in American woods in late summer and autumn when the Golden Rods and Asters are seen in bloom together. It is one of numerous aspects of the vegetation of other countries which the 'wild garden' will make possible in gardens" (3). The whole idea of the "wild garden," as developed by Robinson, was "placing plants of other countries, as hardy as our hardiest wild flowers, in places where they will flourish without further care and cost" (4).

In Germany after 1800, more and more horticultural societies were established, which spread the knowledge of and the interest in foreign plants within different groups of society. The most important of these societies was the Association for the Promotion of Horticulture in the Royal Prussian States (Verein zur Beförderung des Gartenbaues in den Königlich-Preußischen Staaten), which was licensed by the Prussian king Friedrich Wilhelm III in 1822. One of its goals was "the promotion of horticulture in the Prussian state, the cultivation of fruit-trees in all its branches, the growth of vegetable and commercial herbs, the cultivation of ornamental plants, of plant forcing, and of visual garden art" (5).

These societies introduced numerous foreign plants into agriculture, horticulture and also into the field of garden design (Figure 2). Plants, such as wheat, corn and potato, which had proven their economic value for agricultural and horticultural purposes, characterize today's German landscapes. And also today's gardens contain numerous foreign ornamental trees, shrubs and flowers.

But during the late 19th and early 20th centuries this international approach to horticulture was called into question by a number of garden designers and opposed by nationalistic ideas about garden design. In particular during the first half of the 20th century, increasing nationalistic tendencies influenced ideas about garden design and plant use in Germany.

The most important of these garden architects was Willy Lange (1864-1941) (Figure 3). From 1900 until the 1930's he published a series of articles and books in which he described and propagated his concept of the nature garden, which was clearly hostile to internationalism; it was founded on nationalistic and racist ideas.

The political trends in Germany in the time of Lange can be characterized very roughly as follows. In 1871 Germany won the war against France and the German Reich was founded. German ambitions to become a world power promoted ever-increasing nationalistic tendencies.

FIGURE 2. A new cultivated dahlia, named Queen Luisa, on display at an exhibition in Germany 1907 (Möllers Deutsche Gärtner-Zeitung, *22,* 1907, 41, p. 482).

FIGURE 3. The garden architect Willy Lange (1864-1941) and his wife (about 1910). Photograph from W. Lange, *Gartenpläne* (Leipzig, 1927), p. 401.

Not international development, but nation and region became important reference points. At the same time, social and political changes produced by the process of industrialization furthered a middle-class longing for nature and landscapes which had not yet been influenced by industrialization; this longing was manifested in movements such as the nature- and the home-protection movement as well as the so-called *Wandervogel* (6).

The rapid progress of science in the second half of the nineteenth century, an international phenomenon, assumed a specifically national form in Germany. Darwin's book, *The Origin of Species*, in which he developed his theory of evolution, received particular attention. In Germany Darwin's theory of evolution was applied not only to natural, but also to social phenomena and gained considerable ideological influence. The principles of the survival of the fittest and of natural selection became fundamental parts of Volk-nationalistic ideology and were transferred to all spheres of social life. Small circles of people who claimed to follow specific German ideals and searched for true 'Germandom' sprang up.

By referring to Darwin, not necessarily to his book, an increasing number of Volk circles tried, for example, to justify social inequalities which existed in Germany. On an international level such social-Darwinistic assertions were invoked in attempts to prove the cultural superiority of Germany over other nations. The course of evolution which Darwin had described was used to postulate an inevitable course of social evolution from lower to higher cultures. Pure race and, thereby, pure culture were deemed prerequisites for the further evolution of Germany. In no other country of Europe "did the ideas of Darwinism develop as seriously as a total explanation of the world as in Germany" (7).

What then were the nationalistic and social-Darwinistic aspects of Lange's ideology which were worked out in a period of such social development? How did garden theory fit into National Socialist ideology?

Some of the main aspects of Lange's concept of the nature garden can be described briefly as follows. The designer of a nature garden à la Lange should not use geometric and architectural forms; the design of the garden should be informal. The purpose of Lange's nature garden is not primarily to serve mankind; nature, especially plants and also animals, have equal rights. Native plants are preferred to foreign ones. Moreover, Lange interprets the garden as a part of the surrounding landscape, to which it is to be subordinated. The cutting of trees, shrubs and hedges is rejected in the nature garden as anthropocentric, a sign of human hegemony over nature. Instead, the laws of nature should be followed and should be spiritually heightened, in order to produce an artistic display of nature in the garden (Figure 4).

Lange's concept of the nature garden is remarkably similar to the concept of the "wild garden," which the Irish landscape gardener William Robinson had developed in the final quarter of the nineteenth century (8).

The nationalistic tendency of Lange's ideology goes hand in hand with his idea that garden art is a constituent of national culture–indeed, in his thinking culture could only be national. He vehemently rejected the idea that "art could be international," declaring: "Let us find the national style for our gardens, then we will have art, German garden art. As long as there exist different nations, there must exist different national styles" (9). The alleged evolutionary superiority of the German people over other nations should be proven in the field of garden architecture. This nationalism was also oriented against the internationalism of the industrial workers movement and its fight for more rights and democracy.

For Lange garden art became a matter of *Weltanschauung* (ideology). The process of cognition in science required a new *Weltanschauung* which, Lange believed, was characterized by love of nature and consideration of its laws. He therefore saw his concept of the nature garden in opposition to the architectonic garden style which was the dominant tendency at that time in German garden design. According to Lange, the architectonic garden style expressed the anthropocentric and unnatural attitude of other cultures and of lower stages of cultural evolution. For Lange the nature garden was the highest stage of garden culture: "The highest development of garden design is consequently based on the scientific *Weltanschauung* of our times and is reflected in the artistic nature garden" (10).

Lange understood his ideas about the nature garden as a contribution to the *Kulturkampf* in the field of garden art. He claimed that the ability to implement the highest stage of garden art was a racial characteristic of the Germanic or Nordic people. He explained, for example, the contrast between the formal French and the informal English garden style by "different *Weltanschauungen* and these again by the differences between the souls of the two races." In the "formal garden," according to Lange, the Nordic race "spiritually perished in the race-morass of the South" (11). In 1933, on the coming to power of National Socialism, he defamed the formal garden in the National Socialist magazine *Deutsche Kulturwacht* as characteristic of the "South-Alpine race" and as an attempt of this "un-Nordic race" to weaken the "Nordic race" and to strengthen international, anti-German forces (12) (Figures 5, 6).

FIGURE 4. The house of Willy Lange (*Gartengestaltung der Neuzeit*, 6th edition, Leipzig, 1928) plate XII.

FIGURE 5. Garden at the Monte Pincio; according to Willy Lange this formal garden design would have been a sign of cultural inferiority. Photograph from A. Hogenberg, *Hortorum Viridario Rumque Nouiter In Europa* . . . (1655).

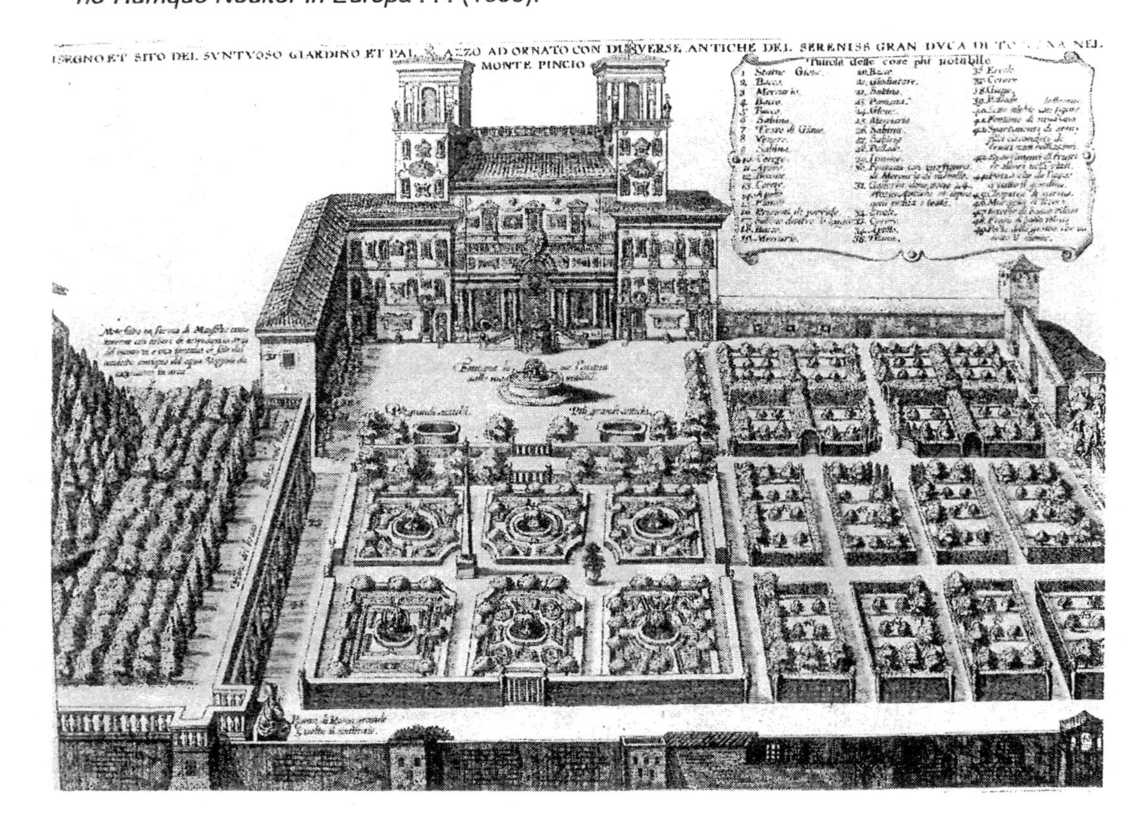

FIGURE 6. Carpet of *Sedum spurium* in the garden of Willy Lange, designed as a nature garden (W. Lange 1928: *Gartengestaltung der Neuzeit*, 6th edition, Leipzig 1928, p. 264).

Lange furthered the idea of the superiority of the German people which was accepted by many Germans during the Imperial period and the Weimar Republic and which, finally, became a State doctrine under National Socialism. The following quotation gives an idea of what he stood for and at the same time of his understanding of cultural development as an evolutionary process: "History will call this new stage of garden style, which is firmly based on its precursors, the stage of the German garden style. Germany has been chosen to lend its name to this style in the history of gardens and to become once again 'an improver of the world' " (13).

A fundamental constituent of Lange's conception of culture, and of the National Socialist Blood-and-Soil-Ideology, too, was the idea that societies on a high cultural level are "rooted in the soil" (*im Boden verwurzelt*), that they live in a close union with the soil and, therefore, the landscape. This idea clearly had a racist and antisemitic background. For example, the Jewish people were often described by Volk groups in Germany as homeless and as nomads, such terms indicating that Jews, Sinti, Roma and other people could not be rooted in the soil and, therefore, could not have a cultural standard.

A consequence of the idea that the German people were rooted in the soil was that every German was thought to require an appropriately designed spatial environment. If the Germans wanted to preserve their power and their cultural standard, then, Lange asserted, all artistic activities had to take as an example the local landscape–and this included the garden and its plants.

The subordination of the garden to the landscape and the use of indigenous plants thus became an essential criterion of nature garden design. Scientific evidence for this garden ideology was sought in such emerging disciplines as phytogeography and plant sociology. By such means an attempt was made to define the kind of vegetation which was considered natural for specific regions or, as it was then preferred, landscapes in Germany. Thus, Lange wrote: "The German nature garden carries the biological idea of design further toward artistic freedom: Our feelings for our homeland should be rooted in the character of domestic landscapes; therefore it is German nature that must provide all ideas for the design of gardens. They can be heightened by artistic means, but we must not give up the German physiognomy. Thus, our gardens become German if the ideas for the designs are German, especially if they are borrowed from the landscape in which the garden is situated" (14) (Figure 7).

A closer look at the social and spatial diversity in the German Reich, however, could have revealed contradictions of the close union of a people with a specific landscape. But it was argued that the German people consisted of a multitude of different tribes which, at last, were united in 1871 in the German Reich. This idea, for example, is also reflected in the constitution of the Weimar Republic, established in 1919. In the preamble of the constitution the German people is characterized as "united in its tribes."

Thus national identity in garden art was not endangered by diversity. Quite the contrary, it could be made legitimate by referring to regional diversity in garden design. This aspect correlated in a specific way with German history, which lacked a united national tradition until the nineteenth century and was marked by a tradition of regionalism, particularism and provincialism (15).

But some experts reacted skeptically towards the increasing infiltration of nationalistic thought into the field of garden design and pleaded for an international and scholarly orientation. For example, in "German Trees in Garden Art" (1911) Curt Schuerer argued as follows: "No wonder, with the increasing amalgamation of the interests of particular countries and peoples internationalism flourishes more and more and as a prompt reaction, emphasis on nationalism is good. Yet, it is questionable if it is better for a nation to keep its distance from what is foreign, to avoid interbreed-

FIGURE 7. "The border between garden and landscape disappears." Garden designed in the natural style by the garden architect Hermann Mattern (H. Mattern 1938: Freiheit in Grenzen, Kassel).

ing with it or to integrate with it. History has shown sufficiently that only those people were really powerful who showed a strong capacity of absorption and that even the fear of being swallowed is the best sign of increasing weakness. . . . Therefore, let us not make the beauty dependent upon national concerns and let us be guided in the selection of trees and shrubs for our parks and gardens only by suitability, beauty and by unrestricted love for nature. Fortunately plant formations in their struggle for survival do not take into consideration political borders" (16).

It should be mentioned that Lange's concept of nature-garden design in adaptation to the landscape allowed the use of foreign plants. In his nature garden it was permitted to complement those plants, which were considered to be characteristic of a local landscape, with foreign plants. The only prerequisite was that foreign plants matched the local plants physiognomically, in order "to express the laws of life, the manifestation of the plant world in a deepened, characteristic way" (17).

In the period before the National Socialists came to power, Lange's ideology of the nature garden received new impetus from the garden architect Alwin Seifert. Seifert (Figure 8), one of the leading garden architects of National Socialism, published his concept of a nature garden in 1929 without, however, any reference to Willy Lange. Seifert introduced the category of "rootedness in the soil" (*bodenständig*), by which he was obviously referring to the National Socialist "blood-and-soil ideology." More doctrinaire than Lange, Seifert required that the design of the garden should match the surrounding landscape. For him this included, for example, avoiding the use of foreign plants where they could be seen by those wandering by.

Like Lange, Seifert saw his concept of the "soil-rooted" (*bodenständigen*) garden explicitly as a contribution to a nationally oriented culture which would help to fight international tendencies. He wrote in 1930 that he introduced the category of "rootedness in the soil" very consciously into the art of gardening: "I wanted to bring garden art into the struggle in all living spaces which has broken out in our days between 'rootedness in the soil' and 'supra-nationality'" (18).

According to Seifert, this struggle was "a fight between two opposing *Weltanschauungen*: on one side the striving for supranationality, for equalization of huge areas, and on the other the elaboration of the peculiarities of small living spaces, the emphasis of which is rooted in the soil" (19). This shows evidence of nationalistic thinking and how it was reflected in the seemingly apolitical field of garden architecture. Seifert wanted to substantiate his plea for "rootedness in the soil" (*Bodenständigkeit*) by science; primarily, however, he attempted to adapt garden architecture to the aims of National Socialism.

With the advent of National Socialism many garden architects tried to create a typical National Socialist garden style. Leaving aside the old tradition of the so-called *Bauerngarten*, they remembered Lange's ideology of the nature garden. They wanted to show the way to a "race-specific," "blood-and-soil-rooted," "homeland-oriented" garden design, as wrote Hasler, one of Lange's students (20). Headlines such as "Plant-Sociology and the Blood-and-Soil-Rooted Garden" and "German Garden Art," under which such ideas were propagated, can serve here as examples. Hasler, with his book *German Garden Art*, claimed "to help to propagate these [that is, Lange's] most German ideas of garden design . . . to help to give the German people its characteristic garden and to help to guard it from unwholesome alien influences" (21).

FIGURE 8. Alwin Seifert (to the right) is introduced to Adolf Hitler, Photograph from *Die Strasse*, 3 (1936), p. 642.

Finally, the subordination of the garden to the landscape by the use of native shrubs and trees became an ideological doctrine. Some garden architects tried to legitimize the elimination of foreign plants with racist and nationalistic, as well as with 'ecological,' arguments. The garden architect Kraemer made plant sociology the main foundation of his racist garden concept. He wrote: "But we still lack gardens which are race-specific, which have their origins in nationality and landscape, in blood and soil. Only our knowledge of laws of blood and of spiritually inherited property, and our knowledge of the conditions of the home soil and its plant world (plant sociology), enable and oblige us to design blood-and-soil-rooted gardens" (22).

In the landscape, German garden architects, nature preservationists and others were even more militant. There they invoked a landscape design which they thought would be typical for the ancient Germanic tribes (Figure 9). The unmerciful extirpation of foreign plants in the German landscapes became essential. For example, in 1942, a team (*Arbeitsgemeinschaft*) of Saxon botanists equated the fight against foreign plants with the fight of National Socialists against other nations, especially "against the plague of bolshevism." The team demanded in a publication "a war of extermination" against *Impatiens parviflora*, a small forest plant. *Impatiens parviflora* was seen as a stranger, spreading and even competing with the somewhat larger *Impatiens noli tangere*, considered native. Presumably, *Impatiens parviflora* endangered the purity of the German landscape. The call ends, "As with the fight against bolshevism, our entire Occidental culture is at stake, so with the fight against this Mongolian invader, an essential element of this culture, namely, the beauty of our home (*heimisch*) forest (is at stake)" (23).

During the period of National Socialism garden architects like Hasler and Seifert were successful because of their ideological conformity with National Socialism. At the time there were only a very few who dared oppose such doctrinaire ideas by referring to more than 1000 years of the international history of garden culture. One of these critics was the writer Rudolf Borchardt. He knew exactly what he was arguing: of Jewish origin, he died in 1942 trying to escape the National Socialists.

Let us end here with Borchardt's criticism of Seifert and other doctrinaire advocates of the nature garden, in as much as it is at the same time a plea for an international garden culture, founded on democratic ideals. In 1938 Borchardt wrote: "If this kind of garden-owning barbarian became the rule, then neither a gillyflower nor a rosemary, neither a peach-tree nor a myrtle sapling nor a tea-rose would ever have crossed the Alps. Gardens connect people, times and latitudes. If these barbarians ruled, the great

FIGURE 9. A typical 'Teutonic' landscape design one can still see today in the so-called Sachsenhain (Grove of the Saxonians), a meeting place designed by the garden architect Wilhelm Hübotter in 1934 for the Reichsleader SS, Heinrich Himmler. Photograph from the Municipal Archives of Verden/Aller.

historic process of acclimatisation would never have begun and today we would horticulturally still subsist on acorns . . . The garden of humanity is a huge democracy . . . It is not the only democracy which such clumsy advocates threaten to dehumanize" (24).

NOTES

1. See O. Sourek, *Antonin Dvorak. Sein Leben und sein Werk* (Prague, 1953), p. 53.
2. M. Hadfield, *A History of British Gardening* (Harmondsworth, Middlesex, 1985), p. 361.
3. W. Robinson, *The Wild Garden or the Naturalization and Natural Grouping of Hardy Exotic Plants with a Chapter on the Garden of British Wild Flowers* (London 1894, reprint of the fourth edition, 1977), p. 156.
4. W. Robinson (note 3), p. 33.
5. "Statuten fuer den Verein zur Befoerderung des Gartenbaues im Preussischen Staate," *Verhandlungen des Vereins zur Befoerderung des Gartenbaues in den Koeniglich Preussischen Staaten*, Erster Band, 1824, p. 7.
6. For more detailed information about the nature- and the home protection movement see G. Gröning and J. Wolschke-Bulmahn, *Die Liebe zur Landschaft*, Part I, *Natur in Bewegung. Zur Bedeutung natur- und freiraumorientierter Bewegungen der ersten Hälfte des 20. Jahrhunderts für die Entwicklung der Freiraumplanung*, U. Herlyn and G. Gröning (eds.), *Arbeiten zur sozialwissenschaftlich orientierten Freiraumplanung*, Vol. 7 (München, 1986); for more detailed information about the bourgeouis youth movement and the "Wandervogel" see J. Wolschke-Bulmahn, *Auf der Suche nach Arkadien. Zu Landschaftsidealen und Formen der Naturaneignung in der Jugendbewegung und ihrer Bedeutung für die Landespflege*, U. Herlyn and G. Gröning (eds.), *Arbeiten zur sozialwissenschaftlich orientierten Freiraumplanung*, Vol. 11 (München, 1990).
7. D. Gasman, *The Scientific Origins of National Socialism. Social Darwinism in Ernst Haeckel and the German Monist League* (London/New York, 1971), p. XIII.
8. For a comparison of the garden ideologies of Willy Lange and William Robinson see J. Wolschke-Bulmahn, "The 'wild garden' and the 'nature garden'–aspects of the garden ideologies of William Robinson and Willy Lange," *Journal of Garden History*, 12 (1992).
9. W. Lange, "Garten und Weltanschauung," *Gartenwelt*, 4 (1900), p. 364.
10. W. Lange, "Meine Anschauungen über die Gartengestaltung unserer Zeit," *Gartenkunst*, 7 (1905), p. 114.
11. W. Lange, *Gartenpläne* (Leipzig, 1927), p. 5.
12. See W. Lange, "Deutsche Gartenkunst," *Deutsche Kulturwacht* (1933), p. 8.
13. W. Lange, *Gartenbilder* (Leipzig, 1922), p. 27.
14. W. Lange, *Gartengestaltung der Neuzeit* (Leipzig, 1907), p. 358.

15. See M. Eksteins, *Rites of Spring. The Great War and the Birth of the Modern Age* (New York/London, 1990), p. 65.

16. C. Schuerer, "Deutsche Bäume in der Gartenkunst," *Gartenwelt*, 15 (1911), p. 34.

17. W. Lange, *Der Garten und seine Bepflanzung* (Stuttgart 1913), p. 48. The idea that the garden architect should express in the garden the local nature by using fitting foreign plants made it possible for Lange to use, for example, *Iris interregna germanica* near wet places; but at the same time he pointed out that this plant "prefers dry places, though it looks wet" (W. Lange, *Gartengestaltung der Neuzeit*, Vol. 2 (Leipzig, 1912), p. 136).

18. A. Seifert, "Randbemerkungen zum Aufsatz 'Von Bodenständiger Gartenkunst,'" *Gartenkunst*, 43 (1930), p. 166.

19. A. Seifert, "Bodenständige Gartenkunst," *Gartenkunst*, 43 (1930), p. 162f.

20. H. Hasler, *Deutsche Gartenkunst* (Stuttgart, 1939), p. 175.

21. H. Hasler 1939, p. 8.

22. A. Kraemer, "Pflanzensoziologie und der Blut- und Bodenverbundene Garten," *Gartenkunst*, 49 (1936), p. 43.

23. Arbeitsgemeinschaft Sächsischer Botaniker, "1. Jahresbericht, Aufruf von Max Kaestner," (Dresden, 1942).

24. R. Borchardt, *Der leidenschaftliche Gärtner* (Nördlingen, 1987; reprint of the 1938 edition), p. 240f.

PLANTS AND THE COMMUNITY

Chapter 10

Cultivating People-Plant Relationships in Community and Cultural Heritage Gardens

San Jose, California (1977-1992)

John Dotter

SUMMARY. This paper documents important people-plant relationships to be found in San Jose's ethnically diverse Community and Cultural Heritage Gardens. Unpublished university research, local magazine articles and newspaper stories are cited which show how partnerships with grassroots community groups have contributed to the garden programs sustainability. Preliminary findings, with investigative surveys, suggest that more use could be made of the non-English speaking gardeners who are skilled horticulturalists.

Several case studies show how cooperative development of the San Jose gardens produces the results identified by the California

Council for Community Gardening in 1977 which stated: *Community gardening improves the quality of life for all people by: beautifying neighborhoods; stimulating social inter-action; producing nutritious food; encouraging self-reliance, conserving resources; and creating opportunities for recreation and education.*

Examples drawn from the San Jose urban gardening experience document local community pride and horticultural accomplishment. A comparison is made of San Jose's largest and oldest community garden (Mi Tierra) with 90% Hispanic plotholders and a smaller garden (Wallenberg), entirely Caucasian.

The Cultural Heritage Garden program is recognized nationally as a new urban gardening program. San Jose's success with its Japanese and Chinese Cultural gardens provided the basis for program expansion to include partnerships with the Vietnamese, Mexican and Filipino communities. A serious effort to have an Indo-American cultural garden is now underway. Results of networking between these projects has unified various factions within each group, as well as created a new sense of belonging to the larger community. Interviews with selected horticulturists and community leaders detail this type of development.

This paper raises familiar questions and long-standing problems, connected with the operation and maintenance of user-developed gardens. Recommendations are made for the effective management of public gardens with volunteers. San Jose Community Gardens are contrasted with other Parks Division maintained gardens in San Jose to ascertain differences in the felt sense of "ownership" and responsibility to care for public landscapes. This research paper closes with suggested methods for starting similar programs elsewhere. An annotated bibliography is provided.

The Recreation, Parks, and Community Services Department of the City of San Jose considers gardening a legitimate form of recreation and a superb community development tool. The people-plant relationships to be found in its community garden plots and cultural heritage garden projects are multi-faceted as well as distinctive due to the ethnically diverse population served. Lionel Castillo, former Director of the Immigration and Naturalization Service under the Carter Administration, commented, when he was in San Jose, that the Statue of Liberty now belongs on the West Coast of our country because the West, particularly San Jose, is the place where immigrants come in record numbers. His remark points out a very important fact about San Jose, its cultural diversity.

BACKGROUND

> As the City grew in population during the 1980's it also grew in ethnic diversity. For the first time in recent history, there is no single ethnic or racial majority within the City of San Jose. Of the total population in 1990, 49.6% identified themselves as White, a 10.4% decrease in percent population of the total from 1980. The Asian population in 1990 increased by 178% from 1980, making it 15.4% of the total population of the City. The Hispanic population in 1990 increased by 48% from 1990, making it 26.6% of the total population of the City. *(Demographic Profile City of San Jose,* Planning Department 3/9/92, p. 2)

The immigration of non-English-speaking peoples with horticultural skills continues. There is less land available now for gardening than at the turn of the century, when San Jose was known as a Garden City. The local supply of undeveloped land continues to decrease while Hispanic and South-East Asian immigrants grow in population. Most of these new residents not only know how to grow some of their own food, but do it as a form of recreation.

Seniors continue to be the volunteer backbone of local urban gardening programs. They are an under-utilized resource for the development and management of gardening programs. Interviews with Senior gardeners conducted for this paper demonstrate their exemplary commitment to the world of plants and how their gardening activities cultivate mutually beneficial relationships with the surrounding community.

Public gardens continue to be common ground where people reconnect with one another and the world of plants. Most of urban California is blessed with a mild Mediterranean climate which provides the incentive to grow flowers, fruit and vegetables twelve months of the year. While this paper's research focuses on San Jose and the Northern California San Francisco Bay area, there are many program ideas and policies which can nurture "people-plant" relationships just about anywhere in the world.

Taped interviews were made with the City's Mexican-American Community Gardening Coordinator, community gardening program volunteers, Cultural Heritage Garden project planning leaders, and with the Convention Center Gardener. These persons all demonstrate commitment to their work. At the community garden sites some of these people begin and end their day in the garden. There have been complaints from some family members that certain individuals spend more time at the community garden than at home.

To more fully understand a gardener's interaction with his garden, a visit to the site helps. Words cannot fully describe a garden's complex and dynamic reality. Photographs and slides can portray aspects of the nurturing environment that can be found in San Jose's Community and Cultural Heritage Gardens.

For the purposes of this paper a community garden is unused public or private land subdivided into plots for persons to grow their own vegetables, fruits, herbs and flowers. It is recreational activity in a neighborhood park setting. The average plot size is 20 × 30 or 600 square feet. San Jose Cultural Heritage Gardens are parks, five acre minimum in size, with other amenities which encourage cultural interpretation through artifacts, museums and special garden plantings.

The Cultural Heritage and Community Garden Programs have their existence and continued support due to City sponsorship. The institutional framework of local government has provided valuable continuity and volunteer training which has enabled these programs to be expanded over the years. As an example, the neighboring City of San Francisco has a larger number of smaller gardens which are administered by a non-profit agency. However, there are more permanent gardens with a greater number of gardeners on fewer large sites in San Jose.

Environmental degradation, coupled with the sincere desire that many people have for ecologically responsible living, are some of the reasons local interest in our urban gardening projects is high and sustained. While many may have never heard the term "Deep Ecology," most everyone acknowledges how out of touch many contemporary lifestyles are with the natural world. The late Thomas Church, San Francisco Landscape Architect, author of *Gardens Are For People,* said: "The garden can be a place where Man can recapture his affinity with the soil, if only on weekends. It must be an oasis where memories of the bumper to bumper traffic will be erased."

Church's mention of soil brings to mind what poet/farmer Wendell Berry said in his noteworthy book *Culture and Agriculture:* (The Unsettling of America)

> The soil is the great connector of lives, the source and destination of all. It is the healer and restorer and resurrector, by which disease passes into health, age into youth, death into life. Without proper care for it we can have no community because without proper care for it we can have no life. . . . It is impossible to contemplate the life of the soil for very long without seeing it as analogous to the life of the spirit. (86)

What Berry points out so forcefully is that our actions have consequences. We cannot damage what we are dependent upon without harming ourselves. This is a form of ecological karma at work where we harvest what we sow.

There is a Kenyan proverb that states: "Treat the earth well, for it was not given to you by your parents, but loaned to you by your children." This proverb dramatically conveys our obligation to be responsible caretakers of the earth. In the words of Lester R. Brown, Worldwatch Institute: "A sustainable society is one that satisfies its needs without jeopardizing the prospects of future generations."

The people-plant relationships of gardening endeavors can keep connections to the natural world vibrantly alive. On the subject of how ecological is one's lifestyle there is the quality of one's life to consider. It is being increasingly accepted that a large percentage of costly illnesses are caused by a harmful lifestyle.

Wendell Berry speaks eloquently about the role gardening can play:

> In gardening, for instance, one works with the body to feed the body. The work, if it is knowledgeable, makes for excellent food. And it makes one hungry. The work thus makes eating both nourishing and joyful, not consumptive and keeps the eater from getting fat and weak. This is health, wholeness, a source of delight. (138)

COMMUNITY AND CULTURAL GARDEN PROGRAMS

In San Jose, as of April 1992, there are 15 community gardens with 850 registered plotholders. These gardens are located throughout the City on 27 acres. A minimum of 3500 persons participate in the San Jose community gardening program. New community gardens are funded with capital improvement funds which come from construction and conveyance tax revenues. These can only be used in the City Council district from which they are allocated. There are 10 San Jose Council districts covering 175 square miles. San Jose's population is growing beyond the 800,000 people currently living at an overall density of 4,334 persons per square mile (California State Department of Finance, May 1990).

Personal and nonpersonal expenses for the program are paid out of the Recreation, Parks and Community Services Department general fund allocation in the City budget. The 1991-92 appropriation for Community Gardens was $105,000 and is $65,000 for the Cultural Heritage Garden Program. The allocation for the two programs combined is $170,000. Approximately $35,000 is still available as grant funds to Cultural Garden projects. New community and cultural heritage gardens are a direct result of citizen interest and effort. The request usually is made by an organized group. If there is sustained and significant community interest for the garden project it is presented to the Parks and Recreation Commission. This Commission makes a recommendation which is forwarded to the City Council along with that of the Recreation, Parks and Community Services Department. This process can take a year. Cultural Heritage Garden planning and community organizing can take several years to get approval for "partnership" status and $25,000 matching grant funds for additional planning.

The Japanese Friendship Garden resulted from a carefully coordinated joint venture which began in 1957 with Pacific Neighbors, Inc., a Sister City Program; the Japanese American Community; and the City of San Jose. It was a response to President Eisenhower's plea for the establishment of friendly "people to people" relationships between cities in the U.S. and other nations. The City of San Jose provided a choice six-and-one-half acre site, clearing and contouring, construction and guniting of the ponds, building of the waterfalls and stream beds, installation of an irrigation system, and fencing–in short, everything necessary to bring the garden to the planting stage.

In the spring of 1965, the garden was ready for the planting of 4,000 trees and shrubs donated by 40 nurseries. These were planted by 120 members of San Jose's Japanese-American community under the supervision of the San Jose Landscape Gardeners Association. Patterned after Okayama's world famous Korakuen Park, detailed plans were made which preserved the traditional and symbolic meaning attached to the location of rock and other garden features. The word "preserve" is the important term to remember, especially when it is linked to a cultural heritage.

Support for and preservation of cultural traditions, that have their roots in the gardening experience, play a very important role in San Jose's urban gardening programs. The level of interest shown in the San Jose Cultural Heritage Garden program corroborates the desire that the culturally diverse local population has to preserve its own distinctive identity. The term "melting-pot" that has been used over the years to describe the different ethnic make-up of those of us who are Americans has the real possibility of lumping everyone together in a bland, lack-luster soup. Local groups identify themselves as Americans but: Mexican-Americans, Vietnamese-Americans, etc. Under close scrutiny the gardens and gardening techniques used by these immigrant groups have noticeable differences. However, when it comes to cooperation and mutual aid for the creation of new garden projects among these distinctly different communities, the response has been most encouraging. An example of this bonding has been the very helpful relationship that has developed in the past year between our established Filipino community and the more recent arrivals, the Indo-Americans. The appendices contain the original goals, objectives, policies and procedures for the Cultural Heritage Garden Program.

A Conference on Refugees was held this year in San Francisco. The needs that were identified were: continued support for ESL classes to assist new immigrants with basic language skills, job placement assistance, and a community of friends to share one's experiences and cultural heritage. While the first two concerns deal with the basics of economic survival, cultural traditions speak to the inner if not spiritual dimensions of one's personality. San Jose's gardens aim to foster community pride and goodwill through cross-cultural association. They are planned as the locations for festivals and events to be shared with everyone. Volunteers benefit directly from horticultural training and valuable work experience. Some volunteers become paid City staff as a result of their service. The original Japanese and Chinese Cultural Gardens are maintained by City staff with some community assistance. The Vietnamese, Mexican and Filipino projects are to be maintained in partnership agreements with the non-profit agency sponsor. Some projects have revenue generating components to offset maintenance costs.

PEOPLE-PLANT RELATIONSHIPS IN SIX GARDEN PROJECTS

The Chinese Cultural Garden

In 1973, five acres of Overfelt Botanical Gardens were designated as a Chinese Cultural Garden largely as the result of the efforts of two members of the local Chinese-American community, Frank and Pauline Lowe. They cultivated local community support as well as another Sister City connection with Taiwan, to obtain large donations of artifacts which include a 15 foot high bronze statue of Confucius, a Chinese gate, Sun Yat-sen Pavilion, President Chiang Kai-Shek Pavilion, and a gazebo-like Plum Pavilion.

Despite Frank Lowe's death several years ago, his widow Pauline spends practically every weekend at the Chinese Cultural Garden sharing its meaning and traditions. The garden is a site for Tai-Chi classes with an instructor who drives the 60 miles from San Francisco every Saturday. In an interview with Mrs. Lowe, she continually stressed the traditional Chinese connection to the world of nature. This connection was not expressed solely in Taoist and Confucian thought. The principles she emphasized were "peace and harmony with nature and family values."

When visitors come to the Chinese Cultural Garden they are shown architectural features of the temples which illustrate these basic principles. They relate very well to her response to the question, "*Why do you garden?*" She said:

> Just touching the soil calms me. When I go into the garden, I feel peace, close to nature. A feeling of relaxation comes over me and I let go of tension. Each part of a garden can have a message which is universal, especially when you go beneath the surface appearances.

Pauline Lowe spoke with a definite sense of a mission and outreach for young people. She was aglow when she described how children in the summer months spend a lot of time in the Chinese Cultural Garden and tell others that it is their park even when they have no Chinese background. All has not been "sweetness and light" with the Chinese Garden. It has suffered, like other City facilities, with acts of vandalism and graffiti. The sum of $7,500 was budgeted this fiscal year for building repair and graffiti removal. However, plans are in the making for development of a docent group as well as for a volunteer corps interested in working on the landscape around the main pavilion to realize increased community involvement and presence on the site.

Mi Tierra Community Garden

In 1976, a coalition of interested citizens from Olinder Residents Inc., the Council on Aging, the Food Bank, the San Jose State University Environmental Center, the University of California Cooperative Extension and the San Jose Parks and Recreation Department banded together to establish the City's first Community Garden. It is the largest and oldest garden in the program and located on approximately five acres of land that had previously been the site of a City tree nursery. Corporate donations from Lockheed, IBM, and the Council on Aging paid the material and labor cost for purchase of the irrigation system and water meter. Volunteers installed the piping, as well as helped clear the site of the unwanted, abandoned trees. Cooperative Extension provided the initial insurance coverage which the City assumed when it formally adopted a program in February 1977. Waiting lists at this garden have numbered over 100 persons.

All of the Volunteer Garden Managers in Mi Tierra's 16 year history have been Mexican-American. One could write a book about the achievements of its first Manager, Artemio Carranza, who, in the spirit of Pancho Villa, created quite a stir. Suffice it to say he managed not just to get the

Mayor and members of the Governor's family to visit Mi Tierra but to promise assistance. He was a "Gray Panther" style advocate for gardeners who needed a place to garden and support services.

Several years later, a different but respected leader emerged. His name was Carlos Robles. Wonderful pictures of him and his grandchildren filled the gardening section of the local newspaper in 1983. Carlos spent almost every waking hour at Mi Tierra until an accident brought about his death in 1987. The deep attachment he had to his garden plot and fellow gardeners is shown in pictures taken where his last wish was granted. On the day of his funeral after the church service his hearse, moving slowly, preceded by Mariachi musicians, entered Mi Tierra. It stopped by Carlos' plot while family members gathered soil in a container. This soil was later mixed with the soil surrounding his casket at his burial site in Santa Clara. The pictures taken by the author convey a fraction of the emotional feeling that was present at Mi Tierra that day.

Following in the dedicated footsteps of Carlos Robles is Concepcion Huerta who recently finished service as a Volunteer Garden Manager at Mi Tierra. Unlike many of the plotholders, Concepcion had no real experience with food growing when he first signed up for a plot in 1981. One might say he is self-taught. A recent visit to his plot this year showed healthy lush red potato plants, onions, corn, carrots and tomatoes growing side by side. Concepcion has worked for the Southern Pacific Railroad for 34 years and is one year away from early retirement. He is already planning to spend most of his retirement at Mi Tierra or assisting as an advisor to the Mexican Cultural Heritage Garden project. For the present, he goes straight to his plot after work, about four in the afternoon, and on weekends hangs out with his friends that have weekly garden barbeques. Concepcion's family does not join him at Mi Tierra, but he claims they love the quality fresh produce that he brings home. Gardening for Concepcion is his #1 recreational activity. This is typical of the Mi Tierra gardeners. Gardening is an important part of the lifestyle. Very few Mi Tierra gardeners participate in other major recreations such as tennis, golf or swimming.

Jesse Frey Community Garden

Dr. Jesse Frey, a veterinarian, came to California in 1916 and established the first animal pathology laboratory in Sacramento. For 31 years he was an Executive for Golden State Milk Co. In 1972 he moved to San Jose. Six years later he founded the community garden which now bears his name. A newspaper published by what is now known as the National Gardening Association featured Dr. Frey in its story (May-June 1981 p. 47) on how he started what was then called the "Willows Senior Center Community Garden."

> It doesn't always take a committee to start a community garden. The San Jose Senior Citizens Center tried to do it that way, and it didn't work. Hardly anybody turned up for the meetings. What it took was a Jesse Frey. It was 1978, and he was 84 and rarin' to go. ("I have to garden; it's part of me.") Jesse went to the organizational meetings; four persons attended the first, two attended the second. They said the ground around the Center was too hard to garden. Not for Jesse. He knew how to inspire interest in a community garden–you plant a garden and grow vegetables. People see them and become interested. They want to join, it's the old "Tom Sawyer painting the fence" syndrome and the garden grows bigger. More people stop by, and more join. Pretty soon you have an active community garden with a waiting list. (*Gardens for All News,* May/June, 1984)

Jess Frey passed away at the age of 93. He was legally blind during the last five years of his life. In spite of this handicap he still gardened. His effective garden management skills were a model for other community gardens.

Wallenberg Community Garden

Located in the Willow Glen District, Wallenberg Community Garden has exceptional soil. The topsoil goes 30-40 feet down. It has some of the finest growing conditions in the entire Santa Clara Valley. Most of the residents in this neighborhood are largely upper middle class and white. While only one and one half acres in size, Wallenberg is thought to be the best looking and managed community garden in the City. A good part of the reason it has this reputation is due to the management and enthusiasm for gardening held by its Volunteer Manager Dick Leal. Dick at 68 is recently retired and up until 6 years ago the only gardening he did was routine maintenance around his home. He readily admits that he is "hooked" on gardening and finds it frustrating when other plotholders don't share his zeal. When interviewed, he shared that he started being a Manager of others at age 16. His last position was a Branch Manager for a well known California Bank. Wallenberg Community Garden presented a meaningful way for him to continue as a leader and community organizer. Basic approaches to good public relations that are used in the business world such as: "Praise in public, criticize in private" are used on a daily basis as he keeps community garden affairs in order. When asked about his expertise and formal training in gardening he admits that it is limited. He deliberately avoids telling others how to garden, but he refers them to other "experts" who might offer assistance.

The Mexican Cultural Heritage Garden and Plaza

Fernando Zazueta is a well known and influential San Jose Attorney. He was born in Mexico and at a very early age was part of his uncle's migrant worker fruit-picking crews working in California. Fernando attended 16 schools before he got his High School Diploma. Gardening is an important recreational pastime for him and his family. In his backyard you will find three pear trees, a fig, four prune trees, a peach, an apricot, lemon, navel orange, and tangerine tree. Fernando's work is no longer in the fields, but put a person from the farm in the city, and gardening takes on a special significance.

Fernando is very clear about the importance of preserving the memories that he has of Santa Clara Valley filled with orchards. As President of the Board of Directors for the Mexican Heritage Corporation, he has spoken in support of having fruit trees in the landscape plan along with popular Mexican flowers like carnations and roses. The Mexican Cultural Heritage Garden and Plaza is a partnership venture with the Recreation, Parks and Community Services Department of the City of San Jose and the Redevelopment Agency.

Philippine Heritage Center

In 1935 Jacinto (Tony) Siquig started farming for himself in the Santa Clara Valley in what is known as the Alviso District of San Jose. Less than ten years earlier he had left the Philippines with some friends and came to this country in high hopes for getting a good job. His starting wages were 10 cents an hour. Soon after his arrival he wished he could return to the Philippines. Tony said: "I didn't have enough money to return so I had two choices, make the best of my situation, or kill myself." Once he was clear in his mind about what to do, things started to get better. He decided to build on his Philippine High School training in agriculture and become a truck farmer on a grand scale. In 1969, he retired from 40 years hard work as an urban farmer. Like many other stories told to the author over the years, Tony quickly got bored. His senior citizen friends asked him to organize the local community. The Filipino-American Senior Opportunities Development Council and the Northside Community Center are living testaments to his leadership abilities.

In the mid-seventies Tony took advantage of special government job training funds and enrolled in propagation classes at the Saratoga Horticultural Foundation with Brian Gage, the capable Director of Horticulture. A whole new world of possibilities and understandings opened as he studied and enlarged upon his agricultural experience.

The author met Tony for the first time in spring 1977. His positive and friendly approach to working with people and plants has not changed. Through his persistence garden programs that connect Seniors and youth have been on-going at the Northside Center Greenhouse and Community Garden. Filipino Seniors serve as mentors and supervisors with high school drop-outs, South-East Asian refugees, persons with physical disabilities, Job Corps and the Summer Program for Economically Disadvantaged Youth (SPEDY), to mention a few. It is a subject beyond the scope of this paper to detail the horticultural therapy that has been a by-product of the nursery propagation work. When asked to describe what fascinates Tony about gardening he said:

> The mind is curious like a seed. Life is a miracle full of mystery and surprises. Where did we come from? How have plants evolved? I feel just like a child when I think about these things. I'm so curious. People have forgotten that farming and gardening are basic skills that we all need to learn and use. One should know how to grow their own food in order to survive. Nowadays, we should have more quality control over the food we eat. I would have farmed a lot differently had I known about the long term damaging effect of pesticides. . . . One of the pleasures I enjoy every day is working in my garden. . . . It's part of the livelihood of being human.

Jacinto Siquig, at 87 now, still is lobbying for the needs of many Seniors, that of low-income housing. Currently he and other Board members of the Fil-Am S.O.D.C. are working with younger generation Filipinos for the creation of a Philippine Center.

Gardens are places to spend time. Seniors usually have lots of time to spare. One afternoon the author came by the Northside Nursery/Greenhouse and heard a lot of loud laughter from inside. An investigation found three of the oldest community gardeners with embarrassed smiles on their faces. They tried unsuccessfully to stop laughing. When the author asked what is it they were laughing about, they did not respond. They made a few remarks to one another in Tagalo. After leaving the Greenhouse another gardener explained to the author that these men in their eighties were planning to grow Durian fruit (Durio zibethinus). In the same way Ginseng has been reputed to enhance one's sexual energy the Durian fruit was believed by them able to restore their youthful passions. Working with Seniors can be full of surprises especially when it comes to the role plants play in their lives.

If you are able to visit the Northside Center today you will find Maria Socorro Araneta busy as the capable and caring Director. She supervises gardeners enrolled in the Title V Senior job training program. "Baby," as she is known at the Northside Community Center, admits that she gardened more when she lived in Chicago for ten years than here in California. When asked why this is so, she said that the soil where she lives now is "impossible, it's as hard as a rock." In addition to management of the Community Center, Baby is providing necessary leadership and project management for the Philippine Heritage Center, a project with definite "People-Plant" connections.

MAINTENANCE OF USER-DEVELOPED GARDENS

Research for this paper, coupled with thirty years of experience working with volunteers on garden projects, lead the author to a very firm conclusion, that volunteers need thorough training and competent supervision. Space limitations preclude detailed case histories but the finite resources of time and materials are wasted without good pre-planning of the simplest project. Community gardening presents ample opportunities for individuals to learn how to work effectively together. Disputes, accidents and generally unwanted situations develop without good management at each community or cultural heritage garden.

Seven years ago, Dave Arroyo, fresh from a stint in the Navy, saw an ad in the paper for a Maintenance Assistant with the RPCS Parks Division. He had no real experience as a gardener. His first few years were spent clearing trails with jail work crews in Alum Rock Park. This is a wooded area in the foothills of the Mount Hamilton range. The next assignment was work at City Hall. Richard Perez trained Dave in the basics of gardening and he moved in two years time up to Groundsworker. At the present time Dave is a full-fledged Gardener with a very visible and difficult job in the heart of San Jose, the Convention Center complex. When asked about working with volunteers Dave had few kind words to say about the government assistance referrals that he has had working with him over the years. He agreed with the author that volunteers need training.

Unfortunately, there are a good number of disgruntled gardeners who work for the City and resent the amount of time they must spend on a daily basis picking up trash and cleaning restrooms as opposed to working with plants. Less than fifty percent of Dave's time is spent on traditional gardening work with plants. Before the interview Dave was working with his back-pack blower.

RESOURCES THAT SUPPORT URBAN GARDENING PROGRAMS

In Santa Clara County there are many agencies that contribute to the operation of urban gardening programs. Cooperative Extension of the University of California has been a consistent provider of technical assistance. Retired Senior Volunteer programs, along with Offices on Aging, both County and City, have provided some paid employment. Educational institutions, elementary grades through University, have from time to time, had school gardening programs to enrich their curriculum. In general, most classes and workshops put major drains on staff resources. It is almost impossible to be teaching and managing a community garden at the same time.

One of the most powerful and effective ways to enhance or start new community gardening programs is to locate outstanding projects in your area and visit those projects. Be sure to make advance arrangements to have the persons in charge at the site. It's human nature to enjoy "show and tell." Garden projects can evoke the same feelings of pride and accomplishment. San Jose's Cultural Heritage Garden program is ambitious by any standard. All the designated projects take years and millions of dollars to construct. However, the encouraging aspect of this type of community development work is the willingness shown by program participants to spend countless hours helping one another.

San Jose's Community Gardening Coordinator, Manny Ruiz, was born on a ranch in Brownsville, Texas. Beginning at eight years of age he traveled with his family as migrant farm workers in 30 states (Florida to Michigan). The best money was made thinning sugar beets in Nebraska. When asked why he gardens after ten years of farm work, his reply was that while it is thought of as work, it is relaxing. Manny loves outdoor work.

He is so organized that one must be careful near his desk. Any papers left that are not his he throws away immediately. For community gardens this means that all plots must be kept free of weeds and cultivated. Being Coordinator for 15 garden sites is a real challenge. The program has less staff with more gardens to look after now than it had eight years ago. Manny refers to this situation as: "Me, myself and I" trying to be everywhere at once. In his opinion the ideal staff to gardens ratio is one coordinator per ten community gardens. The size of these gardens ideally would be one to one and one half acres with 48 to 50 plots per garden. In the opinion of other Bay area garden program managers, the forms he has created are a valuable contribution to community gardening program organization anywhere.

SUGGESTIONS FOR STARTING COMMUNITY AND CULTURAL GARDEN PROGRAMS

It is the carefully considered opinion of the author that urban gardening programs have the best possible opportunity to grow and survive under the aegis of a governmental or educational institution. This approach requires patience with bureaucratic agencies. Broad, sustained community interest in a gardening project is almost certain to win the backing of local officials. A partnership approach is strongly recommended.

An excellent source of supervised labor is jail alternative service work crews. Coordination and pre-planning can make this resource most helpful with big cleanup jobs and tasks that would intimidate a typical plotholder. San Jose has used both the State and Local Conservation Corps for community garden work. Fees are charged and sometimes private foundations will partially underwrite important public benefit project costs.

SUMMARY

Research for this paper uncovered a wealth of underutilized talent in the culturally diverse population of San Jose. The author was surprised that the Volunteer Community Garden Managers interviewed had less than ten years experience. They learned most of what they knew about gardening by actual practice with other more experienced gardeners. The hope expressed on the part of the City Community Gardening Coordinator for more staff is a universal need. At the present time the Cultural Heritage Garden program has more organized and vocal participants.

There was a Community Gardening Advisory Committee from 1978 to 1979. Proposition #13 cutbacks brought about its demise. I believe the time has come for the Community Gardening Program to rebuild its participant support base. An example of a very difficult situation for the City Coordinator is that there is not $1 available in the Non-Personnel part of this budget. Basic materials and supplies that could be purchased in years past lack funds. Gardeners traditionally are "resourceful" people; some are called scroungers. One of the pleasures of working with the immigrant populations is they are very creative in doing a lot with very little.

John and Miles Hadfield, famous English garden writers, state in the Introduction to their *Gardens of Delight* that "Fundamentally, all gardening is the transference of a vision into a touchable seeable reality" (p. 5). In both Community and Cultural Heritage Gardens a "vision" of what is possible needs to be articulated. Visits and conversations with other gardeners are very important ways to expand one's horizons. Gardeners, when you get to know them on a personal level, usually confess that the world of plants provides them a useful medium for their own personal growth and development. As an apprentice to Alan Chadwick the author remembers hearing repeatedly that it is the *garden* that makes the person. Those of us who worked with Chadwick experienced his disciplined but reverent approach to gardening. This made us pay attention to the lessons to be learned from good stewardship of plant life.

The best People-Plant relationships are "partnerships." Wolfgang Goethe puts it succinctly in a quote that the author uses frequently in his gardening classes:

> A plant is like a self-willed person from whom you can obtain anything you want if you treat it in its own way. In ordinary language the idea expressed is: meet the needs of the plant first, and it will reward you later.

Fritjof Capra, author of *The Turning Point* and other seminal works, says:

> The change of attitude from *control to partnership* is characteristic of a profound change of values that is implicit in the shift from the mechanistic to the ecological world view. Whereas a machine is properly comprehended through domination and control, the *understanding* of a living system will be more successful if approached through cooperation and partnership. Cooperative relationships are essential characteristics of the web of life.

AUTHOR NOTE

The author wishes to recognize and thank the San Jose Recreation, Parks and Community Services Department (RPCS) support and encouragement in the development of new urban gardening programs. San Jose was one of the last cities in the San Francisco Bay area to develop a community gardening program. Its commitment is long-term.

Dick Reed, Retired Deputy Director, (Parks Resources) in the mid-1970s used Federal CETA funds to hire the author for the development of the City of San

Jose's Community Gardening. Master Planner Harryette Shuell, Parks and Recreation Commissioner, served as an important liason with the City as Chairperson of the Community Gardening Program Advisory Committee. The valuable input and participation of interested community members on this committee helped lay a solid foundation for the described programs.

My first paid employment in the field of community gardening was in 1972 as Manager of the Saratoga Community Garden, which was located on an abandoned farm next to San Jose. This garden, and several others in Northern California, was strongly influenced by English Horticulturist Alan Chadwick. It was a model community garden with a wide range of educational programs. Dick Reed and other City officials were invited to visit the Saratoga Community Garden. They saw school groups of all ages learning, having a wonderful time, taught by local college and university student interns and Senior volunteers. San Jose officials instantly wanted to reproduce a similar demonstration garden for their neighborhoods. In short, first hand experience of the Saratoga Community Garden exerted a powerful influence on the way San Jose programs developed in subsequent years.

Net-working and cooperation between other community gardening coordinators has contributed to the success of the San Jose gardening programs. Mark Westwind, co-founded, with the author, the California Council for Community Gardening, which co-sponsored several large educational events with the San Jose Recreation, Parks and Community Services Department.

Rosalind Creasy, edible landscaping author, has been the un-official program "photo-journalist" documenting the ethnic diversity in San Jose Gardens. Thanks to her professional work, the stories and faces of San Jose community gardeners have been published nationally.

San Jose's Community Gardening Coordinator is Manny Ruiz. He started as a volunteer at Prusch Farm Park eight years ago. He has developed a series of useful program forms.

Jim Wilson, Former Executive Director of the National Garden Bureau, and Past President Garden Writers of America, provided outstanding professional advice for the San Jose program in its early years. He was the keynote speaker at the dedication of Emma Prusch Memorial Farm Park in 1982. This is a 50 acre farm with large community gardens that serve San Jose's culturally diverse population. It was definitely inspired by the Saratoga Community Garden.

My last and most important thanks go to my wife Mary and children, Denise, Karl and Katherine who have supported my work with people and gardens over the past fifteen years.

BIBLIOGRAPHY

Arroyo, David, April 1992, Interview by author.

Araneta, Maria S., April 1992, Interview by author.

Azcona, Carlos, 1986, "A Brief History of the San Jose Community Garden Program," San Jose State University, Environmental Studies Program Research (SJSU).

Brooklyn Botanic Garden, Spring 1979 *Community Gardening: A Handbook.* N.Y.

Bassett, Thomas J., 1979, "Vacant Lot Cultivation: Community Gardening in America 1893-1978," University of California, Berkeley.

Ferguson, Diane C., 1987, "City Sponsored Community Gardening in San Jose, SJSU.

Flynn, Nancy, May-June, 1981 "Gardens For All News" Burlington, Vermont.

Francis, Mark, "The Park and the Garden in the City–A Comparison of Different Meanings Attached Users, Non-Users and Officials to a City Park and Community Garden in Sacramento, CA," Center for Design Research, University of California, Davis.

Hadfield, Miles and John, 1964, *Gardens of Delight*, Little Brown, New York.

Huerta, Concepcion April 1992, Interview by author.

Jobb, Jaime, 1979, *The Complete Book of Community Gardening*, Morrow and Co. New York.

Leal, Dick, April 1992, Interview by author.

Lowe, Pauline, April 1992, Interview by author.

Northern California Home and Garden, November 1988, "A Harvest of Gardens" (70), Redwood City, California.

Otwell, Carol 1990, *Gardening from the Heart: Why Gardeners Garden*, Antelope Island Press, Berkeley, California.

Ruiz, Manuel, April, 1992, Interview by author.

Siquig, Jacinto, April 1992, Interview by author.

Vision Link Education Foundation, 1991, *Who is Who in Service to the Earth*, Waynesville, North Carolina.

Warner, Sam B. Jr., 1987, *To Dwell is to Garden*, Northeastern University Press, Boston.

Zazueta, Fernando, April 1992, Interview by author.

Chapter 11

Gardening's Impact on People's Behavior

Ishwarbhai C. Patel

Rutgers Urban Gardening (RUG) is a unique public service educational program aimed at teaching gardening skills and food production, utilization and preservation to low-income city residents of Newark and surrounding communities. It promotes gardening through the establishment of community gardens on city vacant lots. Community gardens are neighborhood open spaces managed by and for the members of the community. A typical community garden is divided into individual plots and planted with vegetables by landless gardeners. Some families even share plots. RUG is in operation since 1978. By now, it has motivated over 6,000 city residents to turn 267 public vacant lots measuring about 18 acres into more than 1,100 community gardens growing nearly 60 varieties of vegetables, herbs, small fruits and other food crops worth over $765,000 in 1991 season.

The documented evaluation of educational efforts revealed many benefits that the program participants believed resulted from gardening. The majority of benefits reflect the value of horticulture to human well-being.[1] Gardens became places for human interaction and community building. Gardens improved the neighborhoods and provided opportunity to meet neighbors and know others.

This paper describes how gardening influenced and promoted the use of various sources and channels of communication in the transfer of garden-

---

This paper was presented at the National Symposium "People-Plant Relationships: Setting Research Priorities" held at Meadowlands Sheraton, East Rutherford, NJ.

ing technologies. The word "source" is defined as any individual (example: a county agricultural agent, other Extension personnel, a neighbor, a friend, a relative, or a fellow gardener) who communicates information to gardeners. The word "channel" describes any device in oral, written, or visual form which facilitates the flow of information between two individuals, such as an Extension Agent and a gardener. The data were derived from interviews with a stratified random sample of 133 respondent gardeners residing in Newark, New Jersey and the surrounding communities.[2] These gardeners were the city residents who enrolled in the urban gardening program (UGP) and used various sources and channels of communication for learning and adopting ten gardening practices shown in Table 1.

FINDINGS

The gardeners learned and adopted a wide range of gardening practices. The majority of them adopted 8 out of 10 practices; three-fourths of them adopted the 3 most popular practices, namely "insect-pest and disease control," "fertilizers," and "raised bed gardening"; and one-third adopted "transplant production" in their own gardens. Home transplant production is a healthy and desirable situation because good quality transplants contribute significantly in increasing crop yields and are expensive to buy from nurseries or garden centers.

With respect to the sources and channels of gardening information, the data in Table 1 reveal that Urban Gardening Program Staff (UGPS) was the most common source in all cases; the second most common source was "other"; Friends and Relatives (FR) proved to be second most frequent source only in two cases, and tied with "other" for second in one case; and Printed Media (PM) was the second most common source only in one case and tied with FR for third in one case. This finding indicates that there was a two step flow of information for gardening practices from Extension agents to gardeners. First, a few gardeners in a community learned about new practices from Extension agents and the PM and subsequently, they conveyed information to fellow gardeners. Learning how to put new practices in action is a task oriented function. It is expected, therefore, that the role is performed best by sources with technical "know-how" that can communicate effectively. Of all sources and channels, agricultural Extension agents most effectively communicated these skills. It is not sufficient to use PM or rely on FR to disseminate information about new farm practices. PM has played an important supportive role, however, and will continue to perform a similar role in the future.

TABLE 1. Sources and Channels of Communication Used in Adopting Gardening Practices

Name of Gardening Practice	Number of Respondents Who Used Various Sources and Channels				Total (N = 133)	
	UGPS	FR	PM	Other	Number	Per Cent
Insect Pest and Disease Control	69	12	16	19	116	87
Organic Fertilizer	51	19	22	14	106	80
Raised Bed Gardening	62	10	1	27	100	75
Chemical Fertilizer	36	26	11	24	97	73
Trellising	51	8	8	36	93	70
Mulching	48	14	6	18	86	65
Composting	52	10	2	12	76	57
Soil Testing	60	7	2	2	71	53
Cover Cropping	21	8	4	12	45	34
Transplant Production	31	5	3	5	44	33

UGPS = Urban Gardening Program Staff
FR = Friends, Relatives, Neighbors
PM = Printed Media (newspapers, newsletters fact sheets, flyers, sign-boards, slides, posters, pictures and brochures)

OTHER = Demonstration gardens, clinics, workshops Spring gardening school, seedling give-a-ways, group meetings, garden tours and harvest festival.

To determine the cumulative impact of the use of sources and communication channels, the gardeners were asked to state various sources and channels they employed in learning gardening practices. Table 2 presents the frequency distribution of gardeners according to the number of practices adopted and the number of sources and channels used. A total of 840 information sources and communication channels were used by 133 gardeners averaging 6.3 sources and channels per adoptor. This indicates that repetition of information becomes important.

A gardener who adopted nine practices used 23 sources and channels. Those who adopted 8 practices used on an average 14.4 sources and channels. The use of sources and channels decreased with the decrease in the number of practices adopted. This suggests that, the greater the contact with the sources and communication channels, the greater was the adoption of gardening practices.

Further, the average number of sources and channels used per gardener decreased more rapidly than the decrease in the number of practices adopted. For example, a gardener who adopted 8 practices used on an average 14.4 sources and channels or 1.8 sources and channels per adopted

TABLE 2. Sources and Channels Used and Gardening Practices Adopted

Number of Practices Adopted	Number of Adoptors	Total Number of Sources/ Channels Used	Average Number of Sources/ Channels used Per Adoptor
9	1	23	23.0
8	27	390	14.4
7	22	135	6.1
6	51	211	4.1
5	22	59	2.7
4	9	19	2.1
3	1	3	3.0
Total	133	840	6.3

change, while a gardener who adopted 4 practices used on an average 2.1 sources and channels or 0.5 source or channel per adopted change. This indicates that the relative impact of sources and channels decreased with the increase in the number of practices adopted.

CONCLUSIONS

The findings reveal that extension education usually requires a combination of communication channels or teaching methods. One channel or method supplements and compliments another. It is the cumulative effect on people through exposure to an innovation or practice repeatedly over a period of time that results in action. The differential adoption behavior of the gardeners in relation to ten gardening practices suggests that the adoption of one practice is not dependent upon another. Adoption of a gardening practice is a major consequence of communication. Furthermore, the relative influence of sources and channels decreased with the increase in the number of practices adopted. People's behavior in communities and social groups is influenced by the presence of plants and participation in gardening activities. Gardens promoted the use of various sources and communication channels and thereby improved communication between gardeners and Extension agents. This in turn motivated gardeners to adopt gardening practices that resulted in increased food production. The future holds a tremendous potential for urban gardening and for Extension's involvement.

REFERENCES

1. Patel, I.C. "Gardening's Socio-economic Impacts–Community Gardening in an Urban Setting." *Journal of Extension*: Vol XXIX, Winter, 1991.

2. Patel, I. C. et al. *Urban Gardening Program Evaluation*: Rutgers Cooperative Extension, 1989.

Chapter 12

Gardening Changes a Community

Terry Keller

SUMMARY. This paper is a general presentation on the benefits of community gardens in urban environments. The presentation will identify research needs and at the same time support the thesis that community gardens are a positive attribute in urban communities. Specific examples of successes will be presented from ten years of work in the New York City community gardening movement. Most importantly, work from the last three years in the Bronx will be highlighted and used as a backdrop to identify research needs. Some of these areas of research to be identified include the relationship between property values and community gardens, the frequency of crime in neighborhoods with community gardens and the patterns of demographic change in neighborhoods with community gardens. With many accomplishments over a ten year period working throughout New York City, there are also challenges . . . to growth and how to meet the different needs of communities in a state of profound change. Those of us in the field have a mandate to support the people in the communities, not to provide agencies with facts and figures. Our needs are different and therefore we need to reconcile data collection with hands-on support.

For eight years I have been working in the field of community gardening and I am constantly amazed by the strong need and desire of people to make green spaces for themselves. I am also humbled by, and in awe of, those who succeed. For not everyone does succeed and with every success there are problems.

Bronx Green-Up, the outreach program of the New York Botanical Garden, is a community gardening program which empowers the residents of our neighboring communities in the Bronx to claim neglected, mainly rubble-filled, lots for greening. We help Bronx citizens acquire the expertise and equipment they require not only to start beautification projects, but also to maintain their efforts for the long term. Building garden/park-like spaces on one-time garbage filled lots introduces pride and hope into whole communities. I would like to demonstrate the challenge of starting a community garden in the Bronx, one of five boroughs, the poorest in New York City.

Imagine you are a person living next to, or across the street, from a vacant lot . . . A lot where truckers illegally dump the detritus of construction and demolition from other neighborhoods. Since it is already full of garbage, you and your neighbors continue to use it for its 'designated' purpose. One day, because you have seen a garden or parklike area in another neighborhood, or perhaps there was a news story about people getting together in the city to garden or maybe you were raised in an area where you and your family gardened and you miss it . . . It doesn't matter. You want to clear off that space and make a garden.

How do you start? Usually you start with places where you already connect with people in your community. Maybe you belong to a church and talk to the pastor, or you belong to a P.T.A. and you talk to other parents. Maybe you just talk to people walking by the lot or to your neighbors in the building where you live. You find that most people seem to have a fatalistic or apathetic attitude; taking the position that "nothing will ever change so why should I bother." Finally you find one or two others that may give it a try. Often it is on the day that you just decided to start, went out with your shovel, gloves and garbage bag, and, one by one, they began to work alongside you.

Considering where to begin with the bureaucratic aspects poses a whole other set of challenges. Someone says, "Why not go to the community board?" But you don't know what district you live in or which community board represents your district. By calling around, you locate the office of your community district and talk to someone about your plan. They tell you to call Operation GreenThumb, a city agency run out of the department of general services.

GreenThumb is the agency that gives out one year leases for city owned lots to would-be gardeners. You call them. English may not be your first language but if you are lucky they have someone you can talk to. In a week you get the application and the first thing you see is that you need the block and lot number of the space you want to make into a garden. You call GreenThumb for information and they tell you to go back to your community board to get that information and oh, by the way, you need a letter from the board in support of your project. Just to get this far takes some persistence, for as we all know dealing with bureaucracy can kill the spirit. The dream of a beautiful garden is what keeps one going.

O.K. You have finished the application according to all the rules, you have come up with names of others who will help (another part of the application) and you have the letter of support. You breathe a sigh of relief and send off the letter.

Next you wait. Within a month you will get an answer unless of course you missed the time frame for disposition of your request and then you will wait longer. Finally you get an answer . . . Yes or no. If no, it means the city has plans for that lot or it will be up for auction and a developer will eventually buy it. Then you look for another lot. After all, there are 10,000 such lots in the Bronx from which you can choose. But you want one close to home . . . One you and your friends can monitor from your windows, that you can care for and watch over daily . . . One that is near enough to get to, somewhere the kids can play or work with you. At this point many less intrepid souls give up, for to start the whole process over takes more patience than most people have.

So let's say you received a yes. Now you have to go to GreenThumb headquarters in another borough, perhaps you have never been there before. You are to attend a workshop in garden/park design and it means a subway trip downtown with your small children, or getting someone to take care of them. Maybe it means taking a half day off work. You are determined to do it, whatever obstacles you must overcome. You attend the workshop, meet other people who also want gardens and plants, and learn, with the help of a staff person at GreenThumb, how to design your garden. You also learn what others have done and you are shown pictures of other wonderful community gardens. You are fired up with the beginning of the reality of a garden.

Now you spend time walking around the lot and people see you and conversations about gardening ensue. Soon others are excited about the prospect of a garden. GreenThumb gets the Department of Sanitation on the ball and they come with machinery to take out the big junk . . . Cars, old appliances, furniture and of course, a truckload of old tires. You are left with all the debris not hauled away . . . Stones, bricks, paper, lots of plastic, lots and lots of broken glass and of course a bucket of crack vilas and used condoms. Since much has been taken away, other neighbors begin to see the possibilities. Others begin to help clean the lot and the guys hanging out on the corner pitch in and even the kids get interested. You begin to feel pretty good about the world.

Now you call GreenThumb and tell them you need equipment. The city has a warehouse in Queens (another borough) where you can pick up fencing materials, cement for the posts, lumber for raised beds, a wheelbarrow and tools. There is a problem . . . You have no car, never mind the truck needed to pick up all the material. You find someone with a truck or you give up. Maybe you rent a truck, but if you had enough leftover change to rent one, you would probably be living in the suburbs.

Just for the sake of continuity, let's say you managed to get a truck. You get all the materials and now you have to store them until you get the fence up. You cannot allow the lumber and cement for the fence to remain on the lot for you know it will be stolen as soon as your back is turned. Knowing this, you lined up some people power to put up the fence before you picked up the material. It is cold or it is raining and so no one shows for the fence raising. You store all the materials in your apartment, a fourth floor walkup. Eventually you get the fence up . . . through the church, the P.T.A., because you are a pest and everyone wants you off their collective backs or just maybe because you've managed to get a few others as fired up as you.

GreenThumb says it will deliver soil but they have a long waiting list . . . So you wait. In the meantime you and your would-be gardeners clean the lot periodically. Since not much is going on, people throw junk over the fence. One morning you and your friends manage to put together a few raised beds, buy a few bags of soil and scrape up the little that is on the lot. You plant some seeds and soon something begins to grow. Others come around and ask if they can help. A little gardening goes on, people stop by asking if they can help. The next year soil is delivered and in the meantime you have been networking and a nice little group of 10 to 15 people are ready to garden. By the way, you have been at this for three years now.

Factions develop in the garden. You are a democratic person and organize a committee with a president, vice-president and so on. Your garden grows, and some politician touring the neighborhood for votes mentions it to a newspaper. A photographer comes out and takes pictures. The news spreads like wildfire and soon the entire community is in the garden. Others hear about the garden and before you know it, papers like this are written about how gardening brings a community together.

The woman or man two streets over wonders if the garbage filled lot across from his or her apartment is available for a garden. And so it starts over. Tree plantings begin and now new sidewalks are needed.

You along with your neighbors had the vision, the perseverance and interest in a common goal to bring the community together. The focus was a garden. The pride in accomplishing that task was validated by all the publicity and compliments of visitors. The garden has introduced pride and hope into the entire community and everyone who walks by feels better.

Schools in the neighborhood are now sharing the garden, adopting a couple of the plots, and a community organization has formed to advocate for a ball field. A youth organization formed to help with clean-up details in the fall and spring.

A project such as I described is greatly facilitated and often made possible with the help of Bronx Green-Up. In some cases, the process which culminates in a viable community garden may be measured in months rather than years. We coordinate all available services and provide truck transportation. Community gardeners also receive seeds, trees, shrubs, flowers and vegetable plants through our program and we make available and deliver free compost and soil.

Good things have happened because of gardening activities in the Bronx. In one garden, Felix Graham, an elderly retired, African American gentleman along with two friends, William and Bernie, grew pumpkins and collards that were the pride of the neighborhood. Felix grew ill one winter, suffered a stroke and then pneumonia. By the late spring, supported on each side by William and Bernie, he came to the garden and was seen there most sunny mornings. Felix told me: "When I was the sickest I kept thinking of the garden. I couldn't have gotten better without this place; now, it helps just to be able to sit in the sun with my friends and then, when I feel stronger, I'll do my gardening." Felix is once again going strong and still grows the best pumpkins and collards in the neighborhood.

Around the corner, Mrs. Lois Reddick gardens. For years she stayed in her house, only to sit on her porch once in awhile because "I always felt poorly." Now she grows flowers and vegetables, "almost as nice as the ones I grew in South Carolina." So often I have heard her say, "when I dig in my garden, I dig away the blues."

Last spring, a new group of immigrants arrived in the community. One of the gardens made room for them and soon several families were growing tomatoes, tometillos, epasote, lots of peppers and cilentrillo. Through talking to neighbors in the garden, they learned they were being gouged in rent payments. The community association, initially formed because of the garden, worked to help them with landlord problems and was able to get the rent lowered.

We have helped, in various ways, 150 gardening groups in the Bronx.

Some are connected with social service agencies, others with schools and still others with public housing. All have had problems, yet most work things out. When asked, we step in to help. One such time we were contacted by a group of gardeners complaining bitterly about the "leader." Through another city agency we were able to bring in a community organizer who taught leadership skills and eventually the gardeners voted in a president who delegates responsibilities and makes inclusion part of the way the garden operates. Some gardens are run like benevolent dictatorships, others are very democratic, some are chaotic. But that's O.K. It is the gardeners' community and their garden. I don't think I would like to have someone telling me how to run my own garden.

Which brings me to a very important point which we have learned. That is, not to help unless asked, and not to impose a garden upon a neighborhood. It just doesn't work unless people want a garden and then work to get one and finally maintain it over a long period of time. One organization in New York City built several garden/park-like areas for young children adjacent to day care centers. After the spaces were finished, the neighborhood was expected to maintain them. The people of the community had no initial input into the project and had no feeling of ownership. Most of the time the gates are locked and there is no community activity in those spaces.

We are asked to help with more and more community and school gardens. Often people ask why? The Bronx has more parkland than any other borough. The answer is . . . Most parks are not within the immediate areas of the neighborhoods . . . That is, within walking distance. Parents are reluctant to allow children to leave the community because of the drugs and guns.

Also, most of the parks built in the early part of this century were built with different ethnic groups in mind. Communities in the Bronx, as well as in many urban areas in this country, are in the process of changing their ethnic makeup. Relevant open space, designed with those who live in and around them in mind, is necessary. People living in the neighborhoods must be consulted as to what is needed for recreation purposes and should be involved in the entire process. Only then will neighborhoods maintain their open space. The proponents of public housing are only now beginning to understand the importance of this approach. So often explanations for why poor neighborhoods are run down are bound up in racism rather than on focusing on the lack of municipal services coupled with lack of sense of community ownership.

Last year New York City agency budgets were cut by an average of 10.5%. The Department of Parks and Recreation had its budget slashed by

39%. This brings it to the lowest level of funding in its history. Budgeting for staff has been severely cut. In the poorest areas of the city, parks have no staff on duty and no recreation programs for children and it will not get any better in the near future. We are finding that our community gardens are taking the place of park land. Active recreation in the form of gardening goes on, neighbors congregate and on the weekends there are parties and barbecues. Children are always welcome.

Park budgets will continue to be slashed and those of us helping with neighborhood open spaces will continue to be strapped for program funds unless evidence based on experience as I have just given is quantified and qualified. We are plagued by a lack of adequate funding which is all the more difficult to obtain because we don't have the evidence documented on paper to support our claims. To get the funding we need, I must take prospective funders into the gardens to see for themselves and to listen to the people speak. The people are quite eloquent about what those spaces mean to them.

But to be really effective, to obtain the city and state funding which we need and should have, we need concrete data . . . demonstrating the need for both parks and neighborhood garden/parks. Those of us running the programs spend all our time in the communities, assisting communities, doing support work. We do not have the time, the expertise or the inclination to quantify necessary data. I leave it to you, to those in the social sciences committed not just to studying society but to helping to transform it, to come up with a way to help us assist and support communities, particularly those with the most limited resources.

crime down
beautification up
real estate ↑
relationships built

Chapter 13

Down to Earth Benefits of People-Plant Interactions in Our Community

James W. Zampini

One of America's greatest legacies is that it was built by people helping people. Over the last several decades this legacy has seen bumpy roads. However, within the past few years the American people are falling back on more traditional values and are trying to restore the legacy of people helping people to a higher level than ever before.

Trees and flowers are an intricate part of our daily lives. Their presence improves the quality of our lives in many ways: environmentally, economically, socially, culturally and physically through our health and well-being.

Beautification in the community can have enormous benefits. Just think how many times someone has remarked to you, "Isn't that a beautiful street, village or town. The homes and yards are so beautiful." It's all positive; people feel good to be associated with such beauty and love to talk about it with others. Doesn't it make sense for those associated with the green industry to commit themselves to help aid the beautification effort in some way, no matter how large or small?

This past year, Lake County Nursery committed itself to be more involved in shaping the future of our world through beautification. In September we co-hosted a Beautification Stewards Conference along with the Ohio Department of Natural Resources, the Cleveland Electric Illuminating Company and CLEAN-LAND, OHIO at Lake Erie College.

The conference focused on "How to Increase Tax Revenues and Lessen Crime by the Proper Planting of Trees and Flowering Plants." A group of nationally known speakers made presentations to attendees from all across the United States.

Also, we continue to give many slide presentations to groups who are involved in beautification projects throughout the area.

The following will retrace our steps over the past decade of how we became a catalyst of people-plant interactions through beautification in our community.

LAKE ERIE COLLEGE

Even though Lake County Nursery has been involved for more than 25 years with various beautification projects throughout local communities, it was a project at Lake Erie College that was our "awakening" as to just how powerful an impact beautification can have on people and their daily lives.

Lake Erie College is a small liberal arts college that was established in 1856. It is located in Painesville, Ohio, the county seat of Lake County. Its rich tradition and history, coupled with its location just west of the city square, has made it an integral part of the Painesville and Lake County communities for many years.

My involvement with Lake Erie College was actually not my idea at all, it was my wife's. As part of a rehabilitation program following my heart attack and subsequent open heart surgery, my wife and I would take regular walks throughout Painesville. A stroll through the college campus was always part of this hike.

One day my wife commented on the terrible condition of the college grounds and how the once beautiful campus had deteriorated. She asked if there wasn't something I could do to help.

I realized my wife was right. Help was needed. Not only was Lake Erie College in trouble, but it also *looked* like it was in trouble. From 1982 to 1987, enrollment at the college had declined 42%. There were many rumors that this landmark institution might close. As I walked with my wife, it became clear to me that this college would have a very difficult time attracting new students.

The following day I made an appointment to meet with the president of the college. He was immediately enthused with the plans for the beautification project, agreeing that it was becoming increasingly difficult to promote Lake Erie College as a quality educational institution with its decaying appearance.

The beautification project received almost immediate, widespread support. Many civic-minded community and business leaders became involved.

A landscape architect drew a master beautification plan; many area nurserymen donated plants; a roofing company repaired or replaced several roofs; a company donated the asphalting of parking lots and sidewalks; a painting company repainted a fence enclosing much of the campus; a concrete firm donated new steps replacing those located at some building entrances that were closed, and an architectural firm designed and helped fund new, elegant hand-carved entrance signs.

Since the beautification program began, enrollment at the college has increased 55%. The administration believes that much of this increase is directly attributable to the beautification project.

The remark of Clodus Smith, president of Lake Erie College, probably sums it up best; "The beautification reminds us that first impressions are the ones that last."

THE CITY OF PAINESVILLE

The City of Painesville is a typical American rust-belt community. It has been in an economic decline for nearly 25 years, losing several major industries and the jobs that go along with it. A negative attitude has rippled through the community.

One day Les Nero, Painesville city manager, and two councilmen, made a visit to the nursery and asked me how I thought Lake County Nursery could help the city encourage people to take steps to restore the city back to its prominence.

My suggestion was that in most cases it doesn't take many dollars if the "3 P's"–People, Plants, and Paint–are initiated.

We determined a good place to start would be in the heart of the city–Veteran's Park on the town square. A few years ago a number of cherry blossom trees surrounding the park became diseased and were in varying degrees of deterioration.

The trees were removed by city workers and replaced with Sugar Tyme® flowering crabapple trees. This particular tree was selected not only for its spectacular spring display of blossoms, but also for its abundance of fruit that will attract birds throughout fall and winter. Most importantly, it is a low maintenance plant.

The cost of this beautification program was equally shared between the City of Painesville, Lake County Nursery, the Lake County Commissioners and Bank One.

My other suggestion, which is indirectly related to plants, was lights . . .

Christmas lights . . . and lots of them. Light up the park and light up the buildings that surround it was the goal.

The city council went ahead with this project and each year more and more lights are added. This past Christmas it was breathtaking when driving into the city and around the park. People visited the city just to look at the lights.

From this we are now seeing the buildings around the square being restored. Sidewalks have been repaired and new old-fashioned lamp posts have been installed. There's still a long way to go, but for every goal one must crawl before walking. The city is once again proud of Veteran's Park for its beauty and function by providing a passive outlet for downtown workers and residents.

OPERATION M.A.P.

Wendell Walker, a good friend of mine since grade school, is the president of the Painesville city council. Wendell, knowing the value of plants and what it could mean to his community, has had a longtime goal of having a landscaped city.

Together we decided that the place to begin was by creating a "Welcome" planting at the entrance to the city off Interstate Route 2.

A staff member of LCN designed a landscape that would add beauty and require very little maintenance. With nursery stock provided by LCN and labor by the city, a landscape of trees, shrubs and ornamental grasses was installed.

With that successful project under our belt, I was approached again by Wendell. He said, "Jim, in my black community we need help. We have an area that needs a spiritual uplift. We have some drug problems in this area. Several people in this neighborhood claim that if the city could renovate the community recreation center and provide a greenhouse with some space to garden, the slogan 'Painesville City Pride' could be demonstrated to the fullest potential." He also said, "I know these people. They will give their heart and souls."

After several meetings it was decided that the main corridor leading to the downtown from Route 2 and its industrial park should be beautified first. Next we would come back to work on the community center which is located on one of the side streets off the main corridor.

This project was given the name OPERATION M.A.P. M.A.P. stands for Morse Avenue Project.

Once the city manager and city council gave their blessing, time was of the essence. We had only three weeks to implement and complete this project as winter was rapidly approaching.

Our first step was to get a landscape architect firm to donate their time and draw up a blueprint.

After the city officials, economic development director and landscape architect decided the proposed plan would work, our second step was to encourage participation by various civic groups and companies.

Because of time restraints we contacted as many civic groups as possible, a total of thirteen. Every group contacted said yes. I am sure that any others that we could have contacted would also have said yes. All volunteered to physically participate in helping install the needed materials to beautify this area.

We contacted several banks who donated money for food to feed the volunteers. The Salvation Army donated their canteen truck and time to pass out the food.

Before the planting day arrived, there were two organizational meetings to describe to all involved their duties on planting day.

A work schedule was presented that outlined their particular job. Each group was given a specific color code and a package prior to the planting date for a complete understanding of the task at hand.

Prior to the planting day several people told us, "Do not leave your equipment out in the work area because it will be stripped before morning." So, we hauled it back and forth for a couple days before we decided to take a chance and leave it there. Our equipment stayed out for two weeks without a single incident of vandalism.

Upon delivery of the sod, several people told us, "Do not leave the sod out here on the street because it will not be there tomorrow morning." Again, not as much as one blade of grass was touched.

When "The Day" arrived for planting, even with a nasty, cold, fall day, more than 120 volunteers appeared . . . many more than we had hoped for. These people were from all walks of life. They were a major key to the success of this project. As you looked around everyone was working, smiling, having a good time . . . bonding the community within the various cultural groups. They were a true sign of partnership at its finest. And what did they have in common? They were all *just waiting to be asked* to get involved.

The volunteers were recognized with a hand-signed certificate embossed with the city seal along with gift certificates from McDonalds and other local businesses.

Approximately two weeks after the completion of this project someone did drive their car through a small corner section of the lawn. But, rather than calling the city service department and demanding that the city workers come fix it, the people in this neighborhood put on their work boots, picked up their shovels and rakes and walked down to level and replace the turf themselves.

Also, at Christmas, the pines and blue spruce were beautifully decorated with red velvet bows.

From this project an enthusiasm vacuum was created. A group of individuals led by the Reverend Willie Shaw and his wife went into the Morse Avenue Community Center, a building barely used, and scrubbed, painted and polished it from top to bottom. Now people are asking to use the building for many special functions.

In November the church group held their first annual Community Thanksgiving dinner. It was free to the public and approximately 125 people attended. Next year, with increased publicity, an even greater turnout is expected.

I am sure that this project will have numerous spin-off effects for future individual and joint private/public endeavors throughout the city. I am confident after working with the people living in the MAP neighborhood and the city officials that the Painesville City showcase will continue to grow. As in 1991, city officials budgeted dollars for beautification in 1992.

This project already has motivated several other communities to follow Painesville's lead in beautification. The local Kiwanis group made beautification their group goal for 1992.

To me this project was re-establishing America's rich heritage. Hard work and people helping others. The will and hard work . . . more than money, helped build America and the people involved in these projects proved this. If this beautification project was to be bid out to contractors the cost would have been $70,000-$95,000. Instead it cost approximately $31,000 including all the plants, sod, timber, mulch, topsoil and the automatic underground irrigation system.

As our urban communities become more compressed, plants are going to continue to play an even greater part in the therapeutic world of family bonding. In rural areas with open spaces you need not be so fussy in landscape designing, but in a true urban setting you *must* utilize the proper plants to create a feeling of inner goodness.

Those of us in the green industry too often undervalue and don't recognize the full potential of our products. Our products are more than just pretty things to observe.

In addition to all the environmental benefits that plants provide, many studies have shown that proper plantings reward us with many positive psychological feelings about our surroundings. In many areas, it has been shown that besides dramatic increases in property values, there has been a significant decrease in crime where beautification projects have been completed successfully.

GUIDELINES

For those who intend to get involved in a project of this magnitude, leadership is not enough. You must observe certain key principles.

1. Be a facilitator of excellence. Recognize and know the decision makers of the community whether it be city officials, civic and business leaders or any people of influence. It is important to empower them and encourage them to empower those on their work teams. As I think it has been shown, success in this type of project depends on total community involvement.
2. Develop a comprehensive involvement outline for all levels of participation.
3. Show benefits to all donors. Make them aware of the harmony and sense of inner goodness they will create and receive.
4. Give facts about crime reduction, passive recreation, the therapeutics of plants, population growth, environmental benefits, economic benefits, energy conservations and noise abatement.
5. Wiser investments in plants should be the #1 rule–plant only high quality, low maintenance plants to maintain low expenditures for years to come while the beautification area is enhanced naturally. Suggest a qualified plantsman for the beautification team.
6. Have prepared news releases ready to promote awareness about the project you are undertaking. You may receive additional help you weren't counting on.
7. To maintain community spirit, all people involved in the project should be recognized for their efforts. Whether it be award certificates, names published in the newspaper or a party; show your gratitude for their help.

People-plant involvements can add tremendous value to a community. We encourage those in the green industry to become a catalyst by organizing these types of projects in the places where they live and work. Don't wait to be asked; volunteer your services and become a leader by promoting the many benefits of beautification. While your reward may not be monetary, the sense of harmony and inner goodness you will bring to your community and yourself will last a lifetime and for generations to come.

Chapter 14

Human and Plant Ecology: Living Well with Less

Roger E. Ulrich

SUMMARY. This paper addresses a problem that faces all life on Earth. We, as human beings, are almost literally a cancer that is eating away from the resource base for all existing life forms, leaving behind a polluted desert in our over-consumptory wake. For other animals and plants to survive, we must face the fact that our human addictions to doing whatever feels good at the moment must be curbed.

A quarter of a million people are added daily to the Earth's surface, making even more disastrous the potential problems we face. Human beings are more a part of the Earth's problem than a solution; we have met the enemy and it is us. So what shall we do?

At every level of our individual daily personal actions, we must begin to rise to the challenge of living our lives using fewer resources. Each person, starting now, must begin to spend fewer hours with the lights turned on, encourage our children not to have children, build smaller homes, walk and bike instead of driving combus-

---

This paper is a modified version of a presentation given at the Meadowlands Sheraton in East Rutherford, NJ, as a part of the People-Plant Symposium, "Setting Research Priorities," sponsored by Rutgers University, the People-Plant Council, the American Society for Horticultural Science, the American Association of Botanical Gardens and Arboreta, and the American Horticultural Therapy Association. Editorial assistance was provided by Joan Stohrer, 1608 Sunnyside Drive, Kalamazoo, MI 49001.

tion-engine cars, take fewer trips to shopping malls, build fewer shopping malls, etc. In short, we must each of us dedicate our lives toward practicing the necessary art of living well with less.

> *The white man is a stranger in the night who comes and takes from the land whatever he needs. The Earth is not his friend but his enemy, and when he has conquered it, he moves on. He kidnaps the Sky like merchandise, and his hunger eats the Earth bare leaving only a desert. Humankind has not woven the web of life. Whatever we do to the web, we do to ourselves. All things are bound together. All things connect. Whatever befalls the Earth befalls also the children of the Earth.*[1]

My professional life as a psychologist has been dedicated to the exploration of the nature of humankind. I have attempted to find the truth as it relates to people and report what I have found. The words presented above I pass on to you in the spirit of Chief Sealth, a native on this continent respected as an elder among his people. The points they make are even more poignant today than they were at the time the seeds of his message were planted. We human beings are almost literally a cancer on the Earth, eating it bare and leaving only a desert in our over-consumptory wake.

When I was in the eighth grade, I was taught that we can neither create nor destroy the basic ingredients of the resources that are required to support our lives. Later I learned to refer to this statement as being basic to Nature–the first and second laws of thermodynamics. The amount of energy in the Universe has been fixed for all of eternity. It can, however, be changed in form–not destroyed, but changed. God is supposed to have spoken to Moses through a burning bush. Perhaps God speaks to us again today through the words originating from an Indian Chief who told us:

> *What are human beings without animals? If all the animals ceased to exist, human beings would die of great loneliness of the spirit. For whatever happens to animals will happen soon to human beings. Continue to soil your bed and one night you suffocate in your own waste.*

Near the end of 1990, *E Magazine* contained a cover story entitled *Sheer Numbers* (Hardin, 1990). In it, the point was made that population problems are chronic. What we consider news consists of sharply-focused occurrences (Hardin, 1990). What did President Clinton say today, or ex-President Bush, or some other human politician or college president or research professor? What plane crashed or where did the Earth shake and how many were killed? Extra, extra, read all about it: 263,000 people were added to Earth today. Next day, the same news. Next day, the same again, maybe a few more. Then again and again and still again. Boring–turn that thing off; I'm tired of hearing that a quarter of a million new people are added to the planet each day. Increases in population just aren't news.

> *Continue to soil your bed and one night you suffocate in your own waste.*

As I said before, I study the behavior of humans and other animals. Recently I wrote a book. I called it *RITES OF LIFE* and said that it was about the use and misuse of Animals and Earth. I quote from its Preface:

> It's a happy day. Spring is breaking on the Lake Village Co-Op Farm, where I live. I've just returned from the morning chores. The sheep and cows have been fed, and the newborn goat kids have put on their early show.
>
> Robert Joseph, the herd bull, has had his back scratched. The chickens, pigeons, rabbits, peacocks have settled back into doing what they do all day. The horses broke through a gate and went back to the long pasture two days ago and have not been up to the barn since.
>
> Most of the 40 people from the Co-Op have gone to school and work. They drove cars, which helped add to the millions of pounds of toxic waste we in Kalamazoo County spread out each year into our air, water and Earth.
>
> At school, the students will learn how to become an integral part of our society and thus later be able to join their fathers and mothers, as we all earn money to buy the things we think we need. Much of what we buy, of course, will end up as trash, and next year's pound-count of toxic waste will be that much higher. This fact will prompt more meetings at universities around the world by educated experts who, having received millions of dollars to research the problem, will come out with additional reports. Requests for more millions will then be made for further studies.
>
> Trees around the world are coming down at a rate of 23 million acres per year, denuding an area the size of England–trees that provide oxygen so that we can breathe and counter the toxic wastes put out by the autos we drive to work. In our county, we are taking trees down from along our roadsides for fear that they might kill the drivers who hit them.

> Soon, I will drive to work, adding to the pollution, and meet a graduate class. During the semester, they read books such as *Walden Two, Friendly Fascism, Limits to Growth, Entropy* and *When Society Becomes an Addict*. They saw movies–*Never Cry Wolf, The Animals Film*–and visited with some Amish Mennonites, ate with them, and saw how they lived. Students also came to Lake Village and stayed there with me and my extended family . . . and observed us as we live with all our addictions, family feuds, gates left open, recycling attempts and other issues which relate, basically, to the use and misuse of resources.
>
> So . . . now I'll tell you the conclusion of the psychological research that I have been conducting with animals and our environment since 1955 and to which this book relates: Human beings (God love us) are a real problem . . . and unless we surrender to the other animals and watch and listen very carefully to them, as they show us how to live lives that are more in balance with the laws of nature, we will disappear off the face of the Earth sooner than later. We must, as humans, learn how to live using fewer resources and to show greater respect for all animals and all of life.
>
> I hope that you enjoy this book, should you decide to read it. If you can bring yourself to it, however, I suggest that, between chapters, you also spend time planting trees. I know that planting is more difficult than sitting down reading and writing, but if we do not plant trees, and do not stop driving our cars so much, there are not going to be any wild doves around any more, singing . . . as one is doing just now down by the lake. (Ulrich, 1989)

It was about four years ago that I wrote those observations. Since then, the intensity of my concern has grown. I believe today more firmly than I did then that we humans are the Earth's cancer, her AIDS virus, her most critical illness–Mother Earth suffocating in our waste. For reasons unknown to me, we seem to produce more entropic pollution per capita than any other lifeform. Even with the realization of that fact, it's still hard to tell my children not to have children, although I do and to date they haven't, which I doubt is due to anything I have said. I believe that the Earth is overburdened with a destructive lifeform; I have met the enemy and it is me.

Well, there you have it. I have studied human beings closely and the results of my research suggest that the Earth would be better off growing trees and other plants, not more humans. I feel that these results are supported and highlighted by an NBC News report entitled *EarthWatch*, in which Dr. Paul Ehrlich dramatically narrates the global problems we now face such as the human population explosion, species extinction and the greenhouse effect. (The video *EarthWatch* can be obtained by writing NBC Studies, New York, New York.)

As a psychologist, I feel that it is important not only to present findings based upon research but also to point the direction toward new experiments, and for me there is no longer any experiment of importance other than the real situation. And what is that? I feel the real situation is one in which all of us will sooner or later come face to face with the limits to growth that determine and, by their interactions, ultimately define what happens on this planet. These limitations are determined by population increase, agricultural production, nonrenewable resource depletion, industrial output, and pollution generation.

In 1972, an international team of researchers fed data on these five factors into a global computer model and the behavior of the model was then tested under several sets of assumptions to determine alternative patterns for the future. The resulting message of that study was that the Earth's interlocking natural system of global resources could not support the rate of resource depletion that was then occurring. Thus as humans continue to expand and grow, the results will be the same: resource depletion and the ultimate destruction of life-giving habitat will increasingly manifest itself all around us (Meadows et al., 1972) (see Figure 1).

My exploratory travels around the world, my past 20 years living close to the Earth at the Lake Village Experimental Community, and what I taste in the water, smell in the air and touch on the Earth all affirm what many scientists as well as others have finally come to realize and that the spirit of Chief Sealth so dramatically proclaims:

> *We must live more respectfully with regard to our resources and all our relations. We must begin to work toward living well with less.*

The over-consumptory human habits that have evolved ignoring the universal truth that *"there are limits to growth"* are in no small way a direct function of the widening gap between our daily human actions and what we know as the facts of entropy. We say we should use fewer resources, but we don't do it. Exploitation perpetuated in our own lifestyle needs to be addressed. The myth that state and national well-being requires more and more resources being poured into institutional research and researchers that have repeatedly demonstrated allegiance to the ethic of "Big is beautiful," "More is better," and "Money is our savior" must be exposed so that we understand that we ourselves have become over-consumptive addicts just as Anne Wilson Schaef suggests is true for all of society (Schaef, 1987).

Unlike many other lifeforms, human beings seem to be more prone to

FIGURE 1

Reprinted with permission.

respond to short-term contingencies. If it feels good, do it again. The principles of reinforcement have shown this to be true even of non-human animals, especially when confined and forced to exist in unnatural laboratory settings. Humans, however, seem to be unbounded in their ability to overcome limits of resource depletion displayed by other lifeforms who live in closer balance between the intake and output of energy and matter.

Since 1970, the human population has grown by 47%. We add one quarter of a million people to Earth daily. Agricultural production fails to keep pace as people starve in alarming numbers. Resources decline as industrial growth continues to sully the Earth with pollutants.

Neither behavior analysts nor humans as a species are Earth's saviors. We are, as the basic assumptions of the analysis of behavior implied, simply a natural, lawfully-determined part of the ongoing whole. We are no better than any other lifeform and above all need to become humble in this regard.

In his book *The End of Nature*, Bill McKibben laments the harm we humans have done to the Earth and pleads for its restoration. In talking of

Nature's end, he suggests that it is finished as a force independent of human input.

> Whatever we once thought Nature was–wilderness, God, a simple place free from human thumbprints, or an intricate machinery sustaining life on Earth–we have now given it a kick that will change it forever. (McKibben, 1989)

An old copy of the *New Yorker* contained an advertisement from what in 1949 (the year I graduated from high school) was still the Esso Company. It summed up our century to this point: "the better you live, the more oil you use." And we first-world behavior engineers live well. The trouble is that this pattern of behaving to which we are addicted seems not to be making the planet happy. The atmosphere, the forests, the grasses and the water are all less satisfied than we are.

We need behavioral research–research that will show us how to change human behavior from actions that suggest we are more important than anything else. We must replace the Judeo-Christian ethic, which promotes the "humans are best" attitude, with new behaviors that hopefully might slow down the destruction of our Earth. We must begin to think and act more like our brother and sister plants, the trees, like the lakes, the mountains and the wind, and become sensitive to the fact that our nature is identical to the nature of the Universe. I have a friend, Rolling Thunder, who tells me that the Earth, of which we are a part, is a living organism, the body of a higher individual who has a "will" and "wants" to be well, who is at times less healthy or more healthy. He says we must treat our own bodies with respect and it's the same with the Earth–when we harm the Earth, we harm ourselves; when we harm ourselves, we harm the Earth.

Another friend I knew well, B.F. Skinner, suggested in his book *Walden Two,* that we must "experiment with our own life . . . not just sit back in an ivory tower somewhere as if your own life weren't all mixed up in it" (Skinner, 1976).

Some years back a couple of close relatives of mine handed me a book written by Doris Janzen Longacre entitled *Living More With Less.* This book is a pattern for individual living using fewer resources. It holds a wealth of practical suggestions from the worldwide experiences of the Amish Mennonite culture of which my roots are a part. For each of us, the first step toward healing the wounds which have been inflicted upon the Earth begins right where we are at this moment. Look around you; turn something off that is using up a resource that is already needed by other animals and plants as well as by our children's children's children.

NOTE

1. In about 1854, on the occasion of Washington Territorial Governor Isaac Stevens' first visit to that area, a Duwamish and Suquamish Puget Sound Indian named Chief Sealth is said to have made a speech. Henry Smith, a local man, attended and took notes. More than 30 years later, in 1877, Smith was purported to have reconstructed the speech for a Seattle newspaper. Seattle historian David Buerge concluded that although Smith's text reflects at least some of what the chief had to say, the fact remains that the English version has been no doubt greatly changed over the years.

The seeds of Chief Sealth's original thoughts continue to grow, constantly finding new expression. A University of Texas professor allegedly used Smith's reconstruction for his own version which Ted Perry is supposed to-have heard at a 1970 Earth Day celebration. Perry then rewrote still another version which appeared as part of a film script produced for the Southern Baptist Radio and Television Commission. It has been expanded upon by Britain's Prince Philip and Joseph Campbell and is, in fact, the basis of a recent children's book entitled *Brother Eagle, Sister Sky: A Message from Chief Seattle.*

Wayne Suttles, an emeritus professor of anthropology at Portland State University, is said to have branded the speech as a fake. Jack Hart claims that the speech is nothing but a ". . . phony snippet of greeting-card blather . . ." (Hart, 1992) which raises questions about a trend to accept what we want to hear. In my opinion, far too much of all history has been an acceptance of what a powerful few wanted others to hear–and act upon as if it were the truth. Today, people seem to want to hear more and more about how to save the earth. They seem to know that something has gone wrong and they are attracted to the spirit of Sealth's expanding influence. The indigenous spirit that lived respectfully with nature is touching many of today's youth. Some, like Hart, discredit that spirit by suggesting that the words don't count since Sealth didn't really say them. Others also seem frightened by our new-found interest in Nature and natives and spend great energy trying to prove that whites have invented the "good" Indian (Clifton, 1990).

No doubt there is a little bit in each of us who perhaps would like the world to be something that it isn't. We would like it to afford us endless bounty with ever-increasing access to more and more of everything we want. I am not concerned with who really said the above words; I am concerned that we pay attention to their essence.

REFERENCES

Clifton, James (1990). *The Invented Indian*, Transaction Publishers, New Brunswick, NJ.

Day, Bill (1992). *Detroit Free Press*, April 4, Detroit Michigan.

Hardin, Garrett (1990). Sheer Numbers, *E Magazine*, November-December, pp. 40-47.

Hart, Jack (1992). Putting Words in His Mouth: Chief Seattle's Speech Raises Questions About Trend to Accept What We Want to Hear, *The Oregonian*, February 9.

Longacre, D.J. (1980). *Living More with Less*, Herald Press.

McKibben, B. (1989). *The End of Nature*, Anchor Books.

Meadows, D.H., Meadows, D.L., Randens, J., Behuens, W. (1972). *The Limits to Growth, a Report for the Club of Rorie's Project on the Predicament of Mankind*, A Potomac Associates Book: Universe Books, New York.

Rifkin, J. (1980). *Entropy*, A Bantam New Age Book: Viking Press.

Schaef, A.W. (1987). *When Society Becomes an Addict*, Harper and Row.

Skinner, B.F. (1976). *Walden Two*, MacMillan Publishing Co., Inc.

Ulrich, R. (1989). *Rites of Life*, Life Giving Enterprises, Inc., P.O. Box 404, Kalamazoo, Michigan 49005-0404. pp. iii-v.

Chapter 15

Evaluating Horticultural Therapy: The Ecological Context of Urban Jail Inmates

Jay Stone Rice
Linda L. Remy

SUMMARY. The application of horticultural therapy to prison and jail populations has received little evaluation. The results of baseline interviews with 57 San Francisco County jail inmates randomly assigned to a horticultural therapy program are presented. An ecological analysis of the inner city jail population identifies the relationship between job loss, family disintegration, and history of victimization. The role of trauma in precipitating self-fragmentation is explored with respect to the high incidence of drug abuse and illegal activities. The use of nature to promote healing and growth is discussed. The importance of empathic interventions by the horticultural therapist is noted.

INTRODUCTION

The use of nature, natural growth cycles, and gardening in a therapeutic manner has early roots in the United States. The pioneering American

---

This research was supported by grants from the Wallace Alexander Gerbode Foundation and the Bothin Foundation.

physician, Benjamin Rush, utilized farming with the mentally ill in 1798. He maintained working the soil was healing (Olson, 1976). By the early 1900's gardening therapy programs were in American hospitals, institutions, and reformatories (Grossmann, 1979). While there have been numerous anecdotal reports of horticultural therapy's effectiveness, as well as a few statistical evaluations of horticultural therapy programs, more research is required (J. S. Berry, 1975; Francis & Cordts, 1990; Cotton, 1975; Gilreath, 1976; Hiott, 1975; Horne, 1974; Relf, 1981; Tereshkovich, 1973).

Horticultural programs are utilized in many jails and prisons. They run a gamut from formal therapy programs to prison farming industries. In 1990, the authors telephoned 55 prison authorities in 41 states. Seventy-five percent of those contacted reported that facilities under their aegis had either vocational or correctional horticulture industries, 19% had formal or informal horticultural therapy programs, and 6% currently had no horticulture programs. The state authorities reported there had been no evaluations or other formal studies of these programs. This paper presents baseline data from a study assessing the effects of horticultural therapy on the self-development of San Francisco County Jail inmates.

ECOLOGICAL CONTEXT

To assess the impact of a horticultural therapy program on county jail inmates, it is important to understand the inmate's life-space or umwelt (Uexkull, 1957, cited in Lopez, 1986; Omark, 1977), what he perceives or is able to perceive in his environment, and how this shapes his experience. Understanding the ecology of the inner city provides a basis for appraising the efficacy of this treatment modality. An ecological framework also provides essential information for horticultural therapists working with this population, and may aid them in developing an empathic therapeutic relationship.

Bronfenbrenner (1979) believes an ecological understanding of human development enables researchers to recognize and address environmental factors which obstruct or enable humans to express their full potential within an "ecologically compatible milieu" (p. 7). Ambrose (1977) defines an ecological study as one which examines a species by noting "its habitat, way of life, population characteristics, social organization, and relations with other species" (p. 4). The data presented in this paper includes location of residence, education level, employment status, marital status and family history of trauma, substance use, and criminal history.

Inner City Physical Environment

The vast majority of San Francisco County Jail inmates reside in seven low income urban neighborhoods. While no data were gathered for this study regarding physical environmental influences, these are known to affect urban residents unequally. Inner city neighborhoods are noisier, have higher density, and exposure to environmental pollutants. An inverse relationship has been noted between socioeconomic status (SES) and exposure to air pollution (Burch, 1976; Kruvant, 1975; Schnaiberg, 1975).

Bronfenbrenner (1979) and Garbarino et al. (1982) ascribe developmental significance to the quality of the physical environment. Bronfenbrenner notes psychology and sociology emphasize social class and parental characteristics and undervalue the impact of locality on child development.

The quality of our urban environments has been shown to impact the health and functioning of our communities. Milgram (1970) assesses the habitability of modern cities by comparing them with early human environments. He observes overwhelming environmental stimuli contribute to increased stress, reduced attention, and insensitivity. He proposes some of our most distressing human behaviors may be the result of living in environments which counter human needs.

Density has a detrimental impact on group interactions. As density in university housing increased, students were less willing to participate in cooperative group projects (Bickman et al., 1973; Jorgenson & Dukes, 1976). Rodin (1976) identifies a relationship between density and learned helplessness. In a comparative study of groups from densely and sparsely populated areas, he found more problem-solving capacity in people from the sparsely populated area. He concluded increasing density generated expectations that events will be uncontrollable.

Noise adversely affects helping behaviors in laboratory and natural settings (Matthews & Cannon, 1975; Page, 1977). As students were increasingly exposed to noise, they were less likely to respond positively to a request for phone change or to aid someone who dropped their books.

Socio-Demographic Influences

The availability of work and the number of residents who are gainfully employed is another key determinant of the ecology of a community. Employment directly effects family stability and the capacity to support healthy child development. Wilson (1987) explores in detail the relationship between income, location of residence, and family structure in the genesis of the inner city underclass. He cites U.S. census reports which show the increase in single parent families. From 1970 to 1984 families grew by 20% nationally. The number of female-headed families grew by 51%. The number of families headed by women with one or more children living at home increased by 71%. The increase was 63% in families headed by white women, but 108% for African-American women (U.S. Bureau of the Census, 1980, 1981a, 1984a).

Wilson argues recent economic factors are adversely affecting the formation of families in low income inner city communities. Wilson notes the percentage of African-Americans in the labor force has fallen from 84% in 1940 to 67% in 1980 (U.S. Bureau of the Census, 1984b). In 1979, when the overall unemployment rate was 5.8%, the unemployment rate was 34.1% for African-American male teenagers (Duncan, 1984; U.S. Bureau of the Census, 1979; U.S. Bureau of Labor Statistics, 1984, January).

Wilson maintains that male joblessness may be the single most important factor contributing to the growth in the number of unwed mothers among poor women. He observes studies dating back to the depression consistently show a relationship between unemployment, family instability, and female-headed homes (Bakke, 1940; Bishop, 1980; Cutright, 1971; Komarovsky, 1940). It is important to note that single parents or single parent families are not inherently detrimental. However, when coupled with limited financial resources, single parents find it exceedingly difficult to provide the child with the support necessary for healthy development. As the National Commission on Children (1991) reported, children do best when they have the love, care and support of both parents.

Bronfenbrenner (1978) claims conditions under which families live and the resultant decrease in adult time in caring for children contributes to the increase in academic and behavioral problems, school drop-out rates, drug use, and juvenile delinquency. According to the Children's Defense Fund (1985, 1986), youth growing up in inner city families typically endure difficulties as a result of poverty. This includes living in substandard housing in high crime areas, receiving inadequate health care, attending poor schools, and having limited access to public services. These youth are more likely to have health and behavioral problems.

Job losses have occurred in those blue collar industries which require a less educated workforce. This is significant in light of Brown's (1979) findings that many African-American males are poorly educated. For example, 20% were unable to read at the fourth grade level. Twenty-one per cent of 18-19 year olds and 25% of 20-21 year olds did not finish high school (U.S. Bureau of Census, 1981b), and therefore lack a high school diploma which is the minimal certificate of skills for our culture. Freeman and Wise (1982) claim youth unemployment is associated with crime, drug use, suicide, and school violence. Additionally, the underground drug trade provides economic opportunities which are sorely lacking in low income metropolitan neighborhoods (Clark, 1965; Glasgow, 1981; Johnson et al., 1985).

TRAUMATIC LIFE EVENTS

Political debate often addresses how society is traumatized by the high rate of crime. Too often the role of trauma in the genesis of criminal activity is overlooked. Gibbs' (1988) analysis of young African-American delinquent males suggests their criminal activity has sociological and psychological origins. Their behavior may stem from impaired development of the inner controls necessary for reflecting beforehand on the consequences of their actions. This behavior may also represent the displaced expression of rage at a "society which has been unable to protect and nurture them" (p. 240). Wulach (1983) posits delinquency derives from trauma-based arrested development and/or the lack of role models for realistic behavior and goals.

Pierce (1970) views criminal activity as an attempt to bolster a self made fragile by the "micro-aggression" poor African-Americans face living in hostile environments. These are characterized by constant assaults on self-esteem, which include continual reminders of low status. As the psychiatrist in T. S. Eliot's (1950) *The Cocktail Party* said,

> Half of the harm that is done in this world
> Is due to people who want to feel important.
> They don't mean to do harm–but the harm does not interest them.
> Or they do not see it, or they justify it
> Because they are absorbed in the endless struggle
> To think well of themselves. (p. 111)

Traumatic life events are common for low income inner city residents. Myers, Lindenthal, and Pepper (1974) explore the relationship between psychiatric symptomatology and adverse life events among subjects representing different socioeconomic status. They found the higher rate of symptoms in low income areas was correlated with the higher rate of adverse life events. Their data suggest the social and interpersonal forces within a community contribute to the psychological development of its residents.

Remy, Haskell, Glass, and Wiltse (1983) conducted a major ecological study of 336 high-risk San Francisco families. The explicit focus of the research was on relationships between demographic and socioeconomic characteristics, social problems including victimization, and the response of the helping system. Within the highly disorganized families studied, those with the greatest needs received the fewest services, prevention services were virtually never provided before out-of-home placement occurred, and the delivery of all types of services was persistently related to the demographic characteristics of the families rather than their objective needs.

Parents in low income families may not only be less able to secure or provide support in adjusting to traumatic life events, they may also be a source of trauma. Child maltreatment research typically finds correlations to socioeconomic and demographic variables which are valid indicators of families in need (Burgdorff, 1980; Garbarino & Crouter, 1978; Polansky, Chalmers, Buttenwieser, & Williams, 1981; Schorr, 1974; Shapiro, 1979).

Trauma's Effect on Self Development

According to Kohut (1977), the self represents "the center of the individual's psychological universe" (p. 311). A child's stable and cohesive self emerges through empathic attunement, positive mirroring of its essential nature, availability of others for idealization, and twinning, which represents the child's need to feel alikeness with others (Baker & Baker, 1987). Self-psychology suggests development occurs throughout the life cycle.

Kohut (1971, 1977) asserts repeated and traumatic empathic failures contribute to the fragmentation of the self, thereby, impeding the experience of wholeness. For Kohut, empathy represents the capacity to understand another person's experience from their perspective. While Kohut did not specifically address the direct effect of traumatic life events on the formation of self, other clinicians working with trauma victims have found self-psychology relevant to this population (McCann & Pearlman, 1990). They view traumatic life events as extreme failures of empathy in an unresponsive environment, which contribute to fragmentation of the self (Brende, 1983; Parson, 1988; Ulman & Brothers, 1988).

Victims of trauma may develop antisocial behavior patterns. Examples are children who lived in violent homes (Jaffe, Wolfe, Wilson, & Zak, 1986), children who were physically or sexually abused (Kazdin et al., 1985; Tufts New England Medical Center, 1984), and adult male psychiatric patients who had been abused as children (Carmen, Rieker, & Mills, 1984).

Substance Abuse

Substance abuse is greater in victims of trauma. This may be viewed as a manifestation of anxiety. Rogers (1951) notes threats to one's self produces anxiety. Chronic anxiety symptoms are correlated with post-trauma reactions in Vietnam combat veterans (Blanchard, Kolb, Pellmeyer, & Geradi, 1982; Giller, 1990; Kolb, 1984), victims of severe child abuse who develop Multiple Personality Disorder (Braun, 1983), and persons who are exposed to chronic threat (Davidson & Baum, 1986). Drug and alcohol usage may represent an attempt to treat this anxiety through self-medication.

Substance abuse may also represent a maladaptive attempt to repair a fragmented self (Kohut & Wolf, 1978). Tolpin (1971) suggests the drug or alcohol might be used as a transitional object, necessary for experiencing cohesion at times of distress. The use of substances to bolster a diminished sense of self is supported by research linking negative emotional states to relapses in addictive behavior (Cummings, Gordon, & Marlatt, 1980).

Threats to self brought upon by the experience of trauma are found to contribute to higher alcohol and substance abuse in adult incest survivors (Briere, 1984; Peters, 1984), battered women (Stark & Flitcraft, 1981), and Vietnam combat veterans exposed to atrocities and abusive violence (Yager, Laufer, & Gallops, 1984).

RESTORATIVE ENVIRONMENTS AND REFLECTION

Encountering physical, social, and psychological trauma may evoke a desire to use drugs to numb the daily pain generated by our inner cities. It is interesting to note that drugs have also been used historically as a means for humans to transcend their secular daily experience in order to reflect upon the larger purpose and meaning of life (Weil, 1972).

"People are fascinated with issues that pertain to the self on the one hand, and the cosmos on the other" (Kaplan, 1978, p. 90). The exposure to the natural environment found in a horticultural therapy program may be another avenue for healing and reflection. Kaplan (1983) suggests humans need restorative environments which are conducive to reflection and provide a sense of coherence, fascination, and a feeling of being away from one's surroundings and routines.

Kaplan (1983) defines reflection as a means of extracting meaning from the past which affords an opportunity to plan and anticipate future experiences. Reflection is not possible if survival concerns are pressing. A safe place where one is not required to be vigilant is essential for contemplation to take place. A jail horticultural therapy program may provide a safe haven to people who by necessity live their daily lives vigilantly.

DEPARTMENT DESCRIPTION

New Generation Jails

The San Francisco Sheriff's Department has a long-standing emphasis on rehabilitation and training. Their commitment to providing educational, vocational, and life skills programs is evidenced by the construction of a "new generation" program facility in 1989. The new facility currently houses 358 inmates.

The new generation jail model was developed by jail administrators, working with architects and psychologists, to serve the dual functions of custody and treatment (Gettinger, 1984). These jails communicate an expectation of positive behavior through the use of normal fixtures, carpeting, and furniture. While budgetary constraints forced the Sheriff's Department to exclude some features of the new generation jail, the absence of bars and podular dormitory design, evoke the essence of this innovation. Administrators of new generation jails report less destruction of property than in normal jails.

Walking through the hallways of the Program facility, one sees abundant evidence of a commitment to building self-respect and a sense of community. The walls are tastefully covered with framed works of art and poems produced in creative arts classes. Regularly scheduled cultural awareness programs foster identity, pride, and respect for the cultural diversity of the Bay Area. An environmental ethic is expressed through the boldly marked and prominently placed recycling containers.

In addition to program and architectural changes, San Francisco's program facility is developing a modified direct supervision orientation. In a direct supervision facility, members of the custodial staff stay with groups of inmates 24 hours a day. These staff members become leaders of each group, offering protection, support, and positive role modeling for appropriate behavior. The direct supervision model has lowered the incidences of conflict between prisoners and staff in facilities where it has been implemented (Zupan, 1991).

To be transferred to the program facility, inmates must sign an agreement pledging to an exemplar standard of behavior. They agree to nonviolence, no glorification of crime or drug usage, and no verbal expression of racism or sexism. Failure to comply with these standards may result in being returned to the other locked facilities.

The Garden Project

The main jail and a program facility are located outside San Francisco in San Bruno, CA. They are situated on 110 acres of forest and meadow lands, tucked behind a hill separating them from a community college. The original jails at San Bruno had a working farm with inmates growing some of their food and raising cattle. The farm operation was curtailed in the mid-70's to reduce jail operating costs.

A horticultural therapy program, known as The Garden Project, began to farm this land once again in 1984. This program utilizes the French biodynamic method of organic gardening. The intent of this program is to transmit meaningful work skills while cultivating a heightened awareness of self in relation to community and nature. This garden project exemplifies the new generation, direct supervision approach to incarceration with its emphasis on creating an environment suitable for self-evaluation and establishing respectful mentoring relationships aimed at raising self-esteem.

Food and flowers grown by the garden project are distributed to homeless programs such as Martin de Poores, the Episcopal Sanctuary, and the Mayor's homeless shelters. They are also given to Project Open Hand, which distributes food to home-bound AIDS patients. Trees started at the jail are planted in San Francisco's neighborhoods.

The garden project has also branched out to the community through the development of a post-release garden program in Bayview-Hunter's Point, one of the low-income neighborhoods that is well represented in the jail population. Organic produce grown in this garden is sold to a gourmet restaurant and a quality bakery in the area. Proceeds are used to provide housing for homeless inmates. In addition, an urban tree corps has been initiated on a trial basis with the Department of Public Works. This provides time-limited employment to ex-offenders who have worked with the Garden Project.

METHODS

This study compares inmates participating in the Garden Project with a control group of inmates not in the Garden Project. Inmates are blocked on sex, race, and age and then randomly assigned to the two conditions.

Subjects are assessed at baseline, monthly, at discharge, and 3 months post-discharge. This paper presents selected information gathered at baseline with the revised TCU Intake Form (Simpson, 1991) for 57 subjects. The DATAR Project developed the TCU Intake Form for use with addicts entering drug abuse treatment. The Intake Form was revised for use in a study of substance abusing parents whose children are in foster care (Remy, 1991). This revised Form was administered to subjects prior to their random assignment. Other instruments used are not reported here.

RESULTS

Family Structure

Table 1 summarizes demographic information about the subjects, their partners, other adults living with them, and their children. The study was designed to have equal numbers of male and female subjects. In the jail population, about 85% are male. Thus these data do not reflect the normal jail population, since females are over-represented.

Subjects of this study also do not reflect the jail population in terms of ethnicity. San Francisco has a large non-English speaking population which is represented in the jail. Language barriers made it difficult to include other ethnic groups. In addition, some were not represented in numbers adequate to ensure matching. Subjects therefore were drawn from inmates designated African-American or Caucasian in jail records. However, Table 1 indicates that almost 1 in 5 members of the inmates households are of mixed ethnicity.

Age is not indicated on the table; however, all but 9 subjects are less than 30 years old, which fairly reflects the general jail population. While people often begin families or settle into stable relationships during the mid to late 20's, this has not been the case for this predominantly single population. Eighteen inmates were in a couple relationship at the time of the interview.

Limited education and unemployment are predictors of family and community instability. Subjects in this study typically enter jail unemployed, with less than a high school diploma. The high rate of unemployment extends to the other adults in their household. High unemployment in our inner cities is associated with limited access to community resources and a critical shortage of many vital health care and social services.

These factors contribute to the high rates of chronic health conditions reported in Table 1. The chronic illness category includes subjects who

TABLE 1. Demographic Characteristics of Inmate Families (Expressed in Column Percentages)

		Subject	Partner	Adult	Child
Base N =		57	18	39	67
Sex	Male	47.4	55.6	56.4	56.7
	Female	52.6	44.4	43.6	43.3
Ethnicity	African-Amer	38.6	38.9	41.0	46.3
	Caucasian	43.9	33.3	41.0	23.9
	Mixed	17.5	27.8	18.0	20.8
Education	Lt high school	57.9	29.4	18.8	
	High school	22.8	35.3	56.2	
	Some college	19.3	35.3	25.0	
Unemployed		61.9	66.7	33.3	
Health	None	19.3	33.3	30.7	83.6
	Chronic	15.8	22.2	46.2	16.4
	Substance	22.8	33.3	18.0	
	Multiple	42.1	11.2	5.1	
Treatment	None needed	26.3	44.4	56.4	89.6
	In treatment	50.9	38.9	28.2	9.0
	Needs	22.8	16.7	15.4	1.4
Out-of-home living		100.0	22.2	0.0	17.9

have one or more chronic health conditions. The multiple category identifies subjects with chronic health and substance abuse problems. Of particular concern is that only 19% of the inmates report no chronic health conditions.

Incarceration further destabilizes tenuous family structure. Of the 67 children in the inmates' households, 58.2% are 5 years or younger, and 31.3% are between the ages of 6 and 11. These young children are all at significantly elevated risk by virtue of having one parent in jail. Many are living in informal foster care with relatives, particularly grandparents, or with friends of the family. A substantial proportion of children in these families have two parents in custody, and 12 of 67 are living in formalized foster relationships. Nationally, 6.4 per 1,000 children live outside the home of a custodial parent (Children Now, 1991). Almost 1 in 6 children have chronic health conditions. Nationally about 1 in 10 children are so afflicted (National Center For Health Statistics, December 1991).

Victimization History

Standard measures of child abuse and neglect reported by the subjects during childhood are displayed in Table 2. All data reflect the subject's self-definition of the problems, and likely are an undercount of what might be reported by social service agencies working with these families.

For example, the reported absence of food in the house for periods of time while growing up provides no indication of the quality or amounts of food considered normal by the subjects. Crackers in the cupboard might be considered evidence of food in the house. Time spent alone as a child without an adult caretaker does not address the level of care provided by adults, particularly in families where the subject's parents were alcohol or substance abusers.

Physical abuse was determined by the subject's response to being asked if they had ever been hit harder than they deserved. Many subjects reported that they were only hit when they deserved it. When the author asked what they were hit with, some said broom sticks or electric cords. It is likely that violence towards children has been normalized in these families.

TABLE 2. Inmate Self Report of Childhood Neglect and Abuse (Base N = 57)

Indications of Neglect	
No caretakers	42%
No food	28%
Kept from school	28%
Indications of Abuse	
Physical abuse	63%
Sexual abuse	28%
Saw parents hurt each other	49%

Some national studies indicate that about 1 in 6 boys and about 1 in 3 girls are sexually abused (Finkelhor, Hotaling, Lewis, & Smith, 1990) and that rates are higher in multiproblem families. The true rate of sexual abuse histories doubtless is higher than that reported in Table 2. For example, most of the women and a few of the men have been prostitutes. Silbert (1980) found histories of childhood sexual abuse to be universal among San Francisco prostitutes. The underreporting may point to difficulty subjects have admitting to being sexually abused, particularly in a jail population where a premium is placed on strength and where vulnerability can be exploited.

The extent of violence in the inmate's families is portrayed in Figure 1. Subjects were asked the same series of questions addressing the escalation of violence (1) as a child in their families (Child), (2) as an instigator in adult relationships (Adult) and (3) as a target of their partners' anger (Partner). The variables range from yelling to injury with a weapon, such as a knife or gun. Whether injuries required and received treatment is also noted. The families in which these young adults were raised were violent, and they are continuing that pattern.

Illegal Activities

Substance use history is reported in Figure 2. For each substance, subjects were asked to indicate if they had ever used (Ever), used in the

Figure 1: Victimization History

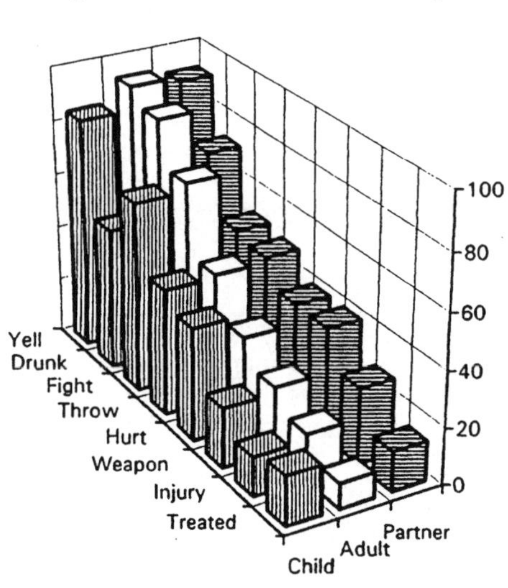

Figure 2: Inmate Substance Use History

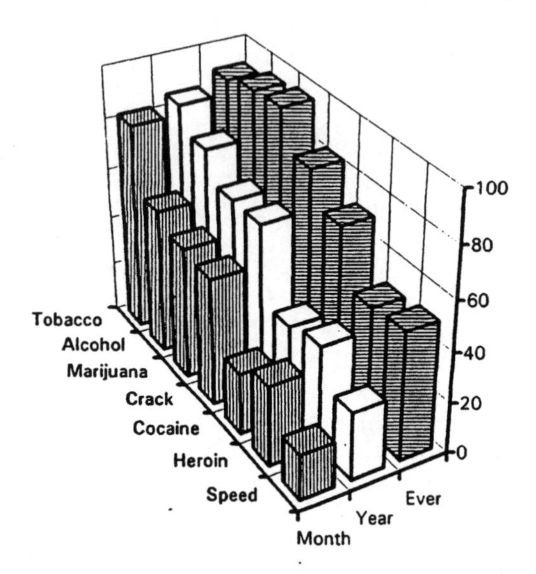

prior year (Year), and within 30 days of incarceration (Month). This population's rate of illicit drug use is considerably higher than Bay Area norms (Swanson, Remy, Chenitz, Chastain, & Trocki, 1992 in press).

Self-reported age of onset of criminal activity, arrest, and incarceration are depicted in Figure 3. Almost all criminal acts were initiated before the age of 18, and most subjects had been locked up by their early 20's. This suggests a path from the initiation of substance use to other illegal activities.

DISCUSSION

The success of the horticultural therapist is predicated upon an empathic understanding of the particular needs of the population being served. The horticultural therapist utilizes adaptive gardening tools when working with disabled clients and raised beds when treating elderly clients. Horticultural therapy must adapt as well to the particular self-needs of inner city jail inmates. Seeds of trauma, victimization, and self-fragmentation are embedded in the level of violence and neglect, social and familial, this population experienced as children. These early experiences vastly impede subsequent psychological, social, and physical development. School fail-

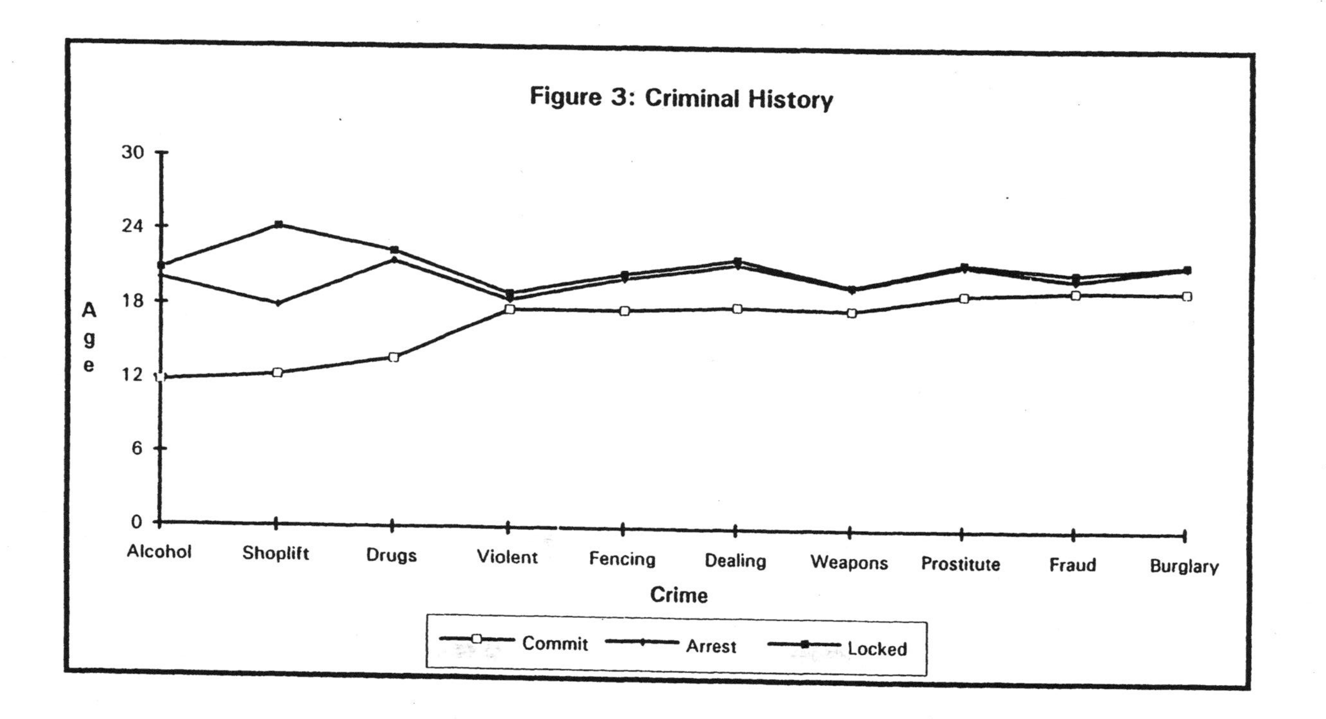

Figure 3: Criminal History
30
24
18
12
6
0
A
g
e
Alcohol
Shoplift
Drugs
Violent
Fencing
Dealing
Weapons
Prostitute
Fraud
Burglary
Crime
Commit
Arrest
Locked

ures, poor health, substance abuse, and criminal activity are the weeds which take hold in traumatized inner city families and communities. County jail inmates are often judged harshly because the "symptoms" produced in our inner cities often manifest in illegal activities. Sensitivity to the childhood trauma and resultant shattered self experience of this population is crucial for effective treatment to occur.

The importance of developing effective treatment programs for this population can not be overstated. County jail inmates are generally under 30 years old and have not settled into stable relationships and family life. Those who have young children frequently do not live with them. Interventions which emphasize nurturing and cultivating life could help to break the generational cycle of childhood neglect and victimization.

The pervasiveness of substance abuse in the inner city is reflected in the data collected. Therapist recognition of the biopsychosocial basis for substance use in this population can facilitate the application of gardening principles to the inmate's life situation. If the therapist is using organic gardening techniques because the plants grow stronger and provide healthier food naturally, this can be related to how the body grows stronger without the use of drugs. Inmates often have an extensive and intimate knowledge of drug intoxication. The natural "high" experienced while working outdoors can be proffered as a viable alternative to the destructive highs they encounter while on drugs. As horticultural therapists teach inmates to weed and compost, both may learn to transform their negative life patterns. They may also learn to recognize and address the adverse social influences which hinder full development.

Horticultural therapy emerged from an appreciation of nature's capacity for promoting healing and growth in human beings. Carl Jung (1960) discerned how the human psyche was akin to nature. Seasonal variations are observed in human development, with periods of dormancy and growth; death and rebirth. Just as the decaying process in nature is integral to the emergence of new life, Jung posited disintegration within the human psyche could lead to reintegration and growth. A therapeutic relationship based on trust, safety, and acceptance can facilitate this process.

As an inmate learns to cultivate life, the resultant increase in self-esteem may lead to the belief that she can be productive in other pursuits. However, it is important to note that a horticultural therapy program operating in a jail facility is only the first step in the long road towards healing individuals and our society. The opportunity for productive work must be cultivated in our communities as well for the county jail inmate to flourish upon release.

REFERENCES

Ambrose, A. (1977). The ecological perspective in developmental psychology. In H. McGurk (Ed.), *Ecological factors in human development* (pp. 3-11). Amsterdam: North-Holland.

Baker, H. S., & Baker, M. N. (1987). Heinz Kohut's self psychology: An overview. *American Journal of Psychiatry, 144*, 1-9.

Bakke, W. E. (1940). *Citizens without work.* New Haven, CT: Yale University Press.

Berry, J. S. (1975). *A hortitherapy program for substance abusers.* (Horticulture Department, South Carolina Agricultural Experiment Station Research Series, 156). Clemson, SC: Clemson University.

Bickman, L., Teger, A., Gabriele, T., McLaughlin, C., Berger, M., & Sunaday, E. (1973). Dormitory density and helping behavior. *Environment and Behavior, 5*, 465-490.

Bishop, J. H. (1980). Jobs, cash transfers, and marital instability: A review and synthesis of the evidence. *Journal of Human Resources, 15* (Summer), 301-334.

Blanchard, E. B., Kolb, L. C., Pallmeyer, T. P., & Gerardi, R. J. (1982). A psychophysiological study of post traumatic stress disorder in Vietnam veterans. *Psychiatric Quarterly, 54*, 220-228.

Braun, B. G. (1983). Psychophysiologic phenomena in multiple personality and hypnosis. *American Journal of Clinical Hypnosis, 26*, 124-137.

Brende, J. O. (1983). A psychodynamic view of character pathology in Vietnam combat veterans. *Bulletin of the Menninger Clinic, 47*, 193-216.

Briere, J. (1984, April). *The effects of childhood sexual abuse on later psychological functioning: Defining a post-sexual-abuse syndrome.* Paper presented at the Third National Conference on Sexual Victimization of Children, Washington, D.C.

Bronfenbrenner, U. (1978). Who needs parent education? *Teachers College Record, 79*, 767-787.

Bronfenbrenner, U. (1979). *The ecology of human development: Experiments by nature and design.* Cambridge: Harvard University Press.

Brown, S. (1979). The health needs of adolescents. In *Healthy people: The surgeon general's report on mental health promotion and disease prevention* (pp. 333-364). (Publication 79-55071A). Washington DC: Department of Health, Education, and Welfare.

Burch, W. R., Jr. (1976). The peregrine falcon and the urban poor: Some sociological interrelations. In P. J. Richerson, & J. McEvoy, III (Eds.), *Human ecology: An environmental approach* (pp. 308-316). North Scituate, MA: Duxbury.

Burgdorff, K. (1980). *Recognition and reporting of child maltreatment: Findings from the national study of the incidence and severity of child abuse and neglect.* Rockville, MD: Westat.

Carmen, E., Rieker, P. R., & Mills, R. (1984). Victims of violence and psychiatric illness. *American Journal of Psychiatry, 141*, 378-383.

Children Now (1991). *What's happening to our children: A county by county guide 1991.* Los Angeles: Children Now.

Children's Defense Fund (CDF). (1985). *A children's defense budget.* Washington, DC: Children's Defense Fund.

Children's Defense Fund (CDF). (1986). *A children's defense budget.* Washington, DC: Children's Defense Fund.

Clark, K. B. (1965). *Dark ghetto: Dilemmas of social power.* New York: Harper and Row.

Cotton, M. (1975). Effectiveness of horticultural therapy in lowering aggressiveness of institutionalized delinquent adolescents. *National Council for Therapy and Rehabilitation through Horticulture: Lecture & Publication Series, 1*(5).

Cummings, C., Gordon, J. R., & Marlatt, G. A. (1980). Relapse: Prevention and prediction. In W. R. Miller (Ed.), *The Addictive Behaviors.* Elmsford, NY: Pergamon Press.

Cutright, P. (1971). Income and family events: Marital instability. *Journal of Marriage and the Family, 33,* 291-306.

Davidson, L. M., & Baum, A. (1986). Chronic stress and post-traumatic stress disorders. *Journal of Consulting and Clinical Psychology, 54,* 303-308.

Duncan, G. J. (1984). *Years of poverty, years of plenty.* Ann Arbor: Institute for Social Research, University of Michigan.

Eliot, T. S. (1950). *The cocktail party.* New York: Harcourt, Brace.

Finklehor, D., Hotaling, G., Lewis, I. A., & Smith, C. (1990). Sexual abuse in a national survey of adult men and women: Prevalence, characteristics and risk factors. *Child Abuse and Neglect, 14*(1), 19-28.

Freeman, R. B., & Wise, D. A. (1982). *The youth labor market problem: Its nature, causes, and consequences.* Chicago: University of Chicago Press.

Francis, M., & Cordts, C. (1992). A research agenda for the impact of urban greening. In D. Relf (Ed.), *The role of horticulture in human well-being and social development: A national symposium* (pp. 71-74). Portland: Timber Press.

Garbarino, J., Abramowitz, R., Benn, J., Gaboury, M., Galambos, N., Garbarino, A., Grandjean, P., Long, F., & Plantz, M. (1982). *Children and families in the social environment.* New York: Aldine.

Garbarino, J., & Crouter, A. (1978). Defining the community context for parent-child relations: The correlates of child maltreatment, *Child Development, 49,* 604-616.

Gettinger, S. (1984). *New generation jails: An innovative approach to an age-old problem.* Washington, DC: National Institute of Corrections, U.S. Department of Justice.

Gibbs, J.T. (1988). Health and mental health of young black males. In J.T. Gibbs (Ed.), *Young, black, and male in America: An endangered species* (pp. 219-257). Dover, MA: Auburn House.

Giller, E. L. (Ed.) (1990). *Biological assessment and treatment of posttraumatic stress disorder.* Washington, DC: American Psychiatric Press.

Gilreath, P. R. (1976). A hortitherapy program for the visually handicapped. Unpublished master's thesis, Clemson University, Clemson, SC., In Olson, A. P. (1976). *The development and implementation of an evaluated study of the effect of horticultural therapy on certain physically disabled patients.* (Horticulture Department, South Carolina Agricultural Experiment Station Research Series No. 175). Clemson, SC: Clemson University.

Glasgow, D. (1981). *The black underclass.* New York: Vintage Books.

Grossmann, R. S. (1979). Horticultural therapy programs in Britain and the United States. *HortScience, 14*(6), 690-691.

Hiott, J. A. (1975). *A hortitherapy for the mentally handicapped.* Unpublished master's thesis, Clemson University, Clemson, SC.

Horne, D.C. (1974). *An evaluation of the effectiveness of horticultural therapy on the life satisfaction level of aged persons confined to a rest care facility.* Unpublished master's thesis, Clemson University, Clemson, SC.

Jaffe, P., Wolfe, D., Wilson, S., & Zak, L. (1986). Family violence and child adjustment: A comparative analysis of girls' and boys' behavioral symptoms. *American Journal of Psychiatry, 143*, 74-77.

Jaffe, P., Wolfe, D., Wilson, S., & Zak, L. (1986). Similarities in behavioral and social maladjustment among child victims and witnesses to family violence. *American Journal of Orthopsychiatry, 56*, 142-146.

Johnson, B. D., Goldstein, P. J., Preble, E., Schmeidler, J., Lipton, D. S., Spunt, B., & Miller, T. (1985). *Taking care of business: The economics of crime by heroin abusers.* Lexington, MA: D. C. Health.

Jorgenson, D. O., & Dukes, F.O. (1976). Deindividuation as a function of density and group membership. *Journal of Personality and Social Psychology, 34*, 24-39.

Jung, C. G. (1960). Stages of life. In R.F. C. Hull (Trans.), *The collected works of C. G. Jung: Structure and dynamics of the psyche* (Vol. 8). Princeton, NJ: Princeton University Press, Bolligen Series XX. (Original work published in (1947/54).

Kaplan, S. (1978). Attention and fascination: The search for cognitive clarity. In S. Kaplan & R. Kaplan (Eds.), *Humanscapes: Environments for people* (pp. 84-90). North Scituate, MA: Duxbury Press.

Kaplan, S. (1983). A model of person-environment compatibility. *Environment and Behavior, 15(3)*, 311-332.

Kazdin, A. E., Moser, J., Colbus, D., & Bell, R. (1985). Depressive symptoms among physically abused and psychiatrically disturbed children. *Journal of Abnormal Psychology*, 94, 298-307.

Kohut, H. (1971). *The Analysis of the Self.* New York: International Universities Press.

Kohut, H. (1977). *The Restoration of the Self.* New York: International Universities Press.

Kohut, H., & Wolf, E. S. (1978). The disorders of the self and their treatment: An outline. *International Journal of Psycho-Analysis, 59*, 413-425.

Kolb, L. C. (1984). The post-traumatic stress disorders of combat: A subgroup with a conditioned emotional response. *Military Medicine, 149*, 237-243.

Komarovsky, M. (1940). *The unemployed man and his family.* New York: Octagon Books.

Kruvant, W. J. (1975). People, energy, and pollution. In D. K. Newman & D. Day (Eds.), *The American Energy Consumer* (pp. 125-67). Cambridge: Ballinger.

Lopez, B. (1986). *Arctic dreams: Imagination and desire in a northern landscape.* New York: Scribner's Sons.

Matthews, K.E., & Cannon, L.K. (1975). Environmental noise level as a determinant of helping behavior. *Journal of Personality and Social Psychology, 32,* 571-577.

McCann, I. L., & Pearlman, L. A. (1990). *Psychological trauma and the adult survivor: Theory, therapy, and transformation.* New York: Brunner/Mazel.

Milgram, S. (1970). The experience of living in cities. *Science, 167,* 1461-1468.

Myers, J.K., Lindenthal, J. J., & Pepper, M. P. (1974). Social class, life events, and psychiatric symptoms: A longitudinal study. In B. S. Dohrenwend & B. P. Dohrenwend (Eds.), *Stressful life events: Their nature and effects* (pp. 191-205). New York: Wiley & Sons.

National Center For Health Statistics. (1991, December). Current estimates from the national health interview study, 1990, *Vital and Health Statistics.* Washington, DC: U.S. Department of Health and Human Services, Public Health Service Centers: Centers For Disease Control.

National Commission On Children, (1991). *Beyond Rhetoric: A New Agenda for Children and Families.* Washington, DC: U. S. Government Printing Office.

Olson, A. P. (1976). *The development and implementation of an evaluated study of the effect of horticultural therapy on certain physically disabled patients.* (Horticulture Department, South Carolina Agricultural Experiment Station Research Series No. 175). Clemson, SC: Clemson University.

Omark, D. R. (1977). Ecological factors and ethological observations. In H. McGurk (Ed.), *Ecological factors in human development* (pp. 37-46). Amsterdam: North-Holland.

Page, R. A. (1977). Noise and helping behavior. *Environment and Behavior, 9,* 311-335.

Parson, E. R. (1988). Post-traumatic self disorders (PTsfD): Theoretical and practical considerations in psychotherapy of Vietnam war veterans. In J.P. Wilson, Z. Harel, & B. Kahana (Eds.), *Human adaption to extreme stress: From the Holocaust to Vietnam* (pp. 245-282). New York: Plenum Press.

Peters, S. D. (1984). *The relationship between child sexual victimization and adult depression among African-American and white women.* Unpublished doctoral dissertation, University of California, Los Angeles.

Polansky, N. A., Chalmers, M. A., Buttenwieser, E., & Williams, D. P. (1981). *Damaged parents: An anatomy of child neglect.* Chicago: University of Chicago Press.

Pierce, C. M. (1970). Offense mechanisms. In F. Barbour (Ed.), *The black 70s* (pp. 265-282). Boston: Porter Sargent Publications.

Relf, D. (1981). Dynamics of horticultural therapy. *Rehabilitation Literature, 42*(5-6), 147-150.

Remy, L. L. (1991). *Intensive family reunification project: Program evaluation and design.* Unpublished manuscript. Mill Valley, CA: Loring Associates.

Remy, L. L., Haskell, S., Glass, R. G., & Wiltes, K. T. (1983). *View of the past, picture of the future?: A study of San Francisco's high-risk youth and families.* Mill Valley, CA: Loring Associates.

Rodin, J. (1976). Crowding, perceived choice and response to controllable and uncontrollable outcomes. *Journal of Experimental Social Psychology, 35,* 564-578.

Rogers, C. R. (1951). *Client-centered therapy.* New York: Houghton Mifflin.

Schnaiberg, A. (1975). Social syntheses of the societal-environmental dialectic: The role of distributional impacts. *Social Science Quarterly, 56,* 5-20.

Schorr, A. L. (Ed.). (1974). *Children and decent people.* New York: Basic Books.

Shapiro, D. (1979). *Parents and protectors: A study of child abuse and neglect.* New York: Child Welfare League of America.

Silbert, M. H. (1980). *Sexual assaults of prostitutes.* San Francisco: Delancey Street Foundation.

Simpson, D. D. (1991, March). *TCU forms manual: Drug abuse treatment for AIDS-risks reduction (DATAR).* Fort Worth, TX: Texas Christian University, Institute of Behavioral Research.

Stark, E., & Flitcraft, A. H. (1981). *Wife abuse in a medical setting: An introduction for health personnel* (Monograph No. 7). Washington, DC: Office of Domestic Violence.

Swanson, J. M., Remy, L. L., Chenitz, W. C., Chastain, R. L., & Trocki, K. F. (1992). Illicit drug use among young adults with genital herpes. *Public Health Nursing.* In press.

Tereshkovich, G. (1973). Horticultural therapy: A review. *HortScience, 8*(6): 460-461.

Tolpin, M. (1971). On the beginnings of a cohesive self: An application of the concept of transmuting internalization to the study of the transitional object and signal anxiety. *Psychoanalytic Study of the Child, 26,* 316-352.

Tufts New England Medical Center, Division of Child Psychiatry. (1984). *Sexually exploited children: Service and research project* (Final report for the Office of Juvenile Justice and Delinquency Prevention). Washington, DC: U. S. Department of Justice.

Ulman, R. B., & Brothers, D. (1988). *The shattered self: A psychoanalytic study of trauma.* Hillsdale, NJ: The Analytic Press.

U.S. Bureau of Census. (1979). The social and economic status of the black population in the United States: A historical view, 1790-1978. *Current Population Reports.* (Series P-23). Washington, DC: U. S. Government Printing Office.

U.S. Bureau of Census. (1980). Families maintained by female householders, 1970-79. *Current Population Reports.* (Series P-23, No. 107). Washington, DC: U. S. Government Printing Office.

U.S. Bureau of Census. (1981a). Marital status and living arrangements, March 1980. *Current Population Reports.* (Series P-20). Washington, DC: U. S. Government Printing Office.

U.S. Bureau of Census. (1981b). School enrollment: Social and economic characteristics of students, 1980 (advance report). *Current Population Reports.* (Series P-20). Washington, DC: U. S. Government Printing Office.

U.S. Bureau of Census. (1984a). Household and family characteristics, March 1984. *Current Population Reports.* (Series P-20, No. 388). Washington, DC: U. S. Government Printing Office.

U.S. Bureau of Census. (1984b). *Census of the population, 1980.* Washington, DC: U. S. Government Printing Office.

U.S. Bureau of Labor Statistics. (1984, January). *Employment and earnings.* Washington, DC: U.S. Department of Labor.

Weil, A. (1972). *The natural mind: A new way of looking at drugs and the higher consciousness.* Boston: Houghton Mifflin.

Wilson, W. J. (1987). *The truly disadvantaged: The inner city, the underclass, and public policy.* Chicago: University of Chicago Press.

Wulach, J. S. (1983). August Aichorn's legacy: The treatment of narcissism in criminals. *International Journal of Offender Therapy and Comparative Criminology, 27*(3), 226-234.

Yager, T., Laufer, P., & Gallops, M. (1984). Some problems associated with war experience in men of the Vietnam generation. *Archives of General Psychiatry, 41*, 327-333.

Zupan, L. L. (1991). *Reform and the new generation philosophy.* Cincinnati: Anderson Publishing.

PLANTS AND THE INDIVIDUAL

Chapter 16

Plants and the Individual: A Recent History

Virginia I. Lohr

SUMMARY. The benefits that an individual can derive from plants have been discussed for thousands of years, yet historically, reports supporting many of these benefits have been anecdotal. The problem with anecdotal evidence involves its veracity: we cannot know if the implied effect is due to plants or other factors. Anecdotes are valuable, however, because they can give researchers ideas about experiments to conduct. In the last twenty years, studies testing hypotheses about how plants affect individuals have begun to appear. This paper briefly reviews the methods and results from a selection of these studies.

INTRODUCTION

The benefits that an individual can derive from plants have been discussed for thousands of years (Stein, 1990). There are obvious and widely

The College of Agriculture and Home Economics Research Center, Pullman, WA releases this as H/LA Paper No. 92-12, Project No. 0695.

acknowledged benefits such as providing food and medicine (Janick, 1992). Other ancient uses for plants include therapeutic recommendations of gardening and working in the soil to promote mental health (Neer, 1976). Why, in the 1990's, are we still discussing this issue?

Historically, most of the information supporting many of the benefits of plants has been anecdotal (Cotter et al., 1978). People who work with plants or spend time with plants have said that doing so made them feel better. For example, Robert Neese, a prisoner in Iowa State Prison in Fort Madison wrote, ". . . when tempers did start to flare due to the tension of constant confinement, a couple of hours' work in the garden made pacifists of potential battlers" (Neese, 1959, p. 40).

There are both positive and negative aspects of anecdotal evidence. The problem with anecdotal evidence involves its veracity. The method of gathering the information is not rigorous, so we cannot know if the implied effect is really due to some aspect of the plants. For example, the proposed response to plants could simply be coincidental or due to an unrelated cause. Did the prisoners mentioned above become pacifists because they were working with plants or simply because they were working? Could any form of physical exercise have caused the same response?

Anecdotal evidence cannot be taken alone at face value, but it should not be ignored. The value of anecdotes lies in presenting researchers with ideas about potential hypotheses to test. Researchers who want to examine the true impact of plants on people need a place to start. By studying what people have been saying for generations about how plants make them feel, we can get many ideas about experiments to conduct. The prisoner stated that working with plants turned prisoners into pacifists. What does that really mean about how the plants affected people and how could it be tested experimentally?

In the last twenty years, studies testing hypotheses about how plants affect individuals have begun to appear. Researchers have used a wide variety of methods in these studies. In this paper, I review the methods and results from a selection of these studies. A sample of the documented information about the impact of plants on individuals is presented. The types of approaches that researchers have used to address this topic are discussed.

THE USE OF INDIRECT APPROACHES

Researchers testing hypotheses about how plants affect people have used a variety of methods. Some of the methods are quite simple, involving little expense, and others are complex, involving elaborate instrumentation. I group the approaches to examining the impacts of plants into two categories: direct and indirect. The indirect approach involves documenting the environmental changes that result from adding plants to a particular location. Impacts on humans can then be inferred from these data by predicting how the documented environmental changes would affect people. The direct approach involves measuring changes in people as a result of the presence of plants.

The indirect approach involves using experimental measures with which traditional plant scientists are familiar. I used a very simple, indirect approach in two experiments I conducted to document the impact of interior plants on relative humidity (Lohr, 1992a; Lohr, 1992b). Plant scientists frequently measure relative humidity in their experiments. I placed recording hygrothermographs in rooms and then repeatedly added or removed plants from the rooms. Relative humidity in each room without plants was below the range recommended for human comfort. I documented a significant increase in relative humidity in both studies. From the measured response, I inferred that the rooms would be more comfortable for people when the plants were present.

The NASA-sponsored research into the ability of plants to remove air pollutants provides a good example of an indirect approach using elaborate instrumentation (Wolverton et al., 1989). In these experiments, plants were placed in air-tight plexiglass chambers into which air pollutants, such as trichloroethylene or formaldehyde, were injected. Levels of the pollutants were measured periodically using air sampling pumps and specific gas detector tubes. Trace metabolites, including ethylene and terpenes, were measured using a gas chromatograph and a mass selective detector. The researchers documented dramatic reductions in the levels of each gaseous pollutant tested, with negligible levels of plant metabolites detected.

THE USE OF DIRECT APPROACHES

The direct approach has usually been used by social scientists. The earliest study I have found using this approach was conducted by Rachel Kaplan (1973) and involved questionnaire or survey methodology. Kaplan simply surveyed known gardeners and asked them why they gardened. By using statistical methods such as factor analysis, she was able to generate meaningful categories to explain people's responses on her questionnaire. For example, her research showed that people seem to begin gardening for very tangible rewards, such as fresh vegetables, but garden later in life for intangible rewards, such as aesthetic pleasure.

Soon after Kaplan's study was reported, another study using the survey method was published (Ulrich, 1974). In this study, shoppers completed questionnaires regarding their use of two routes to a shopping mall. One route was quicker, and thus more economical, but this route was not scenic. The other route was slower, but landscaped and scenic. On the basis of economic value, one would predict that shoppers would travel the route which took less time, however, Ulrich found that shoppers made 56% of their trips on the scenic route.

Later in the 1970's, preliminary results of a study designed to use controlled conditions was published (Talbott et al., 1976). This study measured the behavioral responses of chronic schizophrenics to the addition of flowering plants to the dining hall of a psychiatric hospital. The authors were able to document positive changes in behavior in the patients as well as improved attitudes in the staff when the plants were present compared to in their absence.

In 1979, a study combining questionnaire methodology with controlled conditions appeared (Campbell, 1979). In this study, photographs were taken of an office that had been arranged in specific ways. Pictures were taken of the office in all combinations of the following conditions: with the desk in one of two locations, with or without art objects, with or without plants and aquariums, and tidy or messy. Students were shown the pictures and completed questionnaires about what they felt about the offices. This study documented a positive response to the presence of living elements.

During the 1980's, studies documenting the impact of views of natural areas or "greenscapes" began to appear. These studies involved the careful comparison of factors in addition to views that could have affected the results. Ernest Moore (1981) examined factors that were correlated with rates of sick call among prisoners. He found that prisoners in cells with a view of surrounding farms and forests made fewer sick calls than prisoners with cell views of the prison. Roger Ulrich (1984) reported on the benefits to hospital patients of having a room with a view of trees rather than a view of a brick wall. Patients recovering from gall bladder surgery spent less time in the hospital, used fewer doses of strong pain relievers, and had fewer negative comments from hospital staff on their charts.

The studies by Moore (1981) and Ulrich (1984) documented medical impacts of views of nature, while other studies in the 1980's were documenting reductions in human stress in response to plants. One of these studies looked at physiological measures as a means of documenting human responses (Ulrich and Simons, 1986). In this study, college students were exposed to a stressful situation, a film about accidents at work. They were then shown a "recovery" videotape of a forest with or without a stream, of a commercial street with light or moderate traffic, or of an outdoor shopping mall with few or many pedestrians. Muscle tension, skin conductance, and pulse transit time were measured during the stressful film and after 3, 6, and 9 minutes of the recovery videotape. Results were consistent across the three measures of stress: students were stressed by watching the accident film and they recovered from that stress while watching the recovery videotapes. Students recovered more quickly and completely by seeing the videotapes with trees than by viewing the pedestrian or traffic-dominated videotapes.

In the 1980's, we continued to gain useful information about an individual's responses to plants from the simple methods such as surveys. Joan Aitken and Rodger Palmer (1989) conducted a survey to examine the impact of plants in a business setting by asking students in a communication course to agree or disagree with various statements about plants. They found, for example, that 70% agreed or strongly agreed with this statement: "An office neatly decorated with live plants gives me the impression of a well-organized and well-staffed institution." Only 10% disagreed or strongly disagreed with the statement; 20% felt neutral. The results of this study indicated that interior plants could be a major influence in the first impression created by any business.

CONCLUSION

In the 1990's, we are continuing to see studies using both simple and elaborate measures with direct and indirect approaches. The proceedings from the 1990 and 1992 symposiums (see: Relf, 1992, and this volume respectively) provide good examples in both categories. There are so many questions that remain to be answered about the impact of plants on people that there is room for all types of research strategies to be used. In fact, people are still discovering major questions to ask about the impact of plants. Most of us are asking about the positive impacts, but we should remember that potential negative ones must be addressed as well.

I hope this paper has demonstrated that we have progressed beyond having only anecdotal evidence for the value of plants to individuals. I also hope it will inspire more people to help document what we are hearing in the anecdotes. Use the anecdotes to generate the questions to study. Examine what you believe about how plants are affecting you. Consider what others are saying about how plants are affecting them. Generate researchable questions. There are important questions that need to be and can be addressed whether your research budget is large or almost nonexistent.

REFERENCES

Aitken, Joan E. and Rodger D. Palmer. 1989. The use of plants to promote warmth and caring in a business environment. EDRS: ED 303 843, CS 506 539.

Campbell, David E. 1979. Interior office design and visitor response. Journal of Applied Psychology 64(6):648-653.

Cotter, D. J., R. E. Gomez, and V. I. Lohr. 1978. Enhancing ASHS efforts at the plant-people interface. HortScience 13:216.

Janick, Jules. 1992. Horticulture and human culture, p. 19-27. In: Diane Relf (ed.). The Role of Horticulture in Human Well-being and Social Development: A National Symposium, 19-21 April 1990. Timber Press, Portland, Oregon.

Kaplan, Rachel. 1973. Some psychological benefits of gardening. Environment and Behavior 5:145-162.

Lohr, Virginia I. 1992a. The contribution of interior plants to relative humidity in an office, p. 117-119. In: Diane Relf (ed.). The Role of Horticulture in Human Well-being and Social Development: A National Symposium, 19-21 April 1990. Timber Press, Portland, Oregon.

Lohr, Virginia I. 1992b. Research on human issues in horticulture motivates students to learn science. HortTechnology (in press).

Moore, Ernest O. 1981. A prison environment's effect on health care service demands. Journal of Environmental Systems 11(1):17-34.

Neer, Kathleen. 1976. Horticultural therapy. The Green Thumb 33(4):115-118.

Neese, Robert. 1959. Prisoner's escape. Flower Grower 46(8):39-40.

Relf, Diane (ed.). 1992. The Role of Horticulture in Human Well-being and Social Development: A National Symposium, 19-21 April 1990. Timber Press, Portland, Oregon.

Stein, Achva Benzinberg. 1990. Thoughts occasioned by the Old Testament, p. 38-45. In: Mark Francis and Randolph T. Hester, Jr. (eds.). The Meaning of Gardens. The MIT Press, Cambridge, Massachusetts.

Talbott, John A., Daniel Stern, Joel Ross, and Cheryl Gillen. 1976. Flowering plants as a therapeutic/environmental agent in a psychiatric hospital. HortScience 11:365-366.

Ulrich, Roger S. 1974. Scenery and the shopping trip: The roadside environment as a factor in route choice. Michigan Geographical Publication No. 12. Department of Geography, University of Michigan, Ann Arbor, MI.

Ulrich, Roger S. 1984. View through a window may influence recovery from surgery. Science 224:420-421.

Ulrich, Roger S. and Robert F. Simons. 1986. Recovery from stress during exposure to every-day outdoor environments, p. 115-122. In: J. Wineman, R. Barnes, and C. Zimring (eds.). The Costs of Not Knowing. Proceedings 17th Annual Conference of the Environment Design Research Association. Washington, D.C.: ERDA.

Wolverton, B. C., Anne Johnson, and Keith Bounds. 1989. Interior Landscape Plants for Indoor Air Pollution Abatement–Final Report. NASA, Stennis Space Center, MS.

Chapter 17

People-Plant Principles from the Past

Jean Stephans Kavanagh

SUMMARY. The work of Frederick Law Olmsted, the "Father of Landscape Architecture," and that of his firm from the 1850's through the first half of this century resulted in many of the most professional and socially influential landscapes in the United States. Thousands of these landscapes have attained national significance as plant-oriented places of enduring popularity.

The long term success and viability of these places can be attributed to a number of "Olmstedian" design principles which influence the nature and quality of human interactions within these intensively vegetated settings. This paper examines these Olmstedian principles for approaches to the layout and organization of therapeutic landscapes which can encourage participation with plants and with the outdoor spaces they shape.

INTRODUCTION

The increased importance of designed plant and garden environments in America has contributed to the establishment of the relatively new field of historic landscape preservation. This awareness of historical precedent has emerged not only as the means by which historic landscapes and gardens are preserved for future generations, but also as a process by which these landscapes are understood. In this process, original objectives, principles and values infused into landscapes by their historic designers are identified and considered for the information they contribute to contemporary design modifications (O'Donnell, 1987).

The guiding purpose of the great 19th century landscape designs of Frederick Law Olmsted was the search for "healing" settings to counteract the grind of daily living, the strain of poor working conditions, and the debilitating effects of crowded, contaminated, and congested living and working environments (Olmsted, 1870). The success and effectiveness of this search has placed these among the most highly valued of historic landscapes in the United States (U.S. Congress, 1985).

Frederick Law Olmsted was paramount among nineteenth century landscape designers. His influence on the development of landscape architecture in the United States was predicated upon a design philosophy to which the therapeutic value of landscapes was central. In essence, Olmsted landscapes were intended to provide benefits that are similar to those sought in horticultural therapy. "He had derived from his childhood experiences a romantic idealism which, like earlier nineteenth century Americans, saw in unspoiled nature and the simple life a great force for the rehabilitation and rejuvenation of the human spirit enervated by the pressures of civilized life" (Fabos, Milde and Weinmayr, 1968).

Olmsted is particularly known for his public parks and gardens including Central Park in New York; Druid Hill and Cylburn Parks in Baltimore, Maryland; Golden Gate Park in San Francisco; Yosemite National Park; and as many as a thousand national, state and local parks which exist today on Olmsted projects (Beveridge and Hoffman, 1987). Central Park, among the earliest of Olmsted's landscapes, was conceived and detailed as a therapeutic landscape for nineteenth century New York City's diverse population and classes (Brown, 1982). As important models for 19th century spaces, Central Park and its landscape progeny are valid models for the therapeutic spaces and places we design today. If such historic public parks are evaluated as therapeutic spaces, that is, as landscapes designed to encourage beneficial physiological, social, moral, and psychological effects on users of its spaces, design principles can be identified.

Three factors tend to indicate that these principles remain valid today: (1) the continuing popularity of these 19th century landscapes, (2) the investment of public and private funds in current historic and restoration efforts, and (3) the healthful and therapeutic purposes of the original landscape designs.

The Olmsted landscapes, of which Central Park is prototypical, have lasted more than 100 years. They have survived the growth of cities, population shifts, checkered maintenance histories, and even neglect. Yet, they remain memorable, healthy and supportive landscapes for the people and the cities in the regions they serve (U.S. Congress, 1985).

Many of these sites have a history of renovation efforts dating to the nineteenth century. Central Park, begun in the 1850's, was the subject of renovation and restoration efforts as early as 1871 (Olmsted and Kimball, 1970). The restoration and preservation of these landscapes attests to a constituency sharing an active and interactive sense of the value of these places. Cities and regions hard pressed for income to meet the needs of city operations still were willing to invest money in restoration efforts that ensure yet another 100 years of access to these valued spaces. Historic preservation is, in effect, today's populist vote for the continuation of these parks into the future.

Olmsted's writings clearly indicate that the original landscapes were designed to promote the moral, social and physical health of a broad variety of classes of people within the regions serviced by these parks. Even today, these landscapes of healing bring people closer to nature as a result of their strong people-plant relationship. This and other qualities of national importance were cited in the Congressional Hearings on the Olmsted Heritage Landscapes Act (1985).

OLMSTEDIAN PRINCIPLES

Principles guiding the historic restoration of Frederick Law Olmsted's landscapes have been derived from a growing understanding of these landscapes and a familiarity with his profuse writings. These principles describe the plant oriented settings Olmsted employed to achieve many of the effects and experiences of his spaces.

One well-defined set of principles identified by historic landscape restoration expert, Patricia O'Donnell (1983), includes the characteristics of: (1) profuse planting, (2) spatial extension, (3) separation of circulation, (4) separation of activities, (5) integrate structures, etc., (6) gathering spaces, (7) natural spaces, (8) active recreation settings, (9) design for harmony. Considering O'Donnell's principles in terms of therapeutic goals logically leads to the extrapolation of guidelines more specifically related to the design and layout of therapeutic landscapes.

Profuse Planting: The great variety of settings, spaces, and qualities of an Olmsted landscape design rely on a plant oriented atmosphere to direct and maintain the interest of the viewer. Restricted viewing corridors or viewsheds, entirely defined by plant materials, focus the participant's experience within areas. The defined landscape settings, spaces and qualities primarily depend on the mass, color, texture and seasonal change of plant materials for their image, logic and development. The specific therapeutic guideline derived from the principle of profuse planting, would be to provide large quantities of a broad variety of plants as both the focus and the medium of the landscape.

Spatial Extension: In an Olmsted landscape design, views are exploited to provide an increased sense of space: the prospect of large, expansive, open views extended the horizons of spaces that were close, contained and narrow. A reliance on the shifting of scales from small to large, and the creation of broad vistas and intimate spaces in contrasting sequence, and series, enlarge the space by virtue of the increased variety in the viewer's experiences. The study of Olmsted's before and after sketches for Central Park clearly indicate an attempt to subdivide the original site to complicate and to extend the perceived size of the park (Beveridge and Schuyler, 1983). The specific therapeutic guideline based on spatial extension, would be to provide a high degree of spatial variety through overlapping spaces and vistas, and with distant views to connect distant places and spaces within the garden.

Separation of Circulation: Within an Olmsted landscape design, conflicting circulation modes were separated. In Central Park, five distinct, non-intersecting circulation systems are provided to maximize the experience of the individual within the park setting: pedestrian system, bicycle paths, equestrian paths, pleasure driving park roads, and normal urban cross-traffic movement (Olmsted and Kimball, 1970 and Beveridge and Schuyler, 1983). Clearly a pedestrian's experience in a landscape would be affected by the need to dodge bicycles, horses, carriages or cars and trucks. Within a therapeutic landscape, appropriate circulation modes encourage movement patterns and systems which improve not only access to the landscape elements, but also the experiences of that landscape for the broadest range of populations. An objective for such systems would be to improve the ability of the visitor to focus on the experience as a critical component of achieving access to all parts of the garden. The specific therapeutic guide for circulation then would be to exploit circulation to maximize the sequence, quantity and quality of interactions between people and plants.

Separation of Activities: Olmsted's parks incorporated many different activities with a high potential for conflict. Such conflicts were avoided by clustering activities sharing similar qualities and needs into common areas while separating conflicting activities by greater distances or by patterns of plants and plant masses which buffered conflicting activities. In a therapeutic landscape, the nature of the conflict between activities might be most appropriately described as a potential for distraction from the focus on plants and from the prescribed plant-related activity. Such distractions eventually would reduce the therapeutic benefits to the participant. The specific therapeutic guideline for separation of activities would be to reduce potential conflicts between activities to minimize distractions and to emphasize prescribed activities.

Integrate Structures and Other Features: Within an Olmsted landscape design, structures and site amenities were integrated into a single landscape by plants. Shelters, path systems, bridges and structures of all types, were unified into a landscape setting which reinforced the importance of the plant material as a critical design influence. With these parks, structures having practical, and even engineering importance or significance, also provided multidimensional opportunities to exploit aesthetic, horticultural, and the seasonal interest and experiential values of plants. Emphasis on such multiple functions encourages the diversity that results from an expansive palette of plant materials, colors and textures. The specific therapeutic guideline based on the integration of structures would be to exploit the range of structures and experiences by encouraging multiple functions including practical, engineering, aesthetic, horticultural, seasonal and experiential values.

Gathering Spaces: Olmsted landscape designs provided formal spaces defined by plant material and serving large numbers of people. Frederick Law Olmsted, himself (1870), described such receptive spaces as sites for "gregarious recreation," as sites encouraging the mingling of all classes with a common purpose. These gathering spaces often included rows or allees of trees as formal or informal elements. In addition to accommodating large numbers of visitors, such places were central to the concept of spatial extension. The great open spaces allowed visual organization and improved orientation for visitors. The specific therapeutic guideline derived from gathering spaces would be to include large, open, spaces to accommodate large groups and to organize and orient participants within the therapeutic landscape.

Natural Spaces: An important objective of Olmsted landscape designs was to expose individuals to the positive influences of natural environments. Plants, water and rocks were composed in a specialized environment to bring people closer to nature. Such natural spaces often were enclosed by landforms and plant masses with an emphasis on the details of natural elements within the landscape. Natural spaces were designed to encourage participants to see the detail in the plants, to perceive the texture, shadow and moisture on rock surfaces, to reinforce the sensation of sun, wind and weather. In essence, natural spaces were developed to further enhance the relationship of participants to the plant's environment.

The specific therapeutic guideline derived from natural spaces would be to focus the personal plant-related perceptions and experiences of the visitor by emphasizing the natural details of plant material and other plant related elements.

Active Recreation Settings: Later Olmsted landscapes were not merely passive strolling environments (O'Donnell, 1987). They were landscapes in which activities were expected to occur. Ample opportunity was provided within these settings for fresh air and exertion as a means of developing physical well-being and improving self-image. Modern attitudes about landscape uses continue to include the need for the physical well-being derived from physical activities. The specific therapeutic guideline derived from this design principle would be to provide opportunities within therapeutic landscape for visitors to engage in some type of physical exercise, to work muscles, and to improve physical strength and dexterity.

Design for Harmony: The overriding principle of an Olmsted landscape design was the conscious consideration of the whole landscape and its effects. The many unique and individual elements, principles and environmental characteristics of an Olmsted landscape design could have resulted in a confused site development had unity and harmony not been a central design theme. Indeed, one of the hallmarks of an Olmsted landscape is a total harmony of effects (O'Donnell, 1983). This unity was achieved by maintaining views across landscapes and by using visually related vegetated masses, lines and themes to relate disparate structures and uses. The strong dependence on vegetation as both background and detail design media further established harmony within the setting. The specific therapeutic guideline would be to unify the garden through shared views, similar plant selections, plant massing, and common elements and materials.

CONCLUSION

Frederick Law Olmsted was a master of landscape design. His work and that of his firm from the 1850's through the first half of this century, resulted in some of the most professionally and socially influential landscapes in the United States. Understanding the principles incorporated in these Olmsted landscape designs leads to an understanding of the factors that improve the relationship of people to therapeutic landscape spaces. These landscapes are recognized as being of national significance as plant-oriented places having enduring popularity. They represent the prototype of landscapes incorporating factors which can influence the nature and quality of human interactions within garden settings.

Principles developed for the historic restoration of Olmsted landscapes can aid the formulation of principles for the design of therapeutic landscapes which encourage participation with plants and with the outdoor spaces they shape. The guidelines derived from historic landscape design principles of Frederick Law Olmsted are:

1. Provide large quantities of a broad variety of plants as both the focus and the medium of the place.
2. Provide a high degree of spatial variety through overlapping spaces and vistas, and with distant views to connect places and spaces within the garden.
3. Exploit circulation to maximize the quantity and quality of interactions between people and plants.
4. Reduce potential conflicts between activities to minimize distractions and to emphasize prescribed activities.
5. Exploit the range of structures and experiences through multiple functions including practical, engineering, aesthetic, horticultural, seasonal and experiential values.
6. Include large, open spaces to accommodate large groups and to organize and orient participants within the therapeutic landscape.
7. Focus the personal and plant related perceptions and experiences of the visitor by emphasizing the natural details of plant material and other plant related elements.
8. Provide opportunities within the therapeutic landscape for visitors to engage in some type of physical exercise, to work muscles, and to improve physical strength and dexterity.
9. Unify the garden through shared views, similar plant selections, plant massing, and common elements and materials.

REFERENCE LIST

Beveridge, Charles E. and David Schuyler, Eds. *The Papers of Frederick Law Olmsted: Vol. 3, Creating Central Park, 1857-1861.* Baltimore: The John Hopkins University Press. 1983.

Beveridge, Charles E. and Carolyn F. Hoffman, Eds. *The Master List of Design Projects of the Olmsted Firm 1857-1950.* National Association for Olmsted Parks in conjunction with the Massachusetts Association for Olmsted Parks. 1987.

Brown, Arthur, Ed. *Frederick Law Olmsted.* Twayne's World Leaders Series. Coral Gables, Florida: University of Miami Press. 1982.

Fabos, Julius Gy., Milde, Gordon T., and Weinmeyer, V. Michael. *Frederick Law Olmsted, Sr.: Founder of Landscape Architecture in America.* Amherst, Massachusetts: University of Massachusetts Press. 1968.

McLaughlin, Charles C. and Charles E. Beveridge, Assoc. Ed. *The Papers of Frederick Law Olmsted: Vol. 1, The Formative Years 1822-1852*. Baltimore: The John Hopkins University Press. 1977.

O'Donnell, Patricia. A Process for Parks. *Landscape Architecture Magazine*, American Society of Landscape Architects, 77:56-61, 1987.

_____. "Historic Preservation as Applied to Urban Parks." *The Yearbook of Landscape Architecture: Historic Preservation*. Edited by Richard L. Austin, ASLA, Thomas Kane, FASLA, Robert Z. Melnick, ASLA, and Suzanne Turner, ASLA. New York: Van Nostrand Reinhold Company. 1983.

Olmsted, Frederick Law. *Public Parks and Enlargement of Towns*. A paper prepared as a contribution to the popular discussion of the requirements of Boston in respect to a public park; read at the request of the American Social Association at the Lowell Institute, February 25, 1870.

Olmsted, Frederick Law, Jr. and Theodora Kimball, Eds. *Frederick Law Olmsted: Landscape Architect, 1822-1903*. New York: Benjamin Blom, Inc., Publishers. [1922]. 1970.

U. S. Congress. House. Committee on Interior and Insular Affairs. Subcommittee on Public Lands. Olmsted Heritage Landscapes Act: Hearings on H.R. 37. 99th Cong., 1st sess., 1985.

Chapter 18

The Evolutionary Importance of People-Plant Relationships

Charles A. Lewis

In the first symposium we considered the effect of plants in creating and maintaining human well-being and social development. To delineate the ways in which humans respond to vegetation the speakers approached the topic from several views: historical-cultural, psychological, physiological and sociological. Researchers presented their findings about the kinds of landscape scenes that people prefer to see. We learned of the restorative qualities of green settings. In hospitals, a view of trees helped patients to recover more quickly than patients without such a view. In a number of diverse situations the presence of plants helped people to feel better about themselves and where they lived. The first symposium clearly established the validity of people/plant interactions and pointed the way to further consideration of human issues in horticulture.

Today we will take a closer look at the human preferences for green settings and see if we can find their origin. Why is green so attractive to people? How did we get that way? Since the expression of likes and dislikes, preferences, are reflections of mental attitudes, we will not only be concerned with the green settings but how those settings echo in our consciousness; how our innate thinking patterns about vegetation might have become set within us.

Much of the information in this paper is taken from a forthcoming book, *Green Nature, Human Nature*, by Charles Lewis published in 1993 by Jeremy P. Tarcher, Inc., member of The Putnam Berkley Group, Inc.

To do this we will have to consciously examine an unconscious part of ourselves, our mind which shapes and interprets everything that we experience–see, hear, taste, smell, or touch. Mind is a personal expression of our brain, a many layered organ which has developed as we have over the millennia of evolution. In thinking about our mind it is interesting to consider some of the ways in which it functions.

We can intentionally direct our thoughts as when we focus our mind on solving a problem, or our mind can operate in a more spontaneous manner as when feelings and thoughts present themselves without our consciously calling them forth as when we experience a beautiful sight. It would appear that from a mental viewpoint we really are two people. One is the obvious physical thinking person who walks about in this world making decisions and controlling what happens around him/her. The other is a hidden self which expresses itself through innate responses to the world which do not come out of conscious thought but seem to be stored someplace else and show themselves in unexpected ways.

We all have had the experience of walking down a street and suddenly a car behind us honks its horn, startling us. Immediately our heart starts to beat more quickly and we gain a new level of alertness. In a lecture hall all eyes are focused on the speaker; however, should someone appear at the side of the stage, all audience eyes would immediately turn to look at the newcomer. Often when seeing spiders, snakes and other creatures, people will experience a revulsion, and in frightening situations feel the hairs on the back of their neck "standing up." These are all innate responses which occur not because we first thought out the situation and then decided on the appropriate behavior, but rather because someplace deep inside us these patterns of behavior were already "built in," programmed, part of the essential human. They lie in our subconscious, triggered, ready to express themselves whenever appropriate circumstances occur.

These innate responses often make no sense in a contemporary context, but they serve to remind us that we did not originate in the contemporary world. A close examination reveals that within our contemporary selves there lies hidden an ancient self that was programmed in primitive times for survival under those conditions. Today it continues to respond to stimuli creating echoes of an earlier time when such signals were important for survival of the individual. Understanding these ancient responses and satisfying them helps to bring our two selves–ancient and contemporary–into harmony and reduce the stress resulting from denial of basic intuitive needs. If we are to understand ourselves today, it is necessary to seek the remote origins of these responses. How might they have helped us to survive in primitive environments?

We can gain a view of this hidden self by considering body knowledge, wisdom stored in each of us which helps us to survive each day. If we cut a finger, we do not have to stop and think of all the steps and materials required to heal the wound, stop the bleeding, create a scab, and begin to form new skin. The body "knows" how to do it. When we see blood, our blood pressure is automatically lowered, reducing the chance of excessive bleeding. Body knowledge is ancient information encoded in our genes. Through eons of evolution this wisdom, essential for continued survival, has accumulated within us and is ready to act the instant it is needed. In every second of life our bodies act without our conscious instructions, performing a multitude of functions to maintain an internal homeostasis. Body knowledge is able to signal important needs through feelings. Pain, thirst, and hunger alert us to be aware, call attention to something important. Body wisdom knows what is needed and responds immediately.

Similarly, we might consider our emotional responses to nature settings as the psychic equivalent of body knowledge, something which helped us survive at a time past, but to which we continue to respond today. Is there an ancient meaning in our love for the blazing display of fall color, the fresh bloom of plants and trees in the spring? Perhaps these emotional responses, too, played an important role in our survival and, like body knowledge, were locked in our genes.

How might we examine this ancient past? Where do we come from, what influences may have shaped our journey from its dim beginnings to today? Archaeologists, anthropologists and historians study physical artifacts to tell us how we looked, what we built, the kinds of societies we developed and the events that moved us through history.

But how can we study the role of our non-physical selves–how we thought, felt and reacted as primitives in a primitive environment? There are no physical artifacts to help us here, but rather, we must look at our contemporary selves to see if we can discern traces of the persona of that earlier life. Psychologists find evidence in contemporary patterns of response which are consistent across social, economic, cultural and racial boundaries. When found, these widespread responses can be assumed to be innate in the human psyche, and not the result of contemporary cultural conditions.

PREFERENCES FOR NATURE

Our ancient innate needs and our culturally influenced responses are like two threads, the warp and woof of the human fabric. Our intuitive likes and dislikes for what we see can be the threads that will lead us back to our ancient selves and provide access to the origin of those innate feelings.

Years of research by environmental psychologists and geographers have clearly established that the presence of green is a strong indicator for preference. When vegetation is added to a previously low ranked urban scene its preference is raised. There seems to be something about the presence of green that is important to people. Researchers find a consistency in preference for vegetation among populations that may differ economically, culturally and geographically. This is an indication that what is being measured is not a function of the culture in which people live but rather is something that is inherent in humans, irrespective of who they are and where they live.

What is there about the presence of green nature that elicits preference? The researchers conclude that these settings must provide a kind of information that is important to us at a deep level. Stephen Kaplan conceptualizes humans as information-processing beings–taking in what we find through our senses, then interpreting and using that information as a basis for decisions.[1] Therefore we can conclude that the expression of landscape preferences must somehow be connected with gaining and processing information.

EVOLUTION: PEOPLE AND PLANTS

To find the origin of the meanings of landscape preferences, we will have to look back to our beginnings as a species and the ways in which survival information was gained from the surrounding environment. We cannot see the beginning and certainly there were no psychologists to test our humanoid ancestors, but we can use the best available information to reconstruct the scene; wherever humans appeared on the planet, what they found, and what they needed to survive and reproduce.

In the grand scale of evolution of life on this planet, humans represent but a blip. Our time on Earth is estimated at 100 million years of evolution as a mammal, over 45 million years as primate, over 15 million years as an ape, and 2 million years as *homo sapiens*.

Biologist Richard H. Wagner places us in a proper perspective. He says, "If you were to consider evolution of life on earth as a 30 minute film, you would see wave after wave of new species evolving, filling the environment with a diversity of life forms, and then receding–sometimes totally, but occasionally leaving a few of the best adapted species behind. It is humbling to note that man's existence on earth would flash by in the last 3.5 seconds of that film!"[2]

Humans evolved in a world already populated with a wide diversity of green plants. Learning to survive must have been a full-time task. Only those who were successful were able to produce offspring and carry on the species. Individuals who were not successful were dead ends in the evolutionary process. In addition to the daily instinctive needs for food and water, early humans had to locate a suitable habitat to serve as a nesting site, to provide shelter and protection for offspring. We continue to see the instinctive nesting and courting behaviors in animals and birds that led to their evolutionary survival. Might there be an analogous remnant of our beginnings?

In addition to instinctive responses, our primitive predecessors were endowed with an enlarged brain and thus gained the ability to think, analyze situations and make decisions–plan for the future. In an environment that could both sustain and threaten, they had to learn how to distinguish settings that offered positive opportunities.

Lacking a guide book, humans had to seek in their surroundings the clues which spelled survival. The green environment itself became the source of information on its suitability for sustaining human life. Those successful at learning joined the company of their predecessors that continued the species.

Making Sense Out of What Is Seen

At some point in time our humanoid ancestors started to move down from trees to live on the ground. Current evidence suggests the African Savanna as locale of this change. They went from a multidimensional boreal life in the trees where they could move freely up, down, or out-to an environment which starts and ends at the ground. The change in habitat required an adjustment to the comparatively limited dimensions of life on the ground.

In a savanna already inhabited, the new arrivals had to depend on their ability to outthink those who already lived there. Early humans had to learn how to fend for themselves against all the odds presented in the natural environment. Though slower afoot and physically smaller, they utilized their larger brain capacity to outwit the animals surrounding them, imagine what might happen before it happened, and plan their response to any possible situation.

They had to become skilled at discerning those settings which offered opportunities for success. How might hunter-gatherers analyze a landscape to determine its suitability for their group? Through trial and error, repeated observation and analysis they learned to recognize and favor features in the surroundings which portend safety and to avoid those which offered danger. Only because our hunter-gatherer ancestors had mental capacities superior to the other life forms could they make judgements which required projecting into the future to predict the results of their actions. Only if they could make sense of what they saw would they be able to move to the next step and evaluate its potential for benefit or harm.

Origin of Landscape Preference

The Roles of Intuition and Cognition in Landscape Preference

Let's consider the mental steps that might be required in appraising a setting for its survival value. In the process two mental qualities, intuition and cognition, are utilized. Intuitive reasoning utilizes what has already been learned to produce a more instantaneous or automatic response. In cognition, a slower process, the mind thinks about and evaluates what is seen, reasoning comes into play, and it produces a projection of future implications of what is being seen. If the situation does not pose an immediate threat which would require an immediate decision, assessment can be a slower response with a larger cognitive component.

How might this work? (1) Initially, one sees the setting and through thoughtful analysis makes sense of it, identifying parts of the landscape. (2) Through further thought processes one decides whether the environment is favorable or unfavorable for one's well-being. (3) The next decision, based on the previous analysis, will be to take any action that might be needed. In the cognitive mode, assessment is slower, building piece by piece on what is learned, which is satisfactory if there is no impending threat that would require an immediate decision. However, if the situation is one with a potential for immediate danger or emergency, an efficient quick intuitive response would be more adaptive.

For example, if a person were to hear a rustling in the bushes and then see a large form such as a mammoth, in a cognitive mode the sequence of thoughts might be: (1) Something in the bushes, it is big, it is a mammoth, it is moving, it is moving toward me–quickly! (2) Does this represent a threat? Yes and I had better do something about it–now! (3) Should I run away, or climb the nearest tree that would place me out of its reach? By the time the poor fellow had gone through this chain of thought to reach a decision, in all likelihood the mammoth would be on top of him and no further decision would be required. Not a very successful way of thinking!

If, however, the whole process were to happen more quickly, (particularly steps 1 and 2), then there might be time enough to make a decision on what to do and execute it effectively. If steps 1 and 2 did not require conscious thought but were automatic, culminating in an emotional response, then the appraisal process would be speeded up. The only consciously thought out decision would be "what should I do about it?" With this more efficient way of appraisal, one would have a better chance of surviving.

Gordon Orians and Judith Heerwagen of The University of Washington in their study, *Evolved Responses to Landscapes*,[3] differentiate between intuition and cognition when considering the steps our ancestors might have taken to sort out the dangers and benefits of a setting to decide if it is appropriate.

When arriving in a new environment, the first decision is to stay and explore or to leave. This decision is based on an initial intuitive response, a generalized (gut) feeling about the place. Since the response is intuitive and does not require conscious thought, it is achieved quickly, almost instantaneously, and is therefore, highly efficient. Thinking processes are not tied up in this initial appraisal but are available for concentrating on other aspects, such as the threat of a challenging animal.

The initial response is a quick automatic appraisal of the liveability of the place which includes assessment of the spatial features. For example, an open setting would not be desirable since it would offer no protection, nor would a closed setting of thick forest be desired since movement would be impeded and one could not easily see what was ahead in the tangle of trees and undergrowth.

If the initial instinctive responses are positive, then the individual goes on to intentionally gather more information about the setting. At this more cognitive stage mental associations become important. The individual automatically compares what is seen with any meanings it might call forth from memory of past experiences to provide a more detailed analysis of the habitat.

The final stage of habitat selection culminates in a decision to stay. In making this decision one needs to be sure that the environment does indeed offer enough resources and protection to sustain life activities. Of particular concern would be food and safety. Is there evidence of animals that might be caught? Plants with fruits or other edible parts? Is the food so far away that it would take too much energy to catch it and bring it back to the group? Where is water?

In assessing safety: are there places for escape if attacked by animals, trees to climb, caves in which to hide? Is the landscape open enough to see danger approaching? Does it offer high points where one can see broadly to determine what might lie ahead?

If our hunter-gatherer ancestors would have been deeply involved in the analysis concentrating on assessing the surroundings to select a favorable habitat; while so involved they could have easily been surprised by unnoticed appearance of dangers which crawled, roamed, or slithered in the surrounding environment. If, however, the entire process of habitat selection were a more automatic response, then one's mental faculties could remain alert to encounter the unpredictable.

We have all had the experience of becoming so deeply involved in a mental activity–reading a book, working out a puzzle or problem–that we lose touch with what is happening around us. We could be unaware of someone who comes close and finally taps us on the shoulder, and when it happens, the interruption is startling, comes as a mild shock.

However, when calling on what has already been learned, our minds can work simultaneously at more than one level. Learning a new skill, such as driving a car, requires intense concentration, particularly at the start. But once a skill is "learned," becomes part of our abilities, it is more automatic and requires less mental effort. All the activities of driving–steering, shifting, accelerating, watching out for other cars–happen almost unconsciously and we are able to attend to other things, glimpse the passing scenery or have a conversation with a passenger. What we have learned or know resides within our intellect and functions automatically, does not require an input of constant attention.

Now, if in a similar way, our hunter-gatherer could somehow "learn" to recognize the characteristics of a liveable habitat, become so adept at selecting appropriate habitats that the qualities would be quickly perceived, then appraisal would become more automatic. Thus, while looking for a place to settle one could also remain alert for unexpected dangers lurking in the bushes. We are fortunate that the human mind has evolved with the ability to make complex decisions such as selection of habitat with a minimum of mental effort. Throughout the process of evolution, those that do it best are the ones who survive.

Role of Emotional Responses

In addition to being informed by our thoughts we are also informed by emotions and feelings. When we feel hungry or thirsty, we eat and drink, when we feel pain, i.e., the water is too hot!–we quickly withdraw our hand. We do not have to think–emotions and feelings are non-cognitive ways of becoming aware.

In the evolution of primitive humans, emotional responses became connected with distinct types of behavior. If success in any activity were rewarded with positive emotional responses–joy, pleasure, "good feelings"–it would encourage the individual to repeat the activity whether it would be selecting a safe habitat, or fun activities–eating, drinking, or sex. Negative emotional responses, such as fear, would discourage repetition of the behavior. If it does not feel good you do not do it again. If successful selection of habitat engendered pleasant feelings then one would learn to select an environment *because* it "felt" good. The reinforcement and reward gained from positive feelings would favor quick intuitive responses rather than slower, more deliberately thought out, responses.

In the initial assessment a favorable setting would "feel" good and an unfavorable setting would not "feel" good. Feeling good about successfully recognizing danger or positive qualities would reinforce those adaptive decisions. If the emotional responses were reliable, then it would be much more efficient to utilize these intuitive responses, rather than slower, more conscious processes. One would prefer the adaptive setting because it "felt" good. The place where conscious, deliberate thought is needed is in figuring out what further action to take in a situation, not in making the initial appraisal and evaluation.

If the ability to assess a setting intuitively were genetically programmed, then it would not have to be learned anew by each person. One would be born with these abilities; much as we are born with physiological information which establishes norms of heart rate of 72 beats per minute, or body temperature of 98.6°F.

However, the underlying ability to assess an environment is not as inflexible as normal heartbeat. Though the genetic component is fixed, its expression is mediated by cultural overlay. Inner-city youngsters were terrified when invited to accompany me into a woodland setting at the Morton Arboretum which for regular visitors is a highly favored spot. In a reverse situation, I could well be afraid of the inner city "turf" which serves as the familiar everyday play area for the same city kids.[4] For each of us, cultural experience determines the degree to which the genetic component may be expressed. Both are at work when we look at a setting and intuitively come to know whether we like it or not.

From this view we can understand the basis for landscape preference which has been measured by psychologists[5] and geographers[6] over the past twenty years. Not only is vegetation itself preferred, but the ways in which it is arranged also creates a hierarchy of preference. An open forest is favored over one with thick undergrowth, the presence of a path or opening on which we could enter and explore the setting also rates high in preference. A highly preferred place is at the edge of a woods where one can peek out to see what might be approaching (prospect) yet at the same time be hidden from the view of those outside the forest (refuge). This would be a very safe place for our primitive ancestors. Geographer Jay

Appleton[7] has developed this concept into his prospect/refuge theory of landscape preference. Over the years researchers have found a number of these dimensions which predict preference. Each can be interpreted as a cue to the information content of the scene. It is not solely the presence of vegetation but the potential to gain information from the setting that influences preference.

The researchers conclude that humans, to operate effectively, need to be able to read the cues in whatever setting is immediately at hand. They must be able to infer what is likely to happen and the prospects for continued ability to make sense of it.

The remarkably similar results of preference tests across cultures leads researchers to see that an underlying biological component is involved. Though its expression may be affected by culture, the motivation and underlying mechanism seems to be constant. This is not to say that all people will prefer precisely the same setting: in determining which landscapes people prefer, there appears ample room for cultural influences as well as for the echoes of early human experience.

It is interesting to note that when questioned, people are not able to explain their reasons for preferred settings. The choice comes not from conscious thought but from an intuitive or "gut" feeling that occurs before thought.

Landscape of Savanna

Since our humanoid predecessors are believed to have originated in the savannas of Africa, Gordon Orians[8] has studied that biome to determine which of its characteristics might have offered the best opportunities for survival. Savanna is an open landscape of scattered trees with grasses and shrubs between. Sources of food are available for both people and grazing animals. The open landscape affords distant views for safety and also allows one to see what is in the neighborhood. The savanna offered what was needed: "nutritious food that is relatively easy to obtain; trees that offer protection from sun and can be climbed to avoid predators; long, unimpeded views; frequent changes in elevation to allow us to orient in space."[9]

Water as a resource is relatively scarce and unpredictably distributed on African savannas. However, tree shape can be an indicator of its presence. Gordon Orians has studied tree types in African savannas, measuring their characteristics. Using *Acacia tortilis* as the example, he finds that the shape of the tree will vary with the availability of moisture. "In high quality habitat, this acacia has the quintessential savanna look–a spreading multi-layered canopy and a trunk that branches close to the ground, an umbrella shape. In wetter, overly moist savannas, the species has a canopy that is taller than it is broad with high trunk, while in dry savanna *A. tortilis* is dense and shrubby looking."[10] A savanna with broad shaped trees branched low to the ground would indicate a habitat with proper amount of moisture.

Becoming observant of the "look" of the savanna and preferring tree shapes which indicated moisture availability would have been advantageous for the primitive humans who roamed the area. This was a survivor's landscape, and the ability to quickly recognize it would have been a powerful asset.

If the inhabitants of the savanna did use the appearance of the landscape to aid in assessing its potential as a survival habitat, they might have developed an innate preference for the distinguishing characteristics of that biome's landscape. Researchers have found that Americans prefer park-like settings with a ground cover of grass, no tangled underbrush, and a open wide spacing of mature trees.[11,12] The preferred typical park setting might be characterized as a savanna. Anthropologists sometimes refer to savanna as "parkland."

Further confirmation of continuing savanna preference is found in the results of a study by John D. Balling, psychologist with John Falk, an ecologist (1982).[13] Participants in this study included a broad spectrum of ages–third graders, sixth graders, college students, adults, senior citizens and professional foresters–who lived in an East coast area characterized by temperate deciduous forest. Each group was shown slides depicting five different biomes: tropical rain forest, temperate deciduous forest, coniferous forest, savanna and desert, and were asked to rate the slides in terms of how much they would like to live in or visit a similar area. The third and sixth graders (8 and 11 years old) showed significant preference for savanna over other biomes. Beyond that age, however, familiar natural environments were preferred equally with savanna environments.

Since none of the youngsters had ever been in a savanna, the authors conclude that they were expressing a preference for savanna that is innate rather than learned. Only after they are older and have an opportunity to experience and become familiar with other environments do they begin to exhibit more diversity in selecting a preferred biome. The researchers state that the preferences of the younger participants indicate that "humans have an innate preference for savanna-like settings that arise from their long evolutionary history on the savannas of East Africa." Children are born with the preference for savanna; only after they expand their knowledge base through conscious experience of other landscape types do they express more diverse preferences. They are born with a preference for landscape of the savanna already encoded in their genes. In two subsequent unpublished studies, Balling and Falk ran the same tests on populations in Nigeria and India. The results strongly corroborated the findings of the U.S. study; youngsters strongly preferred savanna while older populations had a wider range of preference.

This thought is echoed by Richard Leakey, son of famed African anthropologists Mary and Louis Leakey, who, when asked why Africa has so profound an effect on people, replied,

> Genetic memory . . . the vast majority of people who come here feel something they feel nowhere else. It is not the wildlife, it is the place. If, as I believe, it is a memory, almost a familiarity, it is very primitive. It is the capacity homing pigeons have, salmon have, to recognize, to go back. You feel it's home. It feels right to be here.[14]

Gordon Orians and Judith Heerwagen are studying preferences for tree shapes by residents from Seattle, Argentina and Australia. Initial results indicate that all three groups rated as most attractive those trees with moderately dense canopies and multiple trunks which originate near the ground. Trees with high trunks and either skinny or very dense canopies are judged as less attractive. The trees selected as attractive correlate with the shape of *Acacia tortilis* growing in adequately moist soils.

The kinds of trees and shrubs we select for our gardens might reflect innate preferences for tree shapes. Gordon Orians[15,16] has studied the woody plants used in Japan where the art of making gardens is an old one. Since few flowers are utilized, the main appeal comes from the shapes and arrangements of non-flowering plants, trees and shrubs.

He has looked for evidence that the Japanese might have been influenced by innate preferences for tree shapes of the African savanna. He notes that among species selected, maples, *Acer*, are used with great frequency. Wild species of maples chosen for garden use tend to be broader than they are high, with shorter trunks, and smaller more deeply divided leaves than the species not chosen. In gardens, the maples generally are not pruned, but allowed to achieve their natural shapes which are characteristic of the savanna trees. In selection of oaks, *Quercus*, evergreen species with smaller leaves were chosen over large leaved deciduous species.

In Japan no conifer achieves a spreading character in its native habitat except for the Red Pine, which, in windswept locations, develops a layered, more horizontal form. The shapes of conifers in Japanese gardens are highly controlled by rigorous pruning, creating evergreens with a distinctly layered aspect, encouraging the effect of a canopy broader than tall with trunks that branch close to the ground, all echoing characteristics of savanna trees. The horizontal effect is encouraged by supporting long, low spreading branches to permit an unusual extension of growth.

At the Morton Arboretum one can find that preferred shape in crabapples and hawthorns. When Tony Tyznik, the landscape architect, selects and prunes specimens of these trees, their savannah shape becomes more evident. For Mr. Tyznik this shape is aesthetically satisfying.

Time Frames of Decisions

Nature's Cues

While perusing day to day affairs, concentrating on the work at hand, it would be important for primitive humans to recognize cues that portend events of great importance. If the cue indicates danger or a sudden change, an immediate response is needed. This would be much the same as the sound of a fire alarm drawing attention to a potentially imminent threat. The response to such a cue must take place quickly, overriding any preoccupation of the moment. From an evolutionarily adaptive viewpoint, alarm cues would be most effective if they aroused a strong emotional response in those who perceive them. The added power of emotion would draw attention even more strongly to the cue.

Cues which portend a change in weather would be particularly important, in that they may indicate a change for which one would have to prepare. Some cues call for immediate action; clouds, wind or temperature changes, would fall in this classification. They could mean that one might have to move to a more sheltered location to find protection from the impending storm.

Other cues portend changes which occur more slowly and are effective over a longer period of time, would not be of such immediacy as to require instantaneous action. Indicators foretelling the change of season would have been of great importance. They would signal the need for a longer preparation period than is allowed for a passing thunder shower; they may affect behavior for several months as in moving from summer pasture to protective winter habitat. The change from summer to fall to winter is heralded by a variety of signals–a difference in the quality of light as the sun's trajectory moves from overhead in summer to lower on the horizon with the approach of winter. Changes in foliage color in fall would be an important indication that preparations for surviving the long winter should be made. The environmental alarm bells set off behaviors in our primitive ancestors that were adaptive, and helped them to survive.

The alarms still sound within us today. We are sensitive to the onset of afternoon and night. Sunrise and sunsets remain deeply fascinating to us. For an animal with poor night vision, such signals would be cues to seek shelter from the coming darkness, to prepare for a new day. We are endlessly enthralled with the shapes of clouds in the sky: poets and painters use their skill to portray them to us. But our ancient ancestors also read the clouds for information about weather, water and shade.

In Spring, trees start to turn green, grass grows, flowers bloom and we are drawn to witness life's renewal. In autumn, millions of us head for the countryside or mountains to view the spectacle of leaves turning orange, red, yellow, and purple before falling to the ground. It is almost a homing instinct that we see each fall at the Morton Arboretum. For several weeks attendance increases and, at the peak, cars ride bumper to bumper through the brilliant fall display. Interestingly, most of the visitors could observe the change in color of vegetation near their homes, but they feel a strong need to come to the Arboretum, a "natural" place where they can be immersed in the spectacle with few reminders of the constructed world in which they live.

Poets and nature writers have tried to portray the emotions hidden in an experience with nature. In William Wordsworth's *The Rainbow*, the poet exclaims, "My heart leaps up when I behold a rainbow in the sky."[17] The rainbow is in the sky, but the up-leaping heart is in Wordsworth. The visual image is brought inside where it comes to life as personal experience. As Wordsworth experienced the rainbow, Annie Dillard in *Pilgrim at Tinker Creek* tells of a cedar transfigured, "the tree with lights in it."[18] She is awestruck by the sight and tries to find words to describe feelings that arise from her experience with nature.

Other signals that were adaptive for primitive humans still call to us today. Fire, contained in a fireplace or within the confines of a camp circle, is comforting and rewarding. One can stare endlessly into the flames, hypnotized by the sounds and the array of brilliant flickering patterns. The same fire, uncontained, consuming buildings, forests, fields is a threat–but none-the-less a strong source of fascination.

Early humans would have had to be quickly aware of large or fierce animals which could easily be life threatening. One would either have to climb to safety, prepare to capture and kill the animal, or be killed. Today hunters reenact primitive rituals and others of us flock to zoos and safari parks in response to a complex relationship, our fascination with the animals with whom we have always shared the earth.

Our love of flowers could well be of ancient origin. In an otherwise green and brown world, flowers could be colorful indicators of future sources of food. Hybridization of flowers often turns to more striking forms and colors–doubles, large blooms, bicolor and multicolored–all of which make the flower more easily apprehended. Flowers have held a place of importance for a long time. Excavation of Neanderthal burial sites in Iraq reveals an abundance of pollen grains in each grave, a sign that flowers were part of that final act.[19]

Why this digression into evolution and the deeply rooted origin of contemporary landscape preferences? Stephen Kaplan comments, "While the survival requirements of humans differ in many ways from those of our ancestors, in many respects the story has not changed dramatically. One must still negotiate the physical environment, assess lurking threats and dangers, and concern oneself with finding one's way back. Nor have humans ceased to be information-based animals, continuously struggling to make sense of their surroundings and exploring new adventures."[20]

We at this symposium are well aware of the strong positive effects of plants on human well-being. From this evolutionary perspective we can see that these responses are not superficial or of the moment, but rather signal the presence within each of us of a genetically continuing connection with green nature, from which our species learned how to survive. We must become more consciously aware of these encoded meanings of green nature and be reminded of that important green connection. When you stroll through gardens, grow plants, enjoy a walk in a park or forest, know and heed the ancient guardian within each of us that continues to guide us in directions that are beneficial for humanity now and in the future.

REFERENCES

1. Stephen Kaplan, Environmental Preference in a Knowledge-seeking, Knowledge-using Organism, *The Adopted Mind: Evolutionary Psychology and Generation of Culture*, J.H. Barkow & J. Tooby (Eds.), Oxford (In press).

2. Richard H. Wagner, *Environment and Man*, Norton, New York, 1971, p. 5.

3. Ibid.

4. Charles. A. Lewis, Nature City: Translating the Natural Environment Into Urban Language, *Morton Arboretum Quarterly*, 11(2) 17-22, 1975.

5. Stephen Kaplan, Environmental Preference in a Knowledge-seeking, Knowledge-using Organism, *The Adapted Mind: Evolutionary Psychology and Generation of Culture*, J.H. Barkow & J. Tooby (Eds.), Oxford (In press).

6. Roger Ulrich, Robert F. Simons, Barbara D. Losito, Evelyn Florito, Mark A. Miles, & Michael Zelson, Stress Recovery During Exposure to Natural and Urban Environments, *Journal of Environmental Psychology*, 11, p. 201-230, 1991.

7. Jay Appleton, *The Experience of Landscape*, John Wiley, New York, 1986.

8. Gordon Orians and Heidi Heerwegen, Evolved Responses to Landscapes, in *The Adapted Mind: Evolutionary Psychology and Generation of Culture*, J. Barkow, L. Cosmides, & J. Tooby (eds), Oxford, (in press).

9. Ibid.

10. Ibid.

11. Herbert W. Schroeder & Thomas L. Green, Public Preferences for Tree Density in Municipal Parks, *Journal of Arboriculture*, 11(9):272-277, September 1985.

12. Rachel Kaplan, Dominant and Variable Values in Environmental Preference, in *Environmental Preference and Landscape Preference*, A.S. Devlin & S.L. Taylor (eds.).

13. John D. Balling & John H. Falk, Development for Visual Preferences and Natural Environment, *Environment and Behavior*, 14(1):5-28, 1981.

14. Aaron Latham, To a Stranger, Africa Feels Like Home, *The New York Times*, November 10, 1991.

15. Gordon Orians, Habitat Selection: General Theory and Application to Human Social behavior, in J.S. Lockard (ed) *The Evolution of Human Social Behavior*, Chicago, Elsevier, 1980.

16. Gordon Orians, An Ecological and Evolutionary Approach to Landscape Aesthetics, in E.C. Penning-Roswell and D. Lowenthal (eds) *Landscape Meaning and Values*. Allen and Unwin, London, 1986.

17. William Wordsworth, The Rainbow, *The Oxford Book of English Verse*, Quiller-Couch (ed), Oxford University Press, New York, 1940.

18. Annie Dillard, *Pilgrim at Tinker Creek*, Harpers, New York, 1974, p. 33.

19. Rose S. Solecki, Shanadar IV, A Neanderthal Burial Site in Northern Iraq, *Science*, 190, 28: 880-881, 1975.

20. Stephen Kaplan, Environmental Preference in a Knowledge-seeking, Knowledge-using Organism, *The Adapted Mind: Evolutionary Psychology and Generation of Culture*, J.H. Barkow, L. Cosmides, & J. Tooby (eds.), Oxford, (In press).

Chapter 19

Indoor Plants and Pollution Reduction

Margaret Burchett
Ronald Wood

SUMMARY. Preliminary experiments by Wolverton et al., first with NASA, and then for the US Interior Plantscape Division of ALCA, have shown that selected indoor potted plants, in test chambers, can reduce concentrations of volatile indoor pollutants such as formaldehyde or benzene by up to 90%. This paper outlines plans to take this pilot work further in new directions, under Australian conditions, by:

- testing the efficacy of local varieties of indoor foliage plants to reduce concentrations of volatile organics in a 'real-world' situation, namely a selected air-conditioned office building in the Sydney area;
- extending Wolverton's methodology, using Australian pot-plant varieties;
- carrying out a series of studies on the mechanisms of absorption and assimilation by plants and soil micro-organisms;
- initiating selection and breeding programs for the varieties tested.

This work has enormous potential world-wide across the full range of indoor environments, in which the population of modern cities spend most of their time. These include commercial, public utility (e.g., schools, hospitals), and private dwelling environments.

BACKGROUND

In July 1990, at the Australian Interior Plantscape Association National Conference held at Surfers Paradise, Queensland, Dr. Bill Wolverton was the keynote speaker. Dr. Wolverton had carried out work with indoor potted plants, first with the National Aeronautic and Space Administration (NASA) and then with the US Interior Plantscape Division of ACLA.[1,2] This work showed that common indoor foliage plants in test chambers could reduce by up to 90%, concentrations of volatile organic indoor pollutants such as formaldehyde or benzene.

The work that Bill Wolverton did with NASA was part of a program on how to devise self-contained ecosystems for space ships. The work that he later did with the interior plantscapers was in response to increasing public concern over the occurrence of 'sick building syndrome' or 'office-related illness' in air-conditioned buildings. It is perhaps not surprising that after the NASA work he could see the possibilities of using plant materials to combat indoor air pollution in buildings here on planet Earth.

Wolverton's work has been exciting news for the interior plantscape industry not only in the USA but in Australia also. It opens up possibilities of producing and supplying potted plants for their direct health benefits, as well as for their beauty and the psychological satisfaction that they bring.

The promise is there, but we are not there in reality yet, because there is more work to be done before we can launch the campaign properly to the public (in Australia or elsewhere), to building owners and managers, and to public health authorities. So in this paper we look at :

- What Wolverton's work has shown so far
- Work planned and in progress in this Urban Horticulture Unit
- Our long-term program plans
- Benefits of the work, that is, the implications for people/plant interactions in indoor environments, in Australia and internationally.

SUMMARY OF PUBLISHED FINDINGS TO DATE

Wolverton, Johnson and Bounds presented a special report for the Nov./ Dec. 1989 issue of *Interiorscape* (USA), entitled *Interior Landscape Plants for Indoor Pollution Abatement*. The findings detailed in that report can be summarised as follows:

- North American varieties of common indoor foliage plants were tested in plexiglass chambers ranging from 0.5-0.8 cubic metres in volume.
- Plant varieties used in various of the tests included, among others:

  – *Dracaena marginata*
  – *Dracaena deremensis 'Janet Craig'*
  – *Ficus*
  – *Gerbera jamesonii*
  – *Dracaena massangeana*
  – *Spathiphyllum 'Mauna Loa'*
  – *English Ivy (Hedera helix)*
  – *Chrysanthemum morifolium*

- Low concentrations (less than 1 mg per litre of air) of formaldehyde, benzene and trichloroethylene were injected into the chambers, and samples taken out over a 24-hour period.
- The concentrations of chemicals were reduced by 10 to 90 per cent, depending on the plant variety and the chemical.
- When the foliage was removed, the pot contents could still be fairly effective.
- Newly prepared pots were not very effective.

CONCLUSIONS FROM WOLVERTON'S WORK

From the results outlined above, it seems clear that:

- the leaves play a major role in absorption.
- the developed plant/soil system is also important.

Wolverton and his son, in 1991, produced a new report for the US Foliage for Clean Air Council, on the *Removal of Formaldehyde from Sealed Experimental Chambers by Azalea, Poinsettia and Dieffenbachia* (Wolverton Env. Services Report WES/100/01-91/005), which shows similar results.

This work, then, has established that potted plants have the ability to absorb, and apparently dispose of, volatile organic compounds.

What we do not know is:

a. how effective they are in a 'real-world' situation, and
b. how the plants do it.

In order to get maximum community health benefit, and market advantage, from this pollutant-eating capacity of pot-plants, we need answers to both these questions.

WORK PLANNED AND IN PROGRESS IN THIS SCHOOL

We can report some progress on a 1-2 year program for tackling both of the questions posed above.

Test of Effectiveness in a 'Real-World' Situation

With Mr. Stan Wesley, of NSW Public Works Department-Engineering (PWD) as Convener of the project team, and with Mr. Stephen McPhail of the Chemistry Branch of the State Pollution Control Commission, and representatives of the New South Wales (Aust.) Interior Plantscape Association, we have designed a practical investigation of the ability of indoor plants to reduce air pollution. The study is specifically aimed at relating total plant surface area to the reduction of formaldehyde concentrations in a known turn-over volume of air-conditioned atmosphere. With colleagues from PWD and other government agencies, we had inspected a number of business buildings in Sydney. The study needed a building with:

- separate air-conditioning units on each floor
- manageable size of floor areas
- two floors with similar, fairly stable, usage
- co-operative management and occupiers.

The first requirement was the hardest to find. Many office buildings have a central air-conditioning system, which could swamp local effects on each floor. We have found a suitable building–the Ashfield regional office of the NSW Public Works Department. (This is a busy suburb 8-10 km south-west of the central business district of Sydney.) This building has six floors, with separate air-conditioning units on each floor.

The design of the proposed project was to pre-test three of the middle floors, second, third and fourth floors, for levels of formaldehyde, which is the most widespread of indoor volatile organic pollutants. Levels of other volatile organics were also monitored.

The two most similar floors in terms of indoor atmosphere were chosen for further study. One of the two floors was supplied with the maximum plant cover that would be likely to be designed for such a setting. Indoor air quality was monitored on both the floor supplied with plants, and the control floor, for a twelve month period; the plants being maintained and renewed as necessary in the normal way. The NSW Interior Plantscapers Association had agreed to supply the plants and plant maintenance for this project.

We also carried out scientific testing of the performance of the plant materials throughout the project, and we would also do a series of post-harvest investigations on them at the conclusion of the investigation. The proposed tests are outlined below.

At the end of the twelve month period the plants were removed, and monitoring continued on both floors for a further three months.

This is an exciting project, for which we negotiated with sponsors. However, it is one of those projects where positive results would be really satisfying, but where negative results, that is, no differences between the two floors, might be what we get. This is because we know so little about how plants reduce concentrations of pollutants from ambient air, that we have no idea how many plants, and of what sort, would be needed in the test building. However, this was a pilot study, under Australian conditions of building engineering and indoor horticulture.

So, while conducting this project, we will continue on other fronts as well, since whatever the outcome of this particular investigation, we can take it as established from Wolverton's experiments that plants do have this property of pollution abatement. The next task is to find out how they do it, and harness the mechanism. That is the essence of horticultural development.

How Do Plants Lower Concentrations of Airborne Volatile Organic Pollutants?

This question has to be subdivided further, as follows:

- For any one species or variety, what is the relationship between leaf area and absorptive capacity?
- If plants of different varieties have equal leaf areas, will they be equally effective, or will there be genetic differences?
- What is the relationship between absorptive power of the pot plant, and the volume of air surrounding it? What difference would air flow or turbulence make?
- What role do roots play in absorption?
- What is the role of the micro-organisms? Which types?
- Does the potting mix have a direct absorptive effect? To what extent? How does it vary from one mix to another?

When we get answers to these questions we will be able to supply and breed plants of the right sort and size for particular situations. Standard breeding programs could be augmented by biotechnology, so that we could even supply tailor-made plants for particular environments–factory, office, shopping centre, private home and so on. This is not a far-fetched prospect. Look at the increases in yields of cereal and fruit crops, and the transformations among ornamental varieties generally that have resulted from selection and breeding over the years. Pollution absorption is just another characteristic that can be bred for in the future. And so too is the production of super-pollution-chomping soil bacteria.

We planned a program that starts off repeating Wolverton's methods, using similar test chambers, local Australian varieties of the same species, and local potting mixes. The aim of this project was to compare the efficiency of local varieties with North American ones. Our aim was to look at the effects of temperature, light intensity, and day length, on performance, and we looked to answer the list of questions posed above. We had an initial grant to commence this investigation in early 1992.

Meanwhile, using basic internal university funding, and the donation of plant materials from Sykes Indoor Plant Services, we were able to start a very limited study of leaf areas, stomate characteristics, and chlorophyll content in a number of local indoor foliage plant varieties. Preliminary results on leaf areas are shown in Table 1.

The total leaf areas of most of these varieties are surprisingly large–the floor area of a single bedroom is usually about 9-10 square metres.

We have also carried out some preliminary measurements of the chlorophyll content of leaves of these species. Shade plants typically have more chlorophyll per unit area than sun plants (i.e., are usually darker green). In addition there is more chlorophyll b per unit of chlorophyll a, since chlorophyll b can help trap the low light that these plants have to use. In future work we plan to compare the rates of photosynthesis and respiration, and the chlorophyll content and Chl. a/b ratios, in plants that have been 'out to work' with those that have been well treated in the nursery. The effects of pollution can sometimes also be directly monitored by such measurements.

A preliminary examination has also been made of the leaf surfaces and stomate distribution on these plants, using scanning electron microscopy (SEM).

LONG-TERM PROGRAM

We have developed a 3-5 year plan to investigate exactly what is happening to the pollutants. For breeding to be successful, and before any biotechnology could be applied to the development of more efficient varieties, we would need this information.

For the benefit of those participants who do not come from a botanical/

TABLE 1. Leaf characteristics of some common NSW varieties of indoor foliage plants

Plant Type	No. of leaves per pot	Total leaf area (m^2)
Ficus benjamina	552	16.7
F. benjamina exotica	1047	18.4
Dracaena fragrans 'Janet Craig,	74	12.8
D.f. 'Janet Craig'	43	8.8
D. deremensis	74	12.8
D. deremensis	76	8.4
D. fragrans massangeana	71	16.6
D. marginata	98	0.5
D. marginata	250	0.9
Spathiphyllum petite	115	5.2
Kentia	6 (185 leaflets)	12.8

horticultural background, let us summarise from first principles what we know of plant function:

All that plants need, in order to be able to grow and reproduce, are a few basic requirements, namely:

- warmth
- light
- air (carbon dioxide and oxygen)

- water
- mineral salts

Given these requirements, plants manufacture their own food, by the process of photosynthesis, carried out in chloroplasts:

$$\text{Carbon dioxide} + \text{water} \xrightarrow[\text{chlorophylls, enzymes}]{\text{light}} \text{SUGAR} + \text{oxygen (by-product)}$$

From sugar, plants can then make all their other components (e.g., DNA, proteins, vitamins, oils, cellulose, pigments, wood, etc). For this they need many enzyme pathways, that can process a huge variety of compounds–even including perhaps formaldehyde, benzene, and other 'pollutant' compounds.

Land plants evolved from aquatic algae, but land plants face a much more difficult environment than their aquatic relatives and forebears, all of whose basic requirements are supplied via the bathing waters of their habitat. In land environments:

- Warmth, light and air come from above, through the atmosphere, but water and mineral salts come from the soil. That is, there is physical separation among the five basic requirements.
- Air is not a very buoyant medium–plants need some skeletal support.
- Temperatures are likely to fluctuate greatly from day to night, and from season to season.
- There is no surrounding water to carry sperms and eggs about in a sort of soup, as happens in rivers and seas and,
- The atmosphere is dehydrating.

So, land plants have developed:

- *cuticle*, a waxy waterproof layer all over the surfaces to stop dehydration.
- *roots*, that take up water and mineral salts, store food, and anchor the plant;
- *shoots*, with leaves for photosynthesis, gas exchange and transpiration through stomates;
- *flowers*, with pollen for fertilisation, and fruit with seeds;
- *stems*, to link shoots and roots, and give skeletal support;
- *veins (xylem)* that carry water and salts up from the roots
- *veins (phloem)* that carry sugars, etc., away from leaves to tips and roots.

Any of these could play a part in pollution abatement. Organic pollutant molecules could theoretically be absorbed:

- into the cuticle layer, which could accumulate fat-soluble organic compounds
- via stomates, into leaf cells, and thence to other parts of the plant
- via root hairs and root tips where water and salts are absorbed
- once inside the plant, could be detoxified by processing into other compounds
- by micro-organisms in the potting mix
- onto the surface of the organic matter (peat, sawdust, etc.) in the potting mix.

To do any advanced breeding program, or any genetic engineering, we need to know exactly what is happening in the plant-growth medium-micro-organism complex.

BENEFITS OF THE WORK

For air-conditioned buildings in particular, the perceived benefits of the utilisation of living plants, rather than engineering means alone, to improve indoor air quality, include the facts that plant materials:

i. Are cost-effective in comparison to the use of sophisticated air conditioning filter systems;
ii. Offer flexibility of location and relocation according to specific needs;
iii. Are environmentally sound;
iv. Have aesthetic appeal to staff;
v. Represent a solution that could be readily developed and refined for each application;
vi. Do not produce acoustical problems;
vii. Do not interfere with any existing air distribution systems or patterns in a room;
viii. Professionally supplied and maintained, retain their effectiveness with comparatively low maintenance costs;
ix. Offer flexible routine maintenance, i.e., the frequency is not absolutely critical;
x. Help create an harmonious effect on staff morale, and thus increase motivation;

xi. Can help form attractive and acceptable visual breaks and barriers in open-planned offices;
xii. Installation does not entail alteration to the fabric or structure of a building;
xiii. Provide a means for discrete implementation of a solution to an environmental problem;
xiv. Entail relatively minor capital and running costs;
xv. Involve only a very remote chance of sudden breakdown or failure of operation.

Apart from offices and other commercial buildings, we can add special environments such as schools, hospitals, restaurants, retirement villages, etc., and private homes as well, all of which could benefit from properly selected and arranged indoor foliage plant varieties for pollution abatement and indoor air improvement. The field is of enormous potential.

REFERENCES

1. Wolverton, B.C., McDonals, R.C., and Mesick, H.H., 1985. Foliage plants for the indoor removal of primary combustion gases carbon dioxide and nitrogen dioxide. *J. Miss. Acad. Sci. 30*: 1-8.

2. Wolverton, B.C., Johnson, A., and Bounds, K., 1989. Interior landscape plants for indoor air pollution abatement. *Interiorscape, Nov/Dec.* (Special supplement, 26 pp).

Chapter 20

Growing Fear: Home Horticulture and the Threat of Lyme Disease

William K. Hallman
Deborah C. Smith-Fiola

SUMMARY. This paper reports results from a survey of 308 tick bite victims who reported their fears, behaviors and changes in lifestyle in response to the threat of Lyme Disease. As the result of fears of Lyme Disease, many reported giving up: gardening, walking through the woods, sitting on their lawns, visiting public parks, hiking, camping, picnicking and other outdoor activities. Nearly four percent of the sample also reported being afraid to go outdoors at all. The results suggest that for some people, the fear of tick bites has changed the way they relate to nature.

Me imperturbe, standing at ease in Nature

–Walt Whitman, 1892

The authors wish to acknowledge the research support of the New Jersey Agricultural Experiment Station and NIOSH grant CCU 206087-02.

I haven't walked in the woods in two years.
I'm afraid to sit or lie in the grass.
I don't even like to go outside anymore.

–Resident of an area endemic for Lyme Disease, 1992

For many Americans, Whitman's line from *Leaves of Grass* accurately describes their feelings of safety and serenity in the midst of nature. Poets, philosophers and writers have long extolled the virtues of nature and popular culture is replete with images of nature as calm, verdant and romantic. Many Americans seek refuge in nature, some by hiking through the woods or camping in a meadow, some through fishing or boating or hunting, some through gardening or yard work. There is evidence that such communing with nature has beneficial effects, relieving stress and improving one's sense of well-being (Fisher, 1990).

The benefits of experiences with nature have been recognized for centuries. Ancient Egyptian physicians recommended walks in the gardens for disturbed patients (Honey, 1991a). Modern horticultural therapy programs have been set up across the country to take advantage of the therapeutic effects of working with nature. In many such programs, simple gardening techniques are used to enhance the lives for people who are experiencing physical, mental or emotional problems (Honey, 1991b).

In stark contrast to the image many Americans have of nature as a place of tranquility, an increasing number of people are beginning to see nature as a threat. For these individuals, nature no longer represents hidden pleasures but hidden dangers. The catalyst for this change in perspective is Lyme Disease.

Lyme Disease is an infectious disease caused by the bacteria, *Borrelia burgdorferi*. It can progress from symptoms of a simple "bullseye" skin rash (Erythema Chronicum Migrans) to serious arthritis, to acute cardiac disease and chronic neurological conditions (Trock, Craft, and Rahn, 1989). Since it was made reportable to the Centers for Disease Control in 1980, Lyme disease has become the most common anthropod-borne illness in the United States (CDC, 1985).

In the northeastern United States, the disease is most often carried and transmitted by the tiny deer tick, *Ixodes dammini*. In addition to deer, other wild animals such as birds, mice, raccoons, and others are also hosts to the tick. The ticks can also be carried close to home by cats, dogs, horses and cows. While many people seem to associate the tick only with areas of deep woods and underbrush, monitoring studies suggest that deer ticks are particularly plentiful in suburban landscapes next to woodlands, fields and high grass. In fact, landscape features that sustain wildlife increase the risks of coming into contact with ticks that carry Lyme Disease. These include gardens, shrubs, perennial borders, woodpiles, bird feeders, bird baths, ornamental grasses, treehouses, dog houses, and other features. As a result, the well-landscaped "estate" home has, for many, become decidedly less attractive. Although nearly 90% of the Lyme Disease cases in the U.S. have been reported in the Northeast, the Disease has already been reported in at least 43 states and it represents a growing public health problem. As more houses are built near woodlands, as reforestation continues and deer populations explode, Lyme Disease is becoming increasingly common in more areas of the country. As a result, the number of reported cases of the disease has increased yearly.

Evidence suggests that people are becoming increasingly concerned about the threat of Lyme Disease. Lyme Disease can be very painful, debilitating and in extreme cases, life threatening.

People are also concerned because of the many uncertainties surrounding the Disease. The best way to prevent Lyme Disease is to avoid a tick bite, yet the ticks that spread the disease are smaller than the head of a pin and are usually unfelt and unnoticed. In addition, the deer tick may be easily missed because it does not look like other, larger, ticks that are more familiar to people (Harbit and Wills, 1990). So, it is very possible to be bitten by a deer tick and infected with Lyme Disease without ever realizing it. Less than half of Lyme disease victims remember having been bitten by a tick (Marzouk, 1985).

There are also uncertainties inherent in the diagnosis of Lyme Disease. Diagnosing Lyme Disease can be difficult, as many symptoms of the disease mimic those of other diseases. As a result, Lyme Disease has been called "The Great Imitator."

Symptoms of the disease can be ambiguous and may occur any time from days to years following the bite from an infected deer tick. Nearly 30% of patients never develop the tell-tale "bullseye rash" (Erythema Chronicum Migrans), most often associated with Lyme Disease. In its early stages, Lyme Disease is often misdiagnosed as the flu (particularly in the absence of the rash), since it often produces fever, chills, headaches, joint stiffness and lethargy (Finkel, 1988; Harbit and Willis, 1990).

Misdiagnoses may also occur because diagnostic tests now used to detect Lyme Disease in patients' blood are somewhat unreliable and may fail to detect true cases of Lyme Disease. Fluid drawn from inflamed joints may also look similar to the fluid extracted from an inflamed arthritic joint, leading to a misdiagnosis of rheumatoid arthritis (Marzouk, 1985).

There are also uncertainties related to the treatment of the disease. In the early stages of the disease, treatment is usually uncomplicated and usually consists of doses of oral antibiotics (Finkel, 1988). If patients are promptly diagnosed and treated, few complications usually result. However, without proper diagnosis and treatment, the disease continues to progress, and can cause acute heart disease, arthritis and acute and chronic neurological disorders (Marzouk, 1985). Patients who do not respond to oral antibiotics may be treated with a costly regimen of intravenous doses of the drugs. This often creates a financial burden for the families of victims since some insurance companies will not pay for the treatment of the disease beyond a set number of days. Chronic cases of the disease which have not responded to antibiotics have also been reported.

Thus, there are considerable uncertainties about the prevention, transmission, diagnosis and treatment of Lyme Disease. These uncertainties are often reinforced by newspaper and television reports in which those affected by Lyme Disease are often portrayed as the victims of an incurable disease caused by an invisible assailant.

Because of these uncertainties, and because the outcomes of Lyme Disease are particularly feared, Lyme Disease has affected the lives of many who live in endemic areas. This paper reports results from an ongoing investigation of how people's lives have been affected by the threat of Lyme Disease, examining peoples' perceptions of the risks of Lyme Disease, and their beliefs, attitudes and behaviors in relating to the natural world. It focuses on how the fear of tick bites has affected peoples' outdoor behaviors including gardening, and presents data from a survey of 308 tick bite victims who reported their fears, behaviors, and changes in lifestyle in response to the threat of Lyme Disease.

METHOD

To accomplish the goals of the study, five hundred individuals were randomly selected from a sample of approximately 3,200 people who brought ticks to the Rutgers Cooperative Extension of Ocean County, New Jersey, for identification in 1991. Ocean County is a large coastal county in Southern New Jersey with a population of about 400,000. It also has one of the highest incidences of Lyme Disease in the State. People who are bitten by ticks are actively encouraged to keep them alive and to bring them to the extension office to find out if they are deer ticks. By properly identifying the species of tick, the uncertainty about whether the tick bite will result in Lyme Disease can be reduced. If the specimen proves to be a deer tick, the tick bite victim is counseled to look for symptoms of the disease, and should they occur, to seek early treatment. In some cases, the tick may be sent to the local health department for dissection to determine if it carries the bacteria that causes Lyme Disease. This information can be helpful for diagnostic purposes.

A survey instrument was developed to collect information about the tick bite victims, their homes and lifestyles and their attitudes and behaviors regarding Lyme Disease. The survey consisted mostly of multiple choice questions, however, several open-ended questions were included to allow respondents to use their own words to describe their experiences. The survey was pre-tested and mailed to the selected sample in mid-April, 1992. A cover letter accompanied the survey, explaining its purpose and assuring respondents that the surveys were not coded to reveal their identity, that their answers would be kept completely confidential and that respondents would not be contacted in the future. Respondents were given a self-addressed, stamped envelope in which to return their completed surveys.

Of the 500 surveys mailed out, six were undeliverable. One was returned but not completed. Of those remaining, 308 were completed and returned, yielding a comparatively high completion rate of 62%.

RESULTS AND DISCUSSION

Of those responding, 62% were female, and 33% were male. The median age was 40 years. Eight percent reported having less than a high school education, 39% finished high school, 30% attended some college or completed a two year degree, 23% finished four or more years of college. Ninety four percent of those responding were white. Eighty five percent of the respondents were married. The median total family income was between $30,000 and $50,000.

Although the sample was drawn from the names of people who brought ticks in for identification, the respondent was not necessarily the person who was bitten by the tick, although in 96% of the cases, the tick had bitten either the respondent or an immediate member of the family. In the remaining cases, the ticks were found on pets, or on clothing, furniture or other objects.

The ages of those bitten ranged from 6 months to 97 years. Interestingly, nearly 25% of those who were bitten were under the age of seven and nearly 40% were under the age of 18. Equally significant is that 20% of those who were bitten were aged 60 or older.

Since nearly 50% of Lyme Disease victims don't remember having been bitten by a tick, the fact that each respondent found a tick on themselves or a family member makes this group of respondents a special population. Ticks feed for a relatively short time and then drop off. That the respondents found a tick on the body suggests that little time had elapsed between when the victim had picked up the tick and when the respondent had discovered it. Some had probably discovered the tick within hours, most had likely discovered it within a day or two. As a result, all but 11% of the respondents were able to say where they thought the victim of the tick bite had come into contact with the tick.

Nearly 26% believe that the victim came into contact with the tick at home. Almost 17% reported that they picked up the tick while gardening. About 12% said they thought they had come into contact with the tick somewhere in their neighborhood. Seven percent said that the tick latched onto them at a park or ballfield. About three percent thought the tick was picked up while camping. An additional three percent said they picked it up while hiking and two percent thought they came into contact with the tick while hunting. Several other outdoor activities were reported, each accounting for less than 1% of the total. These included: bird watching, fishing, vacationing, mountain biking, picnicking, berry picking, sunbathing, and several job related locations.

As might be expected, where the victims picked up the ticks that bit them is associated with the age of the victim. For those who picked up a tick at a park or ballfield, the median age was 10 years. For those who picked up a tick while gardening, the median age was 66 years.

There is ample evidence that the respondents are very concerned about the threat of Lyme Disease. Nearly 85% of the respondents reported that they knew someone who had contracted Lyme Disease: Thirty-four percent said they had acquaintances with the disease, 43% said they knew friends or relatives with the disease, and 7% said they had contracted Lyme Disease themselves.

Nearly 66% of the respondents reported that thoughts or fears of having Lyme Disease had occurred to them at least sometimes during the past week. Almost 23% of the respondents worried about Lyme Disease frequently or constantly. The respondents were also asked, "Out of one hundred people, how many would you guess will get Lyme Disease sometime in their lives?" Only three percent of the respondents correctly guessed that less than 1 out of every 100 people will get Lyme Disease during their lifetime. Half the respondents said they thought that at least 20 out of 100 people would get Lyme Disease sometime in their life. About 30% thought that at least 40 out of 100 people would get Lyme Disease in their lifetime. Nearly 20% thought that 60 or more of 100 people would be afflicted.

Despite their concerns about getting Lyme Disease, 97% of the respondents said that they thought that people could take actions to reduce their chances of getting Lyme Disease. Not surprisingly, 88% of the respondents said that they take precautions against tick bites when they are outdoors. These precautions included many preventive measures recommended by experts such as: avoiding tick infested areas, wearing light colored long pants and long sleeved shirts, tucking pant legs into socks, checking the body daily for ticks, and wearing tick repellents.

What is disconcerting, however, is that nearly 61% of the respondents said that they had changed their outdoor activities as the result of the threat of Lyme Disease. Many (32%) said that they simply try to avoid likely tick habitats such as tall grassy areas and woods with a large amount of underbrush. While this is a reasonable response to preventing tick bites, many respondents went far beyond simple avoidance. A surprising number of respondents said that they had completely given up favorite outdoor activities, especially horticultural activities.

In an open-ended question, seven and a half percent of the respondents reported that they will no longer sit on the lawns surrounding their homes. A little more than five percent of the respondents reported that they will no longer visit public parks. Nearly four percent of the respondents said that they had given up gardening as the result of their fears of Lyme Disease. Other respondents reported that they had given up hiking, camping, picnicking, berry picking, hunting, fishing, and allowing their children to go on class trips. Two respondents also said that they had cut down trees on their property as a preventive measure. Most surprising, and disheartening, however, was that nearly four percent of the respondents said that because of the threat of Lyme Disease, they were afraid to go outside their homes.

In response to other open-ended questions, other extreme approaches to preventing tick bites were suggested. Several respondents suggested that information about pesticides that could be used to treat lawns and gardens for ticks would be particularly valuable for homeowners. Several respondents advocated the use of helicopters to spray acaricides over large sections of the county. Others suggested cutting down trees in public parks. Others advocated eliminating the deer population in the county.

What is interesting about each of these approaches to Lyme Disease prevention is that each tries to reduce the risks of a tick bite to zero through one extreme action. Each is a response to the uncertainties inherent in the prevention of Lyme Disease. For some, it appears that the only good tick is a dead tick, and the only way to feel safe in the garden is to feel totally unthreatened by an unanticipated anthropod.

It is not clear how many people feel this way. It would not be fair to say the results of this study generalize to the population as a whole. On the other hand, in this one county alone, more than three thousand people a year are concerned enough about Lyme Disease to bring ticks to the Cooperative Extension Service for identification. This study suggests that a significant minority of the population of endemic communities have made fundamental changes in their lifestyle as the result of Lyme Disease. The study also provides some evidence that for this minority, a significant shift in perceptions of nature has taken place. Rather than viewing nature as a place of safety and serenity, many now seem to view nature as a dangerous place. Rather than seeking the solace of nature, many are now seeking to avoid nature wherever and whenever they can.

In addition, many parents in the study reported that they are teaching their children that the woods and even the grass on the front lawn are dangerous places, and the only way to be truly safe is to avoid nature. This is particularly troubling since it is not clear that children taught to fear nature will want to preserve nature.

There is no doubt that the threat of Lyme Disease is spreading quickly to other parts of the country and there is little doubt that fears about the disease are also growing. As such, there is a challenge ahead to help people understand that it is possible to prevent Lyme Disease without completely abandoning their lawns, their gardens and their lifestyles. Given the proven therapeutic benefits of horticultural activities, it would be a tragedy to let the threat of Lyme Disease rob the present generation of their enjoyment of horticulture. It would be equally tragic if, in response to Lyme Disease, the next generation was cheated out of the chance to experience it at all.

REFERENCES

Centers for Disease Control. (1985). Update: Lyme disease and cases occurring during pregnancy-United States. *MMWR*, 34(25):376-84.

Finkel, M. (1988). Lyme disease and its neurological complications. *Archives of Neurology*, 45:99-104.

Fisher, K. (1990, October). People love plants–Plants heal people. *American Horticulturist*, 69:12-15.

Harbit, M. D., and Willis, D. (1990). Lyme Disease: Implications for health educators. *Health Education*, 21(2):41-43.

Honey, T. E., (1991a, August). The many faces of horticultural therapy–Part 1. *American Horticulturist*, 70(8):37-43.

Honey, T. E., (1991b, October). The many faces of horticultural therapy–Part 2. *American Horticulturist*, 70(10):19-23.

Marzouk, J. (1985, June). Tick-borne diseases: Where to expect and how to detect such bite-caused syndromes. *Consultant*, pp. 21-31.

Parke, A. (1987). From new to old England: The progress of Lyme Disease. *British Medical Journal*, 294:525-526.

Trock, D. H., Craft, J. E., & Rahn, D. W. (1989). Clinical manifestations of Lyme Disease in the United States. *Connecticut Medicine*, 53(6), 327-330.

Whitman, W. (1892). "Me Imperturbe." in *Leaves of Grass*. Reprinted in *Whitman: Poetry and Prose*, New York: Literary Classics of the United States, 1982. p. 173.

Chapter 21

Studying the Corporate Garden

Madelaine H. Zadik

SUMMARY. Corporations have been building gardens for a variety of reasons ranging from enhancing their images in the eyes of the public to improving the work environment for their employees. One frequently hears anecdotal evidence of the resulting improved employee morale and productivity. The Channing L. Bete Company recently installed an indoor garden to provide a space for employees to go and relax, for informal meetings, as well as a place for the entire company to meet. The garden won an award from *Interiorscape* magazine, but how does this translate into concrete data showing the value to people of such a garden setting? A review of research that has been done in this area is presented and specific research needs and goals for the future are identified.

Dramatic changes are being predicted for the office buildings of the 1990's (Alpert 1991). The past decade has seen much overbuilding; not only were too many office structures put up, but they were also too big, and while many buildings were striking from the outside, the basic needs of office occupants were not always taken into account. What is expected is a shift away from the glitzy skyscrapers to more sensible buildings that will help employees work better, buildings that will perform "productivity-enhancing functions." The question is, what are these productivity-enhancing factors and what role can plants play in improving the work environment and worker performance? Here is an important opportunity to influence how work environments are designed and how people think about this issue.

The Wintergarden, an indoor garden recently built at the Channing L. Bete Co. in South Deerfield, Massachusetts, won an award from *Interiorscape* magazine in 1989, and this kind of interior garden is no longer unusual in the corporate world. Many large companies routinely incorporate interior landscaping in their plans. Yet, very little research has been completed or is being undertaken on the positive effects of these gardens, even though claims of the great benefits–improved employee morale and increased productivity–are commonplace. Can it be that successful (profit-making) companies, like the Channing L. Bete Co., put their money into these new buildings and plantings simply because they think it would be a good thing to do–that it would make their employees happier and more productive, and that it would give their company a better image–without the hard data to back up these projections? Dana Parker, who has studied corporate gardens from a philanthropic perspective (Parker 1989), found that very few of the companies she studied were even interested in doing any kind of research (Dana Parker, telephone interview March 1992). Would the same company buy a computer system, an employee benefits package, or an advertising campaign without some research to make sure they were making the right choice? And how many other companies would be interested in more interior landscaping if we had the research data behind many of the claims of how beneficial it is?

The term corporate garden can mean a variety of horticultural entities, including outdoor gardening plots provided to employees for their own gardening desires or corporate headquarters and office buildings located in a surrounding garden landscape. However, I will concentrate on atria that are used for indoor plantings and other interior landscaping. These other forms of corporate gardens are equally important and deserving of attention and there is much need for research in these areas as well.

STUDIES OF OFFICE ENVIRONMENTS

In the business world, there has been interest and concern about how offices are designed. Michael Brille and the Buffalo Organization for Social and Technological Innovation (BOSTI) cite some important problems affecting office workers. While there is a growth in the size of the office-based sector of our society, there is declining productivity and job satisfaction, and a backlash against the dehumanizing component of office automation. Additionally, there are problems of indoor air pollution and high stress (Brille 1984). Brille cites companies' frequent indifference to office interiors and overemphasis on external image. A study of fourteen different companies reported that office environments are becoming linked to workers' psychological needs, performance, and well-being (Goodrich 1982). In Canada, air quality has been identified as a critical issue in worker productivity–too little, too hot, too dry, and too polluted (Cannon 1987). In this case, productivity was measured in terms of more mistakes, more accidents, workers taking longer to complete tasks, and employees being sick more often. In terms of air pollution, plants have been shown to reduce indoor air pollutants (Wolverton 1989), yet there has not been any direct correlation to worker productivity.

The BOSTI study, although examining how office design can be used to increase productivity, did not specifically address the presence of plants or indoor gardens (there is a brief mention of individuals bringing in plants as a way of personalizing their space). The study stresses the importance of the work environment, and, citing the high ratio of labor to building expenses even in the case of very expensive buildings, it emphasizes that office design can be a very cost effective strategy for enhancing performance and job satisfaction.

Joseph Laviana (1982) examined the effects of the presence of indoor plants specifically, and found a positive effect on subjects' feelings toward and evaluation of the indoor environment. However, this research has limited application as it was conducted in a laboratory setting. Another study of a business environment (Aitken and Palmer 1989) found that people attach meanings to plants in that setting, and that their presence does affect the attitudes of those visiting a business. Neither of these studies addressed the impact of the presence of plants in the workplace on employee performance or job satisfaction.

Kaplan, Talbot, and Kaplan (1988) studied job stress and job satisfaction in relation to "nearby nature." They found that brief non-work related involvements, notably those that included some contact with the outdoors, relieved some stress. In addition, workers who had window views of the outdoors or worked outdoors had less job stress and higher job satisfaction. They documented how job stress affects both physical and mental well-being and suggest that somehow incorporating easily accessible nature into the work setting can be very beneficial. The question here is whether interior landscaping might also provide a setting for these brief restorative encounters.

The only research that actually studied workers' attitudes towards plants in an office setting to determine the effect on job satisfaction was conducted by Kim Randall, Candice Shoemaker, Diane Relf, and Scott Geller (1990). Unfortunately this research was inconclusive because workers were not dissatisfied with their jobs at the start of the study. The research did not show whether plants affected behavior or performance; however, worker attitudes towards plants were generally favorable. This is a very important study nonetheless because there is much to learn in terms of the methodology, even if the results were not conclusive.

EXAMPLES OF CORPORATE GARDENS

Over the last few decades there has been a resurgence of the use of atria in buildings (Bednar 1986). A milestone in the development of indoor corporate gardens was the Ford Foundation building. Built in 1967, it is a twelve story building with an 8,500 square foot brick terraced garden where they introduced large scale plantings and provided a new kind of public indoor urban space. Almost all the offices in the building have direct contact with the garden atrium, and all personnel see each other across the atrium. The claims made about this building were that it fostered a sense of oneness, belonging, and common purpose among foundation employees. Would that be any different if there were no interior plantings? It has been described as a "serene oasis in the hard urban landscape, a place for respite" (Bednar 1986). One effect of this building was that it generated interest among building designers and caused many to reconsider the use of atria.

In 1971, the Boise Cascade Home Office Headquarters was built with a six story 106′ × 106′ square atrium in the center. Design goals were to promote communication among employees by maximizing the opportunity for visual contact and spontaneous exchange of information (Bednar 1986). However, in this case the interior garden is relatively small and it is questionable how much thought was given to this aspect of the design. There was certainly room to increase the size of the interior garden, but, instead, they chose to create a 72′ wide outdoor pedestrian walkway around the entire building on the ground level. This might make more sense in a warmer climate, but it is unclear what purpose it serves here.

In Sacramento, California, the Gregory Bates Building was constructed with a four story 150′ × 144′ atrium. The design objectives here were to create a humane work environment as well as make advances in energy conservation (Bednar 1986). The central atrium serves as both a public plaza as well as an employee lounge. The question again is what role do the plantings play.

In New York City, a number of buildings now sport interior garden spaces: the Citicorp building built in 1978 with a seven story atrium; the Trump Tower, built in 1983 as a retail space and adjacent to that, the IBM building with its Bamboo Court; and the Chemcourt, also an indoor public park. In Wilmington, Delaware, the Hercules Building, constructed in 1983, includes a 90′ × 110′ thirteen story atrium. There are many more such buildings with interior plantings. They serve as a refuge from a frantic cityscape, and could easily be studied as to how they are used and by whom and what effects they have on the people who use them or work in them.

Probably the most well-known indoor corporate garden is the 11,000 square foot garden court at the John Deere Company in Moline, Illinois. Built in 1978, the stated goals for this building were to create a stimulating work environment for their employees. Here, too, they claim increased productivity and morale, and that the garden is an important factor in recruiting new people, yet once again, no one has measured this.

The Channing L. Bete Wintergarden is an 8,200 square foot garden area, built as part of a $4.5 million office expansion. Their stated objectives were to create a New England style garden that would provide areas for small informal meetings (plantings are used to provide some privacy in addition to fountains which provide for acoustical privacy) and they also wanted the space to function as a place for the entire company to meet. Bleachers were installed that are pulled out for such meetings. The entire building was built around the garden and all floors overlook the garden. A cafeteria was added so that most people use the garden during lunchtime. The building itself is very narrow north to south and very long east to west, so all offices in this building either face out onto the Wintergarden, or face a window on the other side. The initial landscape installation cost $50,000, not to mention the cost of the landscape design firm, maintenance costs and other costs associated specifically with the construction of the garden. The Wintergarden at Channing Bete provides a setting in which to examine a corporate garden–both in terms of how it was designed and the effects it has had. Neither Peter Wells, a principal at the Berkshire Design Group, the landscape architect firm that designed the garden, nor Michael Bete, son of the president of the company and also the architect who designed the building, knew of any research they had done themselves or used in making decisions about the design of the Wintergarden in the new building and the effects it might have on employees. There was also no research done after the installation. From the start they had planned the building with the idea that there would be a garden designed for employees. They did use studies of the effects of lighting to justify the use of Vitalites and they relied on the BOSTI study for design of office layout and furniture (Peter Wells, interview October 1991, Michael Bete, interview February 1992).

Again, the question remains, do employees of this company who use the Wintergarden perform their job better? Do they have a better attitude toward the company? Does the increased employee productivity and improved morale make it worthwhile for this company to have put in this expensive building? Here is a setting ripe with possibilities for research. More corporations are taking a hard look at the bottom line–profits–and are interested in taking a look at employee job satisfaction and productivity and how they can improve the work environment.

RESEARCH NEEDS AND POSSIBILITIES

One of the major stumbling blocks to this kind of research is the difficulty in measuring productivity. The BOSTI study actually got away from that term, even though they used it in the title of the study, and instead focussed on job performance using self-ratings and ratings by supervisors. Some of what they used to study job satisfaction included employee turnover, absenteeism, lateness, grievances, and what they termed other withdrawal behaviors.

The Bete Company is actually a very interesting setting because they still have people working in the old building as well as in the new building with the Wintergarden. There are possibilities for a variety of different kinds of research that may also be applicable to other settings as well.

A study of this setting could include the following:

1. How is the Wintergarden actually used? We can track how people use the space: for socializing, for a break to get away from their offices, for informal meetings as was originally hoped, or something else? Are there certain places that people prefer to sit in the garden and why?
2. Why is the Wintergarden a space that serves these functions? Why do employees choose to hold meetings here? Is there a way to compare how productive those kinds of meetings are in the garden or if they are held elsewhere? Do employees go to the garden because they expect to find someone there?
3. Does everyone use the garden for these purposes, if not, who are the exceptions and why? Do people whose offices look out on to the garden use it more or less than others? How about those whose offices are in the old building?
4. Do people avoid the garden at times for specific reasons?

That is a beginning, but we need to find ways we can look at productivity and performance. Other possibilities would be to compare employees who work in the new building to those whose offices remain in the old building (in partitioned cubicles). Are there differences in absenteeism, job satisfaction, levels of job stress, attitudes toward the company? Perhaps there is a way to go back and measure absenteeism before and after the Wintergarden was built. Given a choice, what offices would people prefer to have? Would they prefer an office with a window to the outside or a view overlooking the garden? Can we use simulations to determine employee preferences? How can we measure job stress and see whether visits to the garden are restorative as in the Kaplan's (1988) research? Perhaps we can find situations where employees are under particular stress at work or even in their personal lives and measure stress levels in relation to visits to the garden, or do they use the garden more when they are under greater stress? Are there ways of measuring how long it takes to complete certain tasks and make comparisons before and after a visit to the garden?

There is also the possibility of measuring visitor responses and making comparisons between those who are taken to the Wintergarden for a meeting or those who are taken elsewhere in the old building. How does the garden affect the company's public image? Are the employees' attitudes toward the company influenced by the perception that the company put in the garden for them, because the company truly cares about its employees? Is this true at other companies?

Yet another problem here is that the Wintergarden is a composite of many factors, and plants are only one part of that. We need to find ways of separating out the influence of the plants themselves. What would the space be like without the plants? Perhaps some research could be done with simulations of places with and without plants. Perhaps in other new locations measurements can be made before and after the plant installations. Are the plants affecting temperature, humidity, pollutants?

There are many more questions still to be asked and researched. This is a multidisciplinary field requiring many different kinds of expertise in order to explore both the direct and indirect effects of plants on people in the work setting. This research could provide the data required for many more companies to justify the installation of interior plantscapes and could provide the landscaping industry with information as to how to make their installations more effective.

REFERENCES

Alpert, Mark, 1991, Office buildings for the 1990's, *Fortune* 124 (November 18):140-150.

Aiken, J. E. and R. D. Palmer, 1989, *The use of plants to promote warmth and caring in a business environment*, Paper presented at 11th Annual Meeting of the American Culture Assoc., St. Louis, MO.

Bednar, Michael, J., 1986, *The New Atrium*, NY: McGraw Hill.

Brille, Michael, 1984, *Using Office Design to Increase Productivity*, Buffalo, NY: Workplace Design and Productivity, 2 vol.

Cannon, Margaret, 1987, Give them air, *Canadian Business* 60(April):58-61.

Crouch, Andrew and Umar Nimran, 1989, Perceived facilitators and inhibitors of work environment in an office environment, *Environment and Behavior*, 21:206-226.

Goodrich, R., 1986, The perceived office: the office environment as experienced by its users., in J. Wineman (ed), *Behavioral Issues in Office Design*, NY: Van Nostrand Reinhold, 109-134.

______, 1982, Seven office evaluations: a review, *Environment and Behavior* 14:353-378.

Fong, Allen, 1987, Creating an oasis for employees, *Buildings* 81(August):54-56.

Kaplan, Rachel and Steven, 1989, *The Experience of Nature*, NY: Cambridge University Press.

Kaplan, Steven, Janet Talbot, and Rachel Kaplan, 1988, *Coping with Daily Hassles: The Impact of Nearby Nature on the Work Environment*, Project Report USDA Forest Service, North Central Forest Experiment Station, Urban Forestry Unit Cooperative Agreement 23-85-08.

Laviana, Joseph, 1982, *Plants as enhancers of the indoor environment*, Masters Thesis, Kansas State University.

Michels, Antony, 1991, Getting ideas by rubbing elbows, *Fortune* 124 (September 9):10-12.

Olsen, Christopher, 1986, Boosting productivity and privacy in the open office, *Building Design and Construction*, 27(August):90-92.

Ornstein, Suzyn, First impressions of the symbolic meaning connoted by reception area design, *Environment and Behavior*, 24:85-110.

Parker, Dana, 1989, The corporate garden, *Environmental Responsibility: Taking Root in Public Horticulture, The 1989 Longwood Graduate Seminars* 21:77-83, University of Delaware, Newark, DE.

Randall, Kim, Candice A. Shoemaker, P. Diane Relf, and E. Scott Geller, 1991, *A Study of Relationships Between Plants, Behaviors, and Attitudes in an Office Environment*, Final Report, Department of Horticulture, Virginia Polytechnic Institute and State University, Blacksburg, VA.

Tarquini, Joseph, 1986, Design and productivity: the two go hand in hand, *The Office* 103(June):124-175.

Wolverton, Bill, A. Johnson, and K. Bounds, 1989, *Interior Plants for Indoor Air Pollution Abatement*, Final Report, NASA, Stennis Space Center, MO.

HORTICULTURAL THERAPY

Chapter 22

Corrections and the Green Industry

Joel Flagler

SUMMARY. The New Jersey Department of Corrections has awarded a $90,000.00 grant to Rutgers University-Cook College to develop specialized training for youth correctional programs. The objective is for participants to gain horticultural skills and ultimately be trained and employable. Potential employers include all facets of the green industry–florists, landscapers, garden centers, nurseries, sports turf managers, parks and shade tree departments, and interior plantscapers.

Using a structured set of training modules, participants will have the opportunity to develop new levels of knowledge, responsibility and achievement. Participants will gain hands-on experience as they put to work the lessons learned in class. On a regular basis, students, instructors and counselors will be surveyed to gauge program effectiveness. Academic and socio-emotional progress will be monitored and documented at regular intervals. Career counseling, internships, and a job placement mechanism will serve as follow-up, helping to steer the program graduates toward success.

In the late summer of 1991 a new linkage was developed between the New Jersey Department of Corrections and Rutgers University. There was strong interest and there were funds available to expand and improve the educational programming for incarcerated youth within the correctional system. Monies dedicated to this end were not being used to their best advantage, for lack of progressive, effective programs.

The planning team assembled included the Rutgers University team, a vocational/agricultural specialist, guidance counselor, grant administrator, and Dept. of Corrections liaison. The Rutgers team consisted of faculty from the department of landscape architecture, a horticultural therapist/extension agent, and teaching assistants, plus a liaison to the Office of Continuing Professional Education. Ultimately a plan was developed and a curriculum proposed, which the Dept. of Corrections found exciting and worthwhile.

By January, 1992, only six months after the project conception, $90,000.00 in grant monies were awarded, contracts were signed, and the innovative program was scheduled to commence. Selected for the introductory class session was the New Jersey Flower and Garden Show, a spectacular display of landscape and garden plants offering myriad opportunities for education, entertainment and personal discovery. This first session set the stage for the topic of horticulture which would be the theme for the nine months to follow.

The new program at Rutgers University represents several things. It points to new directions in education and rehabilitation. It also represents important new linkages within State government that could prove highly productive. One such linkage is between the N.J. Department of Corrections and Rutgers-The State University. Another is between the Rutgers Department of Continuing Professional Education, which administers the grant, and Cooperative Extension, the department with which the author/co-investigator is affiliated.

The N.J. Dept. of Corrections specifically wished to target the training for young offenders, aged 14-17. Groups were selected from several residential and day programs around the State. The participating facilities include: Voorhees, Florence Crittenden, Union and Middlesex, representing all geographical areas of New Jersey.

The overall goal of the pilot program could be stated simply "to help participants gain horticultural skills so they can become employed." Employment often spells freedom, independence, self-sufficiency and may imply responsibility and positive productivity. Potential employers would include all facets of the green industry: florists, landscapers, garden centers, parks and shade tree departments; also, grounds and sports turf maintenance, tree farms, nurseries and interior plantscapers.

Each training module in the twenty-five week program is designed to help reach the stated goal. Each class, while building on previous lessons, stands by itself with its own set of lectures, exercises and tasks to complete. The modules are highly structured to avoid excessive breaks in the day-long program. Yet, enough flexibility is built in to allow instructors to seize a teachable moment, like a greenhouse pest problem or the surprise appearance of a hawk or woodchuck.

The curriculum planning team identified specific skills which were deemed important from an employability standpoint. With these skills the individuals might be better equipped to perform routine responsibilities. The skills we rated as being essential include: pruning, planting and transplanting (of annuals, shrubs and trees) and propagation–both vegetative and sexual. Knowledge of pest control is important, too, for combatting weeds, insects, and plant disease problems. Alternative strategies for control are presented to the students as well as the traditional pesticide approach. Awareness of proper cultural practices is another essential for any apprentice horticulturist, whether in a greenhouse, garden, or golf course. Each student must develop some understanding of the balances that exist between soil, water, oxygen and nutrients. Also needed is a functional knowledge of plant parts and processes.

In addition to horticultural skills it is clear that each individual must develop personal skills in order to be successful. Owing to their backgrounds, where family support may be minimal and temptations for trouble great, it may be necessary to first de-program some of the negative attitudes and destructive behaviors. Then it may become more possible to focus on developing new skills and positive behavior patterns.

The correctional youth have already made some major mistakes in their short lives. Thus, they need a solid character framework and positive role models if they are to access a path toward success. The character attributes identified as being necessary include: self-esteem, sense of responsibility, confidence and self-respect. Without these, the horticultural skills alone may not carry the student toward meaningful and productive employment.

TEACHING METHOD

Each of the twenty-five training modules is structured carefully, keeping in mind the population we are working with. Many of the students, ranging in age from fifteen to eighteen, have third and fourth grade reading levels. Attention spans are often limited or non-existent, and traditional teaching approaches have frequently failed them.

The correctional facilities bring the participants to the Rutgers University-Cook College campus once a week for three weeks of each month. The students arrive on time, carrying their Rutgers loose-leaf binders that are issued when a participant enters the program.

The class day begins at 9:00 am with a 45 minute lecture. Visual aids such as overhead transparencies or projected slides make the lecture more tolerable for the students. This is followed by a one and one-half hour hands-on exercise in classroom or greenhouse. The activity is directly related to the lecture theme. The slide presentation on seeds, for example, was followed by an exercise in sowing seeds of annual flowers and vegetables. The lecture on hand tools was followed by a catalog-ordering activity. With calculator in hand each student had to fill out mail order forms and mail away for $250.00 worth of tools and materials.

Lunch is a one hour break and we dine together in the campus student center. It affords a rare chance for the correctional students to mingle with bona fide college students. It is a benign, non-threatening atmosphere. The correctional youth nearly always exhibit good social behavior. Many have stated that they like feeling they are part of the University. Others express that the relaxed, non-judgmental atmosphere of the student center makes them feel good about themselves.

Following the lunch hour there is a structured outdoor activity or field trip. The theme, once again, relates back to the morning lecture. Outdoor activities include: pruning trees and shrubs, transplanting nursery stock, operating landscape equipment, planting up plots in the Rutgers Display Garden, etc. Field trips bring students to places such as: horticultural production greenhouses, sod farms, historic gardens, botanical conservatories, turf research plots and more.

While homework is not a popular item, it is recognized that the correctional students need continuity to keep them 'corrected' while they are away from the Rutgers Campus. Each of the correctional facilities has either a greenhouse, a garden or both. This allows the counselors and staff to duplicate many lessons and activities on-site.

Light homework assignments are given regularly. These include going over new vocabulary lists to familiarize students with new items and plant names. Lesson outlines are distributed a week prior to each class. This provides an opportunity to discuss and think about the lecture and activity for the upcoming day at Rutgers.

EVALUATION METHOD

The project developed by Rutgers for the State Department of Corrections is an experiment not yet one year old. There are many interested parties watching the progress of the pilot program. Of course it is not enough to develop a successful program. It is critical to document our findings and substantiate any claims for success.

Evaluation is essential to measure the program's success and be able to make any changes in methodology along the way. Ongoing assessment affords opportunities for monitoring and improving overall effectiveness. The evaluation mechanism we are using allows for three levels of feedback; by students, by correctional staff, and by Rutgers instructors.

On a monthly basis students complete surveys which give them a chance to rate the lessons, activities and instructors. Portions of the survey measure retention of new information from recent lessons. Most of it, however, asks students about their feelings and attitudes toward plants, hands-on activities and class lectures.

Other methods for recording student feedback include student journals in which daily entries are made. Also, group discussions during the 'wrap-up' at the end of each Rutgers day gives students an opportunity to speak out on what they liked or did not like about the lessons and activities.

Preliminary findings indicate a growing sense of purpose as many individuals express satisfaction and pride in what they have accomplished. Students experience positive feedback from their horticultural projects; their seedlings, their designs, their bonsai plant. Many express excitement about their new successes. A few have stated a very strong interest in possible careers or hobbies involving garden and landscape plants.

Input from correctional staff and Rutgers instructors is also an important component of the evaluation process. Monthly brainstorm sessions and conference calls allow the professionals involved to assess the program and make necessary adjustments. Further, a bi-monthly think tank including staff, instructors and student representatives affords opportunities to see the overall project from different viewpoints. Then, the team can modify and re-shape deficient aspects of the program.

In order to gauge program effectiveness, participants are monitored according to several parameters. These include: academic progress, attainment of pre-vocational skills, social and group behavior, and emotional progress. By keeping them in focus, program objectives are more likely to be met.

The evaluation mechanism starts when the individual enters the Rutgers program. Often this is shortly after the youth arrives at the correctional facility following sentencing. An entry interview determines important factors like reading levels, previous interest or exposure to horticulture, personal aversions or allergies, etc. Monthly student surveys, previously described, help measure academic and personal progress. Input from correctional staff and Rutgers instructors can work to keep the lessons and activities at the right level of difficulty and challenge. Lastly, an exit interview is conducted. At this point the individual is leaving the program and, most likely, being released from the correctional facility to return home. Each youth is informed of opportunities to stay connected to horticulture.

FOLLOW-UP

The final two pieces of the evaluation mechanism are the post-training guidance and the network of mentors and employers. It is hoped that these will help to keep the 'hook' on the program alumni and offer a course of continuity for those who desire it.

It is a widely known fact that in the United States many who leave the correctional system become repeat offenders and are re-incarcerated. It may be that they fall back into the same ruts and negative programming that landed them in trouble in the first place. The correctional system is not doing its job, if, upon release from prison, former inmates are neither better trained nor rehabilitated. For corrections to 'correct,' individuals need to be given some tools with which they can begin to build a solid foundation.

Post training guidance includes essentials like preparation for a job interview and writing a resume. Those who show interest are steered toward an internship, working as a part-time employee with a landscaper, greenhouse operator or interior plantscaper. The green industry network, now being established, will provide additional opportunities for job placement.

Upon release from corrections the youths typically return to the home of a parent, family member or guardian. They disperse to all corners of the State. Fortunately, since Rutgers is the land grant institution, it has cooperative extension faculty in every county. The post-training network includes extension agriculture and 4-H agents, taking advantage of their geographical distribution and commitment to service. Another resource, vocational-agricultural instructors, will be part of the post-training network. They, too, are available throughout the State. Together, these professionals can be available to program alumni, serving as mentors and advisors to provide support and horticultural direction. The post-training network also includes professionals in the green industry, who can serve as employers and hosts for interns and apprentices.

PITFALLS AND PROBLEMS

A major factor to contend with is the turnover of young people in the correctional facilities. Individuals are entering and leaving corrections on a weekly basis. This can be disruptive for students and instructors alike. Knowing this in advance the curriculum was designed so each class session stands independently, although connected to previous lessons. This enables a new participant to get involved right from the start, without feeling like they are lost and behind everyone else. Still it can be a challenge to orient incoming students while providing some level of program continuity. The use of video tape is being explored as a means of bringing a new student up-to-date, through all prior classes. This may prove to be a useful tool.

A second major challenge is to keep the 'hook' on the students who have come through the Rutgers Program. The positive effects of the horticultural intervention can only be sustained if the individuals can get additional training and hands-on experience. The statewide post-training network of mentors and employers will play a critical role in the follow-up process.

PROGRAM EXPANSION

Although the "Corrections and the Green Industry" program is in its infancy, the N.J. Department of Corrections and Rutgers University have turned their sights toward the next phase. Starting in February, 1993, the horticultural training is to expand and four additional facilities will have participated.

Also, three 'spin-off' training programs will be initiated, including: greenhouse management and nursery operations; animal science (including dairy and poultry management); plus environmental science, with the focus on composting and recycling.

The overall goal for next year will remain the same: to impart to the participants those skills needed to make them employable. Thus far, the project seems to be working very well. We will have to wait for the real results, as follow-up determines how many program graduates go into further studies or employment in the many facets of green industry. The individuals involved are talking about new awareness, new interests and new dreams. The comments they share on their surveys may say it best. The following are some actual student comments:

> "Until now, I didn't know I could get seeds to grow."
> "I want to build some bonsai trees."
> "I like the hands-on stuff, I like pruning."
> "I'm pretty good with the mower machine."

Interestingly, 85% of the participants stated that they were exposed to gardening or landscaping by a grandparent, aunt or uncle or a friend. A full 90% of the students felt they had gained a good amount of knowledge since entering the program. All but two agreed that the Rutgers training program might work for them.

Additional responses include:

> "I believe it is possible for me to become a landscaper, I like to create things."
>
> "When I first started I was very bored, but as I started paying attention it was very cool."
>
> "Each time I come here I gain more and more confidence in the program. It taught me a lot of basic skills that I never knew of. I have gained a lot of knowledge about plants. I'm delighted you all want me here."
>
> "It makes me feel like I can do more than just break the law–it makes me feel really good about myself."
>
> "Before the program I never thought of plants as a part of us. Now I know they are a part of us."

REFERENCES

Relf, D., 1992. The role of Horticulture in Human Well-Being and Social Development, Timber Press.

Moore, B. 1989. Growing With Gardening, The University of North Carolina Press.

Rice, J. 1992. Self Development and Horticultural Therapy In a Jail Setting, in proceedings of 2nd People-Plant Symposium, The Haworth Press, Inc. to be published 1994.

Flinn, N. 1985. The Prison Garden Book, National Gardening Association.

Chapter 23

Use of Sensory Stimulation with Alzheimer Patients in a Garden Setting

RTI, Hawthorne, New York

Maxine Jewel Kaplan

SUMMARY. Using an established, self-contained garden area and improving its accessibility with raised beds, winding hard-surfaced pathways, comfortable seating areas and a garden water fountain, a trained horticultural therapist can direct, assist and encourage Alzheimer patients to recapture some of their pleasant, early life experiences in a naturalized setting.

Old-fashioned flowers and vegetables and fragrant, textured, edible plants help stimulate their tactile, olfactory, gustatory, as well as visual senses. The subtle trickling of water from a nearby garden fountain helps reach these patients on an auditory level. Planting, walking, working the soil with their hands . . . touching, smelling, seeing, listening, tasting . . . provides a serene, non-threatening positive experience that in some way improves quality of life. Short- and long-term goals are to help Alzheimer patients to reduce their periods of agitation and aggression.

Standardized tests, such as the Philadelphia Geriatric Center Mental Status Questionnaire (PGC) that measures cognitive functioning; the Ernst Emotional Problems Questionnaire that measures affective functioning; Nursing-Chart notes reporting behavioral changes and/or number of recorded aggressive behavior incidents (+ or −); author-designed measurements and check lists; along with the use of tape recordings of first, intermediate and final horticultural therapy sessions recording number of social initiatives and/or interactions (+ or −) will be used in a collaborative effort to measure and assess whether horticultural therapy will (1) increase verbalization, (2) increase socialization, (3) stimulate long- and short-term memory, (4) improve orientation and (5) improve overall affect.

It has been said by Aristotle, one of the greatest Greek philosophers of our times, that "fire, earth, air and water" are the four basic elements of life (3:403). Gardening deals with three of these elements: earth, air and water. It has been said by Dr. Benjamin Rush, a signer of the Declaration of Independence, that he was convinced that "digging in the soil had a curative effect on the mentally ill" (10:6). There is a "primal association" between plants and people (1)–a primitiveness–a symbiosis, where one's very existence depends upon the other. This is the basis of my proposed research project: using the sensory stimulation of plants and the natural environs, earth, air and water, with Alzheimer patients to improve their quality of life in an institutional setting.

INTRODUCTION

Ruth Taylor Geriatric and Rehabilitation Institute (RTI) is part of the Westchester County Medical Center, Hawthorne, N.Y. It is a University affiliated facility with the New York Medical College located on its campus. RTI is a 425 bed skilled-nursing facility, presently with two Alzheimer Units, each housing about 35-40 patients. Its entire physical plant has been renovated, modernized and refurbished these past 9 years. The Medical Center campus sits on several hundred acres of landscaped properties.

Included in RTI's immediate outdoor environs is the "Mini-Park," a self-contained court-yard that was designed many years ago for patient use–but unfortunately, has been vastly underutilized.

My proposal for the research project is to utilize this area, since it is already fenced and gated, has some pathways, a modest amount of plant material and develop it into a full-scale sensory garden: stimulating the eyes, the ears, the nose, the fingers–one's total "being." Treatment of patients with any disorder, especially in institutionalized settings, is a multi-disciplinary effort. Good medical treatment includes psycho-social treatment as well. Hoyt said about our health-care system (7) that "The single most important concept for cross cultural studies of medicine is a radical appreciation that in all societies health care activities are basically interrelated. Consequently, they should be studied in a holistic manner as socially organized responses to disease that constitute a special culture system: the health care system." Therefore, if we can help stimulate patients to recall fond memories, fragrances, sounds, give them opportunities for productive and useful physical activity; in general, make life more pleasurable–that this positive mental state-of-being will carry over to their physiological well-being. Physical and mental functioning are known to be interdependent states (Habot and Libow 1980). It has been shown that physical impairment is often responsible for diminished performance on mental tests. Furthermore, some investigators (Ernst et al., 1977 a, b) maintain that "sensory deprivation (such as hearing losses and visual problems) produces symptoms of chronic brain syndrome" (6:71). "A person diagnosed as senile, placed in an environment with other similarly diagnosed (patients) and relieved of decision-making, responsibilities and perogatives, is likely to deteriorate rapidly" (6:72).

Allow me to digress briefly to give you some background material about patients with Alzheimer's disease, named after Alois Alzheimer who in 1907 described this disorder that now bears his name. At present, it is estimated that 40-60% of the nation's nursing home residents suffer from dementia, the majority being victims of Alzheimer's disease (2:29) (see Appendix I).

CAUSES OF DEMENTIA SYNDROME

Microscopic brain abnormalities occur in patients with Alzheimer's disease. Diagnosis is made only through autopsy upon death. It is found that when there are increasing neuritic (senile) plaques and neurofibrillary tangles, that they are in direct correspondence to the increasing severity of dementia (2:34). "Dementia means loss or impairment of mental powers." It comes from two Latin words, which mean AWAY and MIND"–it describes a group of symptoms (8:8). The disease becomes more prevalent when people reach their 80's and 90's but 80% of (the general) population never experience severe memory loss (8:10). Presently, about 4 million Americans are afflicted with Alzheimer's disease. It strikes more than 10% of the population over age 65 and nearly 50% of those 85 years or older (4). The average life expectancy of Alzheimer patients is roughly

5-10 years, but there is much variability in the rate of progression (2:25) (see Appendix II).

Patients with Alzheimer's disease have been classified into levels of senility on the basis of severity (Berger 1980). A six-interval scale extends from the least severe (Class I)–patients can still function but forgetful–to the patient confined to bed or chair and is responsive only to tactile stimuli (Class VI) (6:69-70). There are other functional rating scales used for symptoms of dementia such as the one designed by Hutton, Dippel et al. (1985) where scores would range from zero (no impairment) to 42 (severe impairment) (2:21-23) (see Appendix III).

Oftentimes, depression will play a role in the Alzheimer patient's functioning and orientation (6:107-109). Social functioning is indirectly related to mental status (6:141). Information about the patients baseline personality, prior to the onset of Alzheimer's, is important in explaining some of the patient's health behavior or his response to treatments. "Irritability," "bothersomeness" or "socialability" may be a stable personality trait that can create confusion on measurement scales (6:73-74).

Social function is correlated with physical and mental functioning. Activity levels, socialization patterns or morale can be directly associated with sensory deprivation, depression and diminished ability to perform tasks. Again, this suggests the HOLISTIC approach to treatment–physical, emotional and social approaches that directly relate to diagnosis and treatment (6:133).

It must be stressed that Alzheimer patients' abilities may fluctuate from day to day, hour to hour; personality changes occur due to brain damage beyond their control (8:13). It has been said of patients with Alzheimer's that "life may be like constantly coming into the middle of a movie: one has no idea what happened just before what is happening now" (8:35). Awareness of time, place and self-recognition are forgotten in that order (2:101).

Alzheimer patients are more apt to have catastrophic reactions–that is, an extreme reaction to an insignificant or minor stimulus. In handling and treatment of these patients, it is necessary to keep the tasks and environment simple (8:39)–not to "overload" their system. A garden setting can be a soothing environment for them.

Patients may feel tense, embarrassed or even worried about their clumsiness and forgetfulness (8:50). The most terrible loss of all is memory–losing one's day-to-day connections with others and one's past (8:56). Perhaps a garden can help to recall that connection.

"Muscle weakness or stiffness may occur when a person does not move around much. Exercise is important for memory impaired people" (8:52). "Routine exercise benefits people of all ages, particularly older individuals. What is more important to note that health benefits can be shown on relatively LOW levels of activity" (2:39). Gardening can easily and productively incorporate physical exercise. A quote of Berry N. Squyres from a paper on "Exercise and Aging":

> A major task of caregivers is to encourage activity and autonomy in Alzheimer patients. One care giver was very pleased that her husband continued to enjoy gardening. One day, however, a neighbor's dog dug up the plants he had just put out. As Alzheimer patients will sometimes do, when events happen that they cannot understand, this man came to an erroneous conclusion. He decided that his neighbor had dug up his plants. Using the premise that turnabout is fair play, he went next door and removed his neighbor's plants! Gardening continued to be an important source of exercise and activity for this man, and the neighbors forgave him (2:38). Previously enjoyed activities may remain important and enjoyable even for seriously impaired people. (8:88)

"Everyone enjoys experiencing things through his senses." A brilliant sunset, the smelling of a fragrant flower or the tasting of a favorite food; touching a furry animal, a smooth piece of wood or putting a hand under running water (8:89)–all previously enjoyed activities that are no longer available to patients because of their isolating environment. What better opportunity for these experiences than a sensory garden!

Some patients develop an agitated, determined pacing; can even become violent. This can be their way of communicating: a desire to go home (to the familiar); expressing restlessness, boredom, the need for exercise, to be "doing something" (8:132-133). A garden once again can be the opportunity for them to reminisce about home; to walk; to till the soil and even sit, while still feeling "life" about them.

There is some unexplained "worsening" of behavior in the evenings–perhaps a decrease of light or decreased tolerance for stress at the end of the day. Patients may have a symptom referred to by health care professionals as "sun-downing"–a person who becomes confused late in the day, i.e., when the sun goes down" (2:33). Patients experience sleep disturbances because of damage to their internal clock and will sometimes doze during the day if permitted. This lethargy can also be caused by lack of ample exercise (8:140-141). Mace and Rabins wrote in "The 36-Hour Day" that "once a person gets started doing something, she may begin to feel less apathetic (and lethargic). Perhaps she can peel only one potato today. Tomorrow she may like doing two. Perhaps she can spade the garden. Even if she spades for only a few minutes, it may have helped her to get moving. . . " (8:167). The use of grow-lights can be used in patient indoor areas to increase amount of light received during daylight hours to facilitate "daytime" orientation and adjust that internal clock.

Self-stimulatory behaviors (nose-picking, skin scratching, rocking, masturbation, etc.) are self-reinforcing and usually indicate inadequate environmental stimulation (2:115). Reality orientation, behavior modification, reinforcers, reminiscence therapy are all continual, repetitive reminders that keep patients stimulated and increases orientation (2:94-111).

"Behavior therapy holds no cures but it promises great improvement in the day-to-day management of Alzheimer patients and other dementia symptoms" (2:91).

GOALS AND OBJECTIVES

The goal of this proposed research project is to improve the manageability of patients with Alzheimer's disease in a long-term health care facility by the use of sensory stimulation in a garden setting.

METHODS AND MATERIALS

This proposed research project would be working with patients that have increased memory losses; decreased sensory stimulation; lack of adequate physical exercise; "sun-downing" and loss of orientation. Horticultural Therapy would be a "natural" treatment modality. My idea was born (concept), matured (put into written form) and culminated with this paper presentation–hopefully to be reborn with outside funding to create my Sensory Garden.

When funding does become available, a part-time Horticultural Therapist and Assistant would be hired 2 1/2 days/week specifically for this therapy program. THE CONSTRUCTION PHASE would include: a garden-site with several raised-beds, accessible both to wheel-chair and ambulatory patients; the placement of trellised arbors with seating, along the pathways; and the placement of a garden water fountain with some semi-secluded seating areas adjoining it. THE GARDENING PHASE would include: the plantings of some semi-deciduous and evergreen shrubs to aesthetically enhance the overall appearance of the Mini-Park; in addition, plantings of some old-fashioned perennials such as peonies, hollyhocks, honeysuckle, iris, lilies, lambs ear, herbs, chicks and hens, etc., etc.–would be placed for visual, tactile, olfactory as well as memory stimulation.

THE PROGRAM DESIGN

Initially, only 6-8 patients would be serviced. It is important that patients, especially Alzheimer patients, be closely supervised with their therapy; hence the small ratio of staff to patients. Volunteers would be additionally solicited. The initial group of patients would be carefully selected by the research criteria–individuals with moderate impairment.

The therapist/doctor/social worker would initially test, with standardized tests, each patient to assess general mental status: either the Ernst Emotional Problems Questionnaire (10 questions) (see Appendix IV) or the Philadelphia Geriatric Center (PGC) Morale Scale (22 questions) (see Appendix V) would be used. Both were selected because of their briefness and could be administered by the interviewer. They were not dependent upon the literacy or the ability for the patient to self administer. Further, a base-line of personality and behavior patterns would be obtained from patients' charts.

An outdoor Horticultural Therapy program would operate three times per week, May through October, depending upon weather conditions. Patients would be assisted out to the garden area and perhaps the first several times, would only familiarize themselves with this new setting. With the assistance of Staff and Volunteers, they could walk, touch, feel, smell, see and hear their new surroundings. "What do you see? What do you hear? What do you smell? What do you feel? Do you remember . . . ?" says the therapist. The twittering sounds of the birds, the soothing warmth of the sun, the sweet smell of freshly cut grass, the recollection of pleasurable glimpses into their past–senses that could never be duplicated within the four walls of a building–would be further enhanced by the encouraging words and assistance of staffing. These sessions at first might only last 15-30 minutes but gradually increasing to full sessions of 45 minutes to 1 hour.

Ultimately, planting of plants, cultivating and watering would soon be incorporated into their activities. Each person (patient) would be allowed to participate to whatever level of participation they felt comfortable. Much encouragement and praise would be necessary. The aesthetic quality of their gardening efforts would not be a primary goal, only their heightening awareness of their interaction between themselves and the natural world immediately surrounding them. Some patients might be happy to just walk the encircling paths; others might choose to sit in this newly discovered naturalized setting and "listen," while yet others might choose to "till" the soil.

The Horticultural Therapist would lead the group not only in the gardening aspects of this program, but by talking to them about the flowers, the seasons, the weather . . . and all that horticultural therapy involves, would help to make them more mentally alert, physically stronger and emotionally happier.

After each session, the Horticultural Therapist would be responsible for all record keeping and note charting. She would fill out the "Behavioral Assessment Check List" (see Appendix VI) for each patient and note any specific behavioral or mental changes for later assessment. If possible to utilize tape recordings of sessions (but in its absence, clinical observations) to measure (+) or (−) social interactions and/or initiatives; also to note any increase or decrease in aggressive or other undesirable behavior patterns. The same standardized tests that were used in the initial assessment would be administered at the termination of the outdoor programming (October) to note any changes.

An indoor program can be brought in when weather necessitates to carry through the sensory stimulation theory. Fragrant herbs, perfumes, foods, water and other natural materials can be used. Words, sounds, songs, pictures and games all about nature can be bought or made specifically for this projects use. The same time-frame and staffing would be needed.

It is important for Staffing to treat their confused patients with respect and dignity. At all times, to remember we are dealing with adults whom should not be talked down to or treated as a child. Likewise our activities should be adult activities brought down to their levels of functioning–not children's activities given to adults.

CONCLUSION

Institutionalized patients, particularly such patients as Alzheimer's, who are confined to their Nursing Units, experience a deprivation of all natural sensory stimuli–fresh air, wind, sun; the harmonious sounds of nature–the blends of the birds twittering, the leaves rustling and the trickling of the waters running. At present, there has been little research specifically dealing with Alzheimer patients and a horticultural therapy program. Based on evidence of horticultural therapy treatment of patients with other mental and physical disabilities, I conclude that Alzheimer patients will experience reduced agitation and aggression through the use of sensory stimulation and horticultural therapy.

REFERENCES

1. Conklin, E., 1973 "Minus plants and flowers in a malcontent," Florist & Nursery Exchange (Jan.);1-3, 8.

2. Dippel, Raye Lynne, Ph.D. & Hutton, Thomas J., M.D., Ph.D. 1988 "Caring for the Alzheimer Patient," Prometheus Books, Buffalo, N.Y.

3. Encyclopedia Britannica Inc., 1952 "The Works of Aristotle," Chicago, London, Toronto, Vol. 1.

4. Gannett Suburban Newspapers "Controversial Alzheimer's drug to be available," Dec. 3, 1991.

5. Judd, Mary 1971 "Why bother. He's Old and confused," Winnipeg Municipal Hospital, Manitoba.

6. Kane, Rosalie A. and Kane, Robert L. 1981 "Assessing the Elderly," Lexington Books, Lexington, Mass.

7. Kleinman, A. 1980 "Patients and healers in the context of culture: an exploration of the borderland between anthropology, medicine and psychiatry," University of California Press, Berkeley, Calif.

8. Mace, Nancy L. and Robins, Peter V., M.D. 1984 "The 36-Hour Day," Warner Books, New York, N.Y.

9. Relf, Diane, Ph.D. "Dynamics of Horticultural Therapy," Rehabilitation Literature, May-June, 1981, Vol. 24, No. 5-6.

10. Sullivan, Mary, HTR Journal of Community Gardening, Spring 1985.

APPENDIX I

TABLE I Causes of Dementia Syndrome
Alzheimer's disease
Pick's disease
Multi-infarct dementia
Bilateral subdural hematomas (blood clots on the brain)
Brain tumors, especially involving the frontal lobes
Chronic fungal or tuberculous meningitis
Kidney failure
Liver failure
Electrolyte imbalance
Drug overdosage
Hypo- or hyperthyroidism
Hydrocephalus
Huntington's disease
Late Multiple Sclerosis
Late Parkinson's disease
Post-traumatic brain injury
Depression
Chronic Alcoholism
Pernicious anemia
Neurosyphilis
Creutzfeldt-Jakob disease
AIDS

Reprinted with permission by *Texas Medicine* and Prometheus Books.

APPENDIX II

FIGURE 1. Comparison of rates of progression between two Alzheimer groups whose members differed in dementia severity at onset of study. Average time of nursing home admission for each group is indicated by an X.

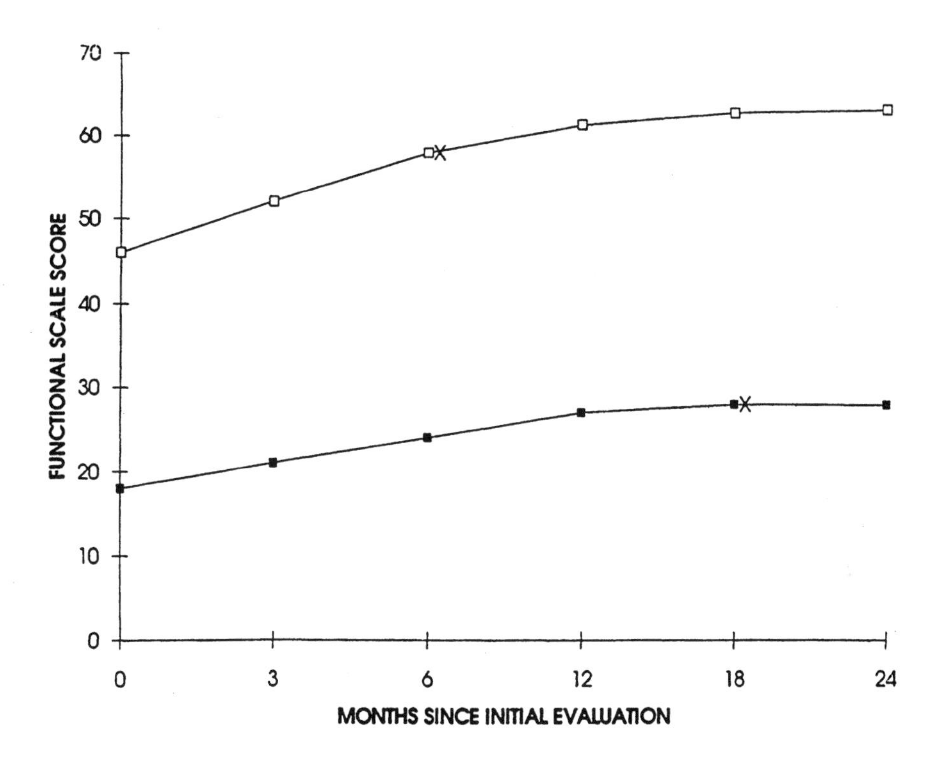

Hutton, Dippel, Loewenson, Mortimer, and Christians, 1985. Reprinted with permission.

APPENDIX III

TABLE I
Functional Rating Scale for Symptoms of Dementia

Instructions

1. The scale must be administered to the most knowledgeable informant available. This usually is a spouse or close relative.

2. The scale should be read to the informant one category at a time. The informant is presented the description for behavior in each category. The informant is read each of the responses beginning with zero response. All responses should be read before the informant endorses the highest number response which best describes the behavior of the patient.

3. Responses obtained for each category are summed to give an overall score for functional rating of symptoms of dementia.

Circle the highest number in each category that best describes behavior during the last three months.

Eating:

0 Eats neatly using appropriate utensils
1 Eats messily, has some difficulty with utensils
2 Able to eat solid foods (e.g., fruits, crackers, cookies) with hands only
3 Has to be fed

Dressing:

0 Able to dress appropriately without help
1 Able to dress self with occasional mismatched socks, disarranged buttons or laces
2 Dresses out of sequence, forgets items, or wears sleeping garments with street clothes–needs supervision
3 Unable to dress alone, appears undressed in inappropriate situations

Continence:

0 Complete sphincter control
1 Occasional bed wetting

Facial Recognition:

0 Can recognize faces of recent acquaintances
1 Cannot recognize faces of recent acquaintances
2 Cannot recognize faces of relatives or close friends
3 Cannot recognize spouse or constant living companion

Hygiene and Grooming:

0 Generally neat and clean
1 Ignores grooming (e.g., does not brush teeth and hair, shave)
2 Does not bathe regularly
3 Has to be bathed and groomed

Emotionality:

0 Unchanged from normal
1 Mild change in emotional responsiveness–slightly more irritable or more passive, diminished sense of humor, mild depression
2 Moderate change in emotional responsiveness–growing apathy, increased rigidity, despondent, angry outbursts, cries easily
3 Impaired emotional control–unstable, rapid cycling or laughing in inappropriate situations, violent outbursts

Social Responsiveness:

0 Unchanged from previous "normal"
1 Tendency to dwell in the past, lack of proper association for present situation
2 Lack of regard for feelings of others, quarrelsome, irritable
3 Inappropriate sexual acting out or antisocial behavior

Sleep Patterns:

0 Unchanged from previous, "normal"
1 Sleeps, noticeably more or less normal
2 Restless, nightmares, disturbed sleep, increased wakefulness
3 Up wandering for all or most of the night, inability to sleep

Verbal Communication:

0 Speaks normally
1 Minor difficulties with speech or word-finding difficulties
2 Able to carry out simple, uncomplicated conversations
3 Unable to speak coherently

APPENDIX III (continued)

Memory for Names:

0 Usually remembers names of meaningful acquaintances
1 Cannot recall names of acquaintances or distant relatives
2 Cannot recall names of close friends or relatives
3 Cannot recall names of spouse or other living partner

Memory for Events:

0 Cannot recall details and sequences of recent experiences
1 Cannot recall details or sequences of recent events
2 Cannot recall entire events (e.g., recent outings, visits of relatives or friends) without prompting
3 Cannot recall entire events even with prompting

Mental Alertness:

0 Usually alert, attentive to environment
1 Easily distractible, mind wanders
2 Frequently asks the same questions over and over
3 Cannot maintain attention while watching television

Global Confusion:

0 Appropriately responsive to environment
1 Nocturnal confusion or confusion upon awakening
2 Periodic confusion during daytime
3 Nearly always quite confused

Spatial Orientations:

0 Oriented, able to find his/her bearings
1 Spatial confusion when driving or riding in local community
2 Gets lost when walking in neighborhood
3 Gets lost when in own home or in hospital ward

APPENDIX IV

Emotional Problem Questionnaire[c]

1. How do you sleep (well, so-so, badly)?
2. How are your bowel movements (constipated, OK, sometimes constipated)?
3. Is it hard for you to get up in the morning?
4. Do you feel better in the afternoon or the morning?
5. How do you feel generally?
6. Do you think people like you?
7. Do you think people hate you?
8. Do you think that people are trying to run your life?
9. Are you sociable?
10. Are you lonely?

APPENDIX V

Philadelphia Geriatric Center Morale Scale

1. Things keep getting worse as I get older. (No)[a]
2. I have as much pep as I did last year. (Yes)
3. How much do you feel lonely? (Not much)[b]
4. Little things bother me more this year. (No)
5. I see enough of my friends and relatives (Yes)[b]
6. As you get older, you are less useful. (No)
7. If you could live where you wanted, where would you live? (Here)[b,c]
8. I sometimes worry so much that I can't sleep. (No)
9. As I get older, things are (better, worse, the same) than/as I thought they'd be. (Better)
10. I sometimes feel that life isn't worth living. (No)
11. I am happy now as I was when I was younger. (Yes)
12. Most days I have plenty to do. (No)[b,c]
13. I have a lot to be sad about. (No)
14. People had it better in the old days. (No)[b,c]
15. I am afraid of a lot of things. (No)
16. My health is (good, not so good). (Good)[b]
17. I get mad more than I used to. (No)
18. Life is hard for me most of the time. (No)
19. How satisfied are you with your life today? (Satisfied)
20. I take things hard. (No)
21. A person has to live for today and not worry about tomorrow. (Yes)[b,c]
22. I get upset easily. (No)

APPENDIX VI

Behavioral Assessment Checklist

Patient's Name:

Treatment of Group:

Date:

Patient participates:

- immediately
- frequently
- moderately often
- seldom
- not at all

Patient focuses on tasks and activities:

- immediately
- frequently
- moderately often
- seldom
- not at all

Patient interacts with group members:

- immediately
- frequently
- moderately often
- seldom
- not at all

Patient's affect:

- bright
- flat
- labile
- depressed
- angry
- improves every session

Special comments:

Chapter 24

Measuring the Quality of Treatment in Horticultural Therapy Groups

Martha C. Straus

SUMMARY. The mental health-care system is in a state of change. Hospitals are finding the need to provide more thorough and more rapid treatment with less resources. Lengths of stay are reduced, and the need to provide effective treatment quickly has increased. As part of the quality assurance program, the Rehabilitative Therapies committee designed and implemented a four month study to measure patients' perception of their treatment groups run by the Rehabilitative Therapies Department. The data is currently under analysis. Horticultural Therapy groups have been separated and are being reviewed and compared with other Rehabilitative Therapies groups.

INTRODUCTION

Friends Hospital, the first non-profit short term hospital, was founded in 1813. The hospital is the first private institution in the country with the specific mission of caring for the mentally ill. Today Friends Hospital adheres to the principles of moral treatment, combining them with a modern medical approach to the treatment of mental and emotional illness. The 192 bed in-patient program and the 24 patient Day Hospital program are accredited with JCAHO.

JCAHO, the Joint Commission on Accreditation of Health Care Organization, accredits hospitals through a standardized system of review based on inspection. JCAHO is a private, voluntary, nonprofit organization comprised of representatives from several medical associations. JCAHO standards have been adopted by other accrediting or licensing organizations for treatment and accreditation purposed (AOTA, 1978). Accreditation helps insure a patient base, reimbursement to the hospital, and a certain level or standard of services to the patients.

JCAHO requires a Quality Assurance (QA) program which addresses effective mechanisms for review and evaluation of patient care. The Quality Assurance plan describes how the hospital will meet its QA objectives. The objectives of a QA program should include assessment of patient programming and the assurance that patients receive appropriate treatment in a safe and timely manner, that staff and services are monitored, and to coordinate information throughout the hospital. For staffing, QA can be used for credentialing, to identify educational needs, and to reduce liability risks (NAPPH, 1987).

Friends Hospital has a multi-departmental Quality Assurance Committee, and as part of the hospital's program to assess and improve the quality of biopsychosocial rehabilitation, each department also has its own Quality Assurance committee. Services are monitored and evaluated in accordance with established procedures. Each department reports to the main multi-disciplinary hospital committee quarterly. It is essential that monitoring and evaluation of information on important aspects of care and clinical activities be coordinated and integrated.

THE SURVEY

This study was designed and based on important aspects in monitoring and program evaluation recommended by NAPPH. They include establishing the indicator or clinical activity to be evaluated, designing measurable criteria for each indicator, planning who will collect the data, for how long, and how the data will be evaluated (NAPPH, 1987).

REHABILITATIVE THERAPIES MONITOR

The QA committee decided to monitor our treatment groups in order to assess therapist accountability and to establish if groups achieve the goals designated in the treatment plan. Therapists were asked to identify the top three parameters that they felt made what they do with patients in their group a success. For example, feeling accepted, having a successful experience, or feeling safe to express feelings. A questionnaire was written by the committee and reviewed by the department and the hospital's QA director. The questionnaire also coincided with the "Level of impairment" study the hospital was participating in. This study used a level of impairment index with 11 impairments (Namerow, 1988) and all treatment goals were to be placed into one of these areas. Questions were designed to measure social interaction, self-esteem, ability to function independently, expression of feelings and the therapeutic impact of the group within the treatment plan. Patients answered 14 questions on a scale of 1–never to 5–always. Six optional questions were to be circled at the end if the patient felt they applied to their progress. Therapists were to note the length of time the patient had been in the group to assess changes in treatment and patient's awareness of their therapy.

The survey was pre-tested on one group. Therapists were to administer the questionnaire after 3 sessions in the group and then before discharge to determine if group experience changed as time in treatment lengthened. Each therapist was to have 20 patients surveyed. This became impossible to collect. Patients were being transferred into and out of groups quickly, discharges were happening without prior notice, and trying to keep surveys separate so they could be matched up, yet confidential, was a difficult task. The committee stopped the survey and reorganized the entire procedure. The survey was to be handed out one day a week by all therapists running groups on that day. The survey now included number of weeks in the group in order to assess the impact of the group over time. It was distributed from June to September. Difficulty remained with patients not filling out the questionnaire completely which made the test invalid, and with therapists not administering them correctly, or at all. Printing was impossible for most geriatrics to read and frequently patients did not have their reading glasses with them. Most patients have 2 groups daily and were tired of filling out the questionnaire after several weeks.[1,2] Some patients filled them out without really thinking through the questions, and some filled them out with all 5's to please their therapist (see Patient Questionnaire).

OUTCOMES

Across the board, scores are lowest when patients are first placed and tested in groups. Their satisfaction peaks around the third week, and then satisfaction goes down from week 5-7. Certain questions scored differently in task oriented versus interactional groups. We are still evaluating the data. With an average of 20 groups per day, and 8 patients per group for 4 months, we have over 1600 surveys to evaluate.

CONCLUSIONS

Since we have not finished examining the data, conclusions are still in the preliminary stages. Several conclusions may be made with the identification that satisfaction peaks around week 3 (see Post-Questionnaire Analysis Form).

1. Our groups are designed to be short term and three weeks is the maximum a patient may improve in a group.
2. After 3 weeks in a group it is time to upgrade the treatment plan and move patients on to another group.
3. Around the 3 week mark, patients begin to get restless and want to go home and start to feel general dissatisfaction in order to separate themselves from the hospital and their groups. Talking about these feelings within the group may help improve the quality of treatment in the later weeks.
4. Groups differ in focus and intent. Some offer more support than others. Some groups are sedentary, others require sharing or parallel participation. In the Horticultural Therapy Skills Group, patients have the opportunity to work independently in any area of the program, greenhouse, garden, or potting shed, and may not have the chance to interact with others.

We hope to have the data analyzed by the end of 1992. The information from these surveys will be used to evaluate patients' perceptions of their Rehabilitation Therapies groups and if we are meeting designated treatment goals and objectives.

REFERENCES

Namerow & Associates, Inc. The Level of Impairment Index. Pilot Testing, 1989. Baltimore, MD.

The American Occupational Therapy Association, Inc. 1978. Manual on Administration. Kendall/Hunt Publishing Company, Dubuque, Iowa.

The National Association of Private Psychiatric Hospitals, January 1987, NAPPH Guide to Quality Assurance Manual. Education and Research Foundation, Washington, D.C.

PATIENT QUESTIONNAIRE

It is the goal of Friends Hospital to provide the best possible care for our patients. By completing this questionnaire, you will be helping us to evaluate the quality of our program.

Thank You.

Today's Date ____

Group Title

(Please circle 1)
Number of Weeks in this group: 0 1 2 3 4 5 ______ (more)

Instructions: Please circle the number that rates your response to question.

	Never				Always
1. I feel I am a member of this group.	1	2	3	4	5
2. I feel accepted by other group members.	1	2	3	4	5
3. I feel able to talk in this group.	1	2	3	4	5
4. I am able to express my opinions in the group.	1	2	3	4	5
5. The group leader's direction of the group helps me understand why I am in the group.	1	2	3	4	5
6. My thinking becomes clearer when I am working in this group.	1	2	3	4	5
7. My outlook is more positive when I am in the group.	1	2	3	4	5
8. I feel more energetic when I am doing something.	1	2	3	4	5
9. I feel less alone when I am in a group like this.	1	2	3	4	5
10. I am able to do things for myself in this group.	1	2	3	4	5

PATIENT QUESTIONNAIRE (continued)

	Never			Always	
11. I can ask for assistance in this group.	1	2	3	4	5
12. I think this group helps me prepare for leaving the hospital.	1	2	3	4	5
13. I feel I can discuss changes in my treatment plan and groups with the therapist.	1	2	3	4	5
14. This group is an important part of my overall treatment.	1	2	3	4	5

Please circle as many of the following as apply:

This group has helped me improve:

1. My self-confidence with other people.
2. My ability to accomplish something.
3. My understanding and expression of my feelings.
4. My attitude towards my work.
5. My control of my moods and impulses.
6. My ability to function every day.

POST-QUESTIONNAIRE ANALYSIS FORM

Directions: *Please enter the rating next to the corresponding question # according to the number of weeks the patient has been in the group. A patient who has been in the group for 1 week and rates question #5 a "3" would be placed under column marked "1" and next to row marked "5" with the #3 for their rating. Enter the name of the group being rated and your name as the rater.*

Group Name: *HORTICULTURAL THERAPY 2:35*

Rater: *Colleen*

Question #	Impairment #	Number Of Weeks In Group				score sheet had 1 low score for 4 & 5			Avg of Ratings
		0	1	2	3	4	5	6+	
1	10	4	4.2	4.5	4.5	5	4		4.3
2	10	4	4.2	4.3	4.5	5	3		4.25
3	5	3.5	4.6	4.1	4.5	5	3		4.17
4	5,10	3.5	4.4	4	4.5	5	5		4.1
5	5	3.5	4.6	4.2	4.5	5	4		4.2
6	5	3	4.2	3.3	5	4	3		3.87
7	5	3.5	4.2	3.8	5	4	4		4.25
8	5	3.5	4	3.8	5	5	4		4.07
9	5	3.5	4.4	3.8	4	4	3		3.92

POST-QUESTIONNAIRE ANALYSIS FORM (continued)

		Number Of Weeks In Group				*score sheet had 1 low score for 4 & 5*			
Question #	*Impairment #*	*0*	*1*	*2*	*3*	*4*	*5*	*6+*	*Avg of Ratings*
10	*10*	*3.5*	*4.4*	*3.8*	*4*	*4*	*3*		*3.92*
11	*10*	*4*	*4.8*	*4.3*	*5*	*5*	*4*		*4.52*
12	*10*	*3.5*	*4.2*	*3.5*	*4*	*5*	*3*		*3.8*
13	*5,8*	*5*	*2.4*	*4.1*	*4*	*4*	*3*		*3.87*
14	*8,10*	*4*	*2.6*	*3.6*	*5*	*5*	*4*		*3.8*
Avg of Ratings		*3.71*	*4.1*	*3.9*	*4.60*				*4.07*
				PART II					
1	*10*								
2	*10*	*3*	*4*	*6*	*1*	*1*	*1*		
3	*5*	*4*	*3*		*1*	*1*	*1*		
4	*10*	*3*	*2*		*1*	*1*	*1*		
5	*5*	*3*	*2*		*1*	*1*	*1*		
6	*10*	*4*	*4*		*1*	*1*			

Chapter 25

Surveying the Therapeutic Landscape: A Quest for Cases of Outdoor Therapy Settings

Jean Stephans Kavanagh
Thomas A. Musiak

SUMMARY. This paper is an initial report on the nationwide survey of outdoor facilities of horticultural therapy programs conducted from Texas Tech University. This report is intended to encourage discussions which explore areas of future research into the optimum physical design of outdoor plant-oriented therapeutic landscapes.

The premise for these discussions is found in the perception of therapeutic landscapes therapy as vernacular landscapes. Yet, as places of personal expression and significance, these ad hoc gardens often are found within institutions and organizations having artful, even expensively, designed and maintained grounds. The nature and forms of the purposefully therapeutic gardens as they are currently and as they might become, pose questions which can profoundly affect the quality and likelihood of plant-people connections.

BACKGROUND

Understanding the nature and extent of both outdoor horticultural therapy facilities and outdoor recreational horticultural settings is an important component in the development of future research into the optimum physical design of plant oriented therapeutic landscapes (Relf, 1992). The frequency and type of horticultural therapy activities in outdoor settings today is relatively undocumented and, to some extent, remains a matter of conjecture.

Site visit observations suggest that existing horticultural therapy facilities are essentially vernacular landscapes. That is to say, they are landscapes derived from the gradual accretion of effects and from adaptations to immediate and perceived program needs. The unpolished nature of these gardens is particularly perplexing as they are often found within institutions and organizations which have stressed intensively designed and maintained grounds in other components of their physical plant. Observation alone cannot provide insight into the purposefulness of these vernacular qualities in therapeutic landscapes, since questions about the relationship between client outcomes and visually unstructured therapeutic landscape settings remain unanswered.

THE SURVEY

A copy of the Therapeutic Landscapes Survey is included in an appendix to this article. This survey design is intended to aid the development of design guidelines for therapeutic landscapes (Kavanagh and Musiak, 1991) by providing information about the forms, genesis and nature of purposely therapeutic gardens in the United States (Kavanagh and Musiak, 1992). The questionnaire (see Appendix) solicits information about the institutional setting, the design process, and plans for future therapeutic landscapes within the institution. This information will provide a basis for future discussions and for research which can profoundly effect the ease with which people interact with plant environments.

The therapeutic landscape survey was originally initiated to identify positive examples of therapeutic landscapes within the United States. As the survey instrument has matured and as research goals have focused, the therapeutic landscape questionnaire has broadened to include three general areas of information. These three purposes have derived from the realization that the ad hoc qualities identified during limited site observations of existing landscapes have indeed evolved over extended periods of time. However, current stress on environmental and social awareness has brought about an increasing recognition of plants and horticulture as contributors to positive therapeutic activity. The demand for such planted therapeutic settings is tending to create a need for "instant landscapes."

These complete gardens are designed consciously, in lieu of landscapes developed in incremental phases over a period of years.

The survey design was revised to elicit information about factors affecting these landscapes. First, as originally planned, the questionnaire inventories the field of horticultural therapy as the initial step in identifying, describing, and documenting positive examples of outdoor horticultural therapy settings. Secondly, another purpose of the questionnaire, to identify nationwide trends in plant related therapy settings, emerged from the desire to define "effectiveness" in the landscape and horticultural therapy program merger. Third, specific data about therapeutic landscapes through the identification of physical facilities and factors continued to be necessary to formulate guidelines for optimum design of therapeutic landscapes.

During the development of the survey instrument, as questions arose about the nature of the field, lack of information led to the observation that few horticultural therapy landscapes have been documented. To date, actual data collection has been minimal. A more rigorous approach to collecting and evaluating data about the sites is required if progress in the field is expected. The survey furthers documentation of therapeutic landscapes as a step in improving the design of such therapeutic landscapes.

The Therapeutic Landscapes Survey questions are organized into three logical blocks. This organization is based upon the landscape architect's five-part design process sequence of (1) programming, (2) planning, (3) design, (4) implementation, and (5) post-occupancy evaluation. This design process, by including several phases or stages, serves as a systematic way of developing proposed landscape projects (Simonds, 1983).

For the purposes of this survey, only the first three stages of the design process have been examined. In the first stage, the development of a program, the program–written or unwritten–lists all requirements, desires, resources, and limitations anticipated in the therapeutic landscape. This conscious identification of the desired objectives, clients, therapies, facilities, and psychological effects is necessary if effective treatment is to be achieved through horticulture in the landscape. Therefore, the first division of survey questions seek to identify common program factors and influences.

In the second stage of the design process–the planning and site selection stage–design team participants select and employ the criteria used in the facility planning effort. Often those factors affecting the location of proposed facilities are particularly influential. In this stage, seeking optimum locations for gardens, garden structures and amenities, and avoiding poor locations and interrelationships are major objectives. As a result, the second division of survey questions explores the relationship between the therapeutic landscape and its site.

The third phase of the design process–the design phase–includes layout and shape or form-giving by emphasizing those components and arrangements which result in organization of the texture, color, and scale of elements incorporated into the landscape. The third division of the survey attempts to identify these as general design characteristics.

The two other design process phases not evaluated in this study are the implementation and the post-occupancy phases. While guidelines can be developed for optimum construction and evaluation, study of these phases was postponed in favor of a distinct focus on the three most significant design and form-giving stages.

CONCLUSION

This survey, as originally defined, has expanded from one in which the development of design guidelines was of primary concern to one in which further identification of needs and problem-solving for therapeutic landscape settings can be encouraged through the research process. At this time, the study continues to explore status of therapeutic landscapes in the United States. The survey (see Appendix) has been mailed to AHTA members. Further phases of the survey are anticipated for related health professionals' organizations. As the responses are returned, it is expected that accumulated knowledge of existing therapeutic landscapes will result in substantial information which can be assimilated into guidelines for the planning and design of horticultural therapy landscapes.

REFERENCE LIST

Kavanagh, Jean S., and Musiak, Thomas A. *Design and Planning of Horticultural Therapy Landscapes*. Paper presented to the AHTA at its annual meeting in Philadelphia, PA. 1991.

Kavanagh, Jean S., and Musiak, Thomas A. *Therapeutic Landscapes Survey.* Questionnaire. Lubbock, TX. 1992.

Relf, Diane, ed. *The Role Of Horticulture In Human Well-Being And Social Development*. Portland: Timber Press. 1992.

Simonds, John O. *Landscape Architecture* Second Ed. New York: McGraw-Hill Book Co. 1983.

ADDITIONAL SOURCES

Bennett, Corwin. *Spaces for People: Human Factors in Design*. Englewood Cliffs, NJ: Prentice-Hall, Inc. 1977.

Cooper Marcus, Clare, and Francis, Carolyn, eds. *People Places: Design Guidelines for Urban Open Space*. New York: Van Nostrand Reinhold. 1990.

Diffrient, Niels, Tilley, Alvin R., Bardagjy, Joan C. *Humanscale 1/2/3*. Cambridge: The MIT Press. 1974.

Kaplan, Stephen and Kaplan, Rachel. *Humanscape: Environments For People*. North Scituate, MA: Druxbury Press. 1978.

Laurie, Michael. *An Introduction to Landscape Architecture*. New York: American Elsevier Pub. Co. 1975.

APPENDIX

AHTA

THERAPEUTIC LANDSCAPES SURVEY

Please help us in our survey of exterior landscapes used in therapies. This survey attempts to identify the type and location of these outdoor horticultural therapy settings (therapeutic landscapes) in the United States. Questions are asked about the respondent, the institution within which the therapeutic landscape functions, the process of design and planning leading to the eventual development of the therapeutic landscape, and the landscape elements provided for the therapy setting. Completing this survey should require less than 15 minutes. Please try to return it by September 15, 1992.

We hope that you will complete as much of the questionnaire as you can, **even if your institution offers no such therapy or therapeutic landscape.** Thank you for your kind assistance in our task!

Jean Stephans Kavanagh, ASLA and
Thomas A. Musiak, ASLA
Department of Landscape Architecture
Box 42121
Texas Tech University
Lubbock, Texas 79410

* **Does your program employ any outdoor horticultural therapy facilities?** Yes No*

*** If you answer "No", please provide the information below and answer questions 1 - 5 before returning this survey. If "Yes", please complete the survey. Thank you for your help!**

General Information:

Name of respondent ______________________ Position ______________

Type of therapy or allied health practice ______________ Office Phone ____/__________

Name of Institution ______________________________________

Address of Institution ______________________________________

____

I. HT PROGRAM BACKGROUND

Institutional Data

1. Please circle the most appropriate description of your institution:

Critical care hospital	Convalescent/hospice
Rehabilitation hospital	Outpatient "clinic"
Psychiatric hospital	Residential care
General purpose hospital	Other (Please specify:) ____________

2. Please identify the disabilities and life stage(s) your institution serves:

DISABILITY	AGE RANGE	DISABILITY	AGE RANGE
Altzheimers	___ to ___	Psychiatric	___ to ___
Dementia	___ to ___	Terminal illness	___ to ___
Head injuries	___ to ___	Vision	___ to ___
Learning	___ to ___	Other: (specify)	
Movement	___ to ___	____________	___ to ___
Pre& post operative	___ to ___	____________	___ to ___

3. **Does your institution provide horticultural therapy, recreational horticulture, or other garden therapy activities?** yes no

4. **Is a horticultural therapy program expressly incorporated into the mission of your institution?** yes no

5. **If you offer horticultural therapy and your answer to question number 4 was "No", is your program a direct outgrowth of the mission of your institution?** yes no

Program Information

6. **How many patients does your program serve monthly?** _____

7. **Does your program utilize outdoor facilities for horticultural therapy or horticulturally based recreation?** yes no

APPENDIX (continued)

8. **Do your clients physically garden in your outdoor horticultural therapy (therapeutic landscape) facilities?** yes no

9. **Does your institution plan to enlarge your therapeutic landscape facilities in the future?** yes no

10. **Have your therapeutic landscape facilities been enlarged within the past two years?** yes no

11. **If you answered "No" on questions 9 and 10, does your institution plan to improve or renovate or have your therapeutic landscape facilities been improved or renovated within two years?** yes no

II. FACTORS IN DEVELOPMENT — Early Goal-setting

12. **Was a written description of the horticultural therapy program's goals, objectives and needs developed prior to the design and construction of the therapeutic landscape?** yes no

13. **Please indicate who participated in writing this description?**

Client	Landscape Architect
Administrator	Horticulturist
Program Director	Horticultural Therapist
Maintenance Personnel	Other therapist: (specify)
Other:____________	____________________
________________	____________________

14. **If no written program description was prepared, who was responsible for setting goals for the design and construction of the therapeutic landscape?**

____

15. **Are clients permanent residents of your institution?** yes no

16. **Please list five major horticultural activities normally included in your program and list the associated tasks for each activity.**

Activity	**Tasks**
1.____________________	______________________________
2.____________________	______________________________
3.____________________	______________________________
4.____________________	______________________________
5.____________________	______________________________

17. **Where is the therapy garden located in reference to major indoor facilities?**
Adjacent Within "walking" distance (but not adjacent) Distant

18. **Who initially was selected to maintain the garden?**

Therapy staff	Custodial staff
Client	Contracted maintenance firm
Volunteers	Other (Specify):____________________

19. **Who currently maintains the garden?**

Therapy staff	Custodial staff
Client	Contracted maintenance firm
Volunteers	Other (Specify): ____________________

20. **When was the garden construction initially begun?** ______________

21. **When was it completed?** ______________

22. **If not completed presently, please indicated expected date of completion.**

____

APPENDIX (continued)

Planning and Site Selection

23. Please rank (1 highest through 9 lowest) the site selection criteria employed?

___ Topography
___ Slope
___ Land availability
___ Utilities availability
___ Ability to define (fence)
___ Type of Soils
___ Proximity to other facilities
___ Sun orientation
___ Other(Specify):______________________

24. Which of the following best describes the accessibility goals defined for your therapeutic landscape?

Total accessibility
Access for selected populations
No accessibility goals formulated
Access to selected areas
Access only with assistance

25. Rank the importance (1 highest through 6 lowest) of the following factors in planning for your therapeutic landscape.

___ Group size
___ Age of Clients
___ Individual clients
___ Group scheduling
___ Number of concurrent groups
___ Gender

26. What specific site facilities were proposed and which currently exist? (P = proposed; E = exist)

___ Canopy/shade structure
___ Equipment storage
___ Materials storage
___ Fountain/aquatic garden
___ Irrigation
___ Raised plant beds (Estimated _____% of planting area)
___ Ground level beds (Estimated _____%of planting area)
___ Other (Specify):______________________

Design

27. Please describe how the paths and walks are constructed:

Material (s) ______________________________________

Width of major path ______ft.
Width of minor path(s) ______ft.
Maximum slope of paths ______%
Minimum slope of paths ______%

28. Please describe the size and quantity of space(s) in this therapeutic landscape:

	Group size	Approx. area (s.f.)	Quantity
Maximum group-size capacity	______	__________	_____
Minimum group-size capacity	______	__________	_____
	Total number of spaces		_____

29. Please circle the typical perimeter definition(s) of spaces and work areas within the therapeutic landscape:

Wall	Fence	Hedge
Raised bed	Plant mass	Overhead canopy
Tree line	Paving change	Path
Roof	Arbor	Seating
	Other (Please explain:) ____________________	

____

30. Please estimate the size of planting areas within the therapeutic landscape:

Maximum width of beds _____ ft.	Minimum height of raised beds _____ ft.
Minimum width of beds _____ ft.	Maximum height of raised beds _____ ft.
Beds are irregular -- width varies	Maximum length of beds _____ ft.

31. Please describe any specialized accessibility adaptations to therapeutic landscape beds:

____

____

32. Please list the types of plants most often selected for the therapeutic landscape:

Woody plants	Full season/perennials	Partial season/ annual plants
__________	__________	__________
__________	__________	__________
__________	__________	__________
__________	__________	__________
__________	__________	__________

APPENDIX (continued)

33. **Please circle the description which best explains how the spaces within the therapeutic landscape are organized and related to each other?**

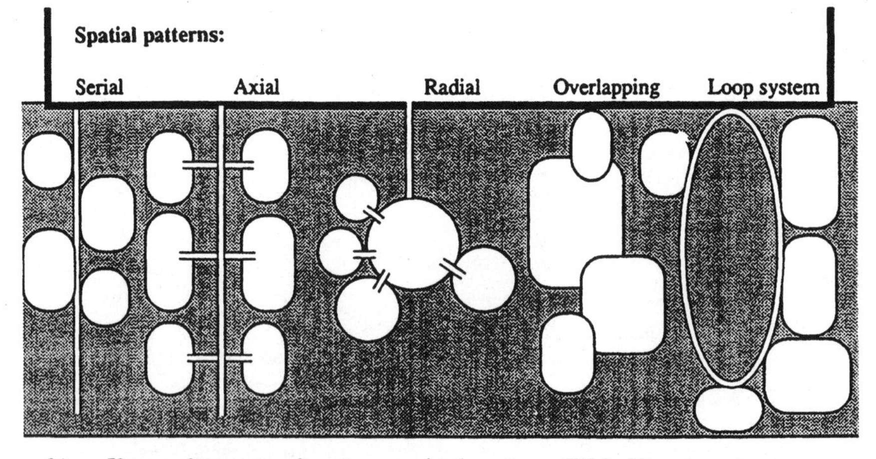

34. **If none of these describes the organization of your HT facility, please try to draw its organization below.**

Thank you very much for your help with this survey.

Chapter 26

Combining Phototherapy with Horticulture Therapy

Dan Greenlee

I have been self employed in the greenhouse business for 20 years, with specialization in hydroponics since 1985. After assisting in the establishment of a commercial hydroponic greenhouse for Community Mental Health Services, that employs and provides rehabilitation for their clients, I was asked to become the rehabilitation facility's representative for Cropking Inc. of Medina, OH in 1989. Since that time I have been developing new horticulture therapy programs for various facilities.

Let me stress that I am not a clinical researcher, but rather am a person with an inquiring mind who cares about people. Growing interest in horticulture therapy has allowed me to develop new programs and concepts. In studying some of these new programs, I started looking at the light levels needed for plant growth, and began wondering what effect light had on people. I will be sharing some of my observations and conclusions.

I have learned that recurrent fall-winter depression can be caused by insufficient levels of light and is treated with supplementary light. The syndrome is identified as Seasonal Affective Disorder (SAD) or Light Deficiency Syndrome (LDS) (Rosenthal et al., 1989). With the help of Dr. Paul Brown I learned that some of the programs in place have plant lighting that would also be of sufficient quantity and quality to be useful in the treatment of SAD.

In early December, 1991, two programs were selected for further study. Wes Harris at Lewis Wetzel Care Center agreed to make observations at their facility, which has residents with disabilities including: MR/DD, mental handicap, substance abuse, and a geriatric population. John Obloy agreed to make observations at the School of Hope, which is a MR/DD school. Time and lack of diagnostic methods did not allow positive diagnosis of SAD in either facility.

At Lewis Wetzel Care Center, a large hydroponic unit is operated in the basement with high pressure sodium lighting. Three smaller hydroponic units with fluorescent lighting are located in residents' rooms. Harris reports that a resident who is receiving treatment for substance abuse has experienced less depression since working with the unit in the basement and has taken the initiative to produce plants to transplant to an outdoor garden. One geriatric resident with a unit in the room appears to have experienced less depression. It has been concluded that diagnostic and evaluation methods are needed for continued research. It was also concluded that additional research was needed to explore varying degrees of SAD.

At the School of Hope, the classroom program is lovingly referred to as "John's Jungle." The program includes a large hydroponic where the students are growing tomatoes and cucumbers using high pressure sodium lighting. Two smaller units grow lettuce, herbs and foliage plants using fluorescent lighting. Obloy reports that all students are calmer and more at ease in the lighted environment. A student with Downs Syndrome was undergoing evaluation by a behavioral modification specialist for disruptive acting out behavior and chronic absenteeism in the previous school year and the beginning of the current year. Changes in behavior this year were narrowed down to the lighting. The plant lights are controlled by a timer. When the lights were on, acting out stopped, when the lights cycled off the acting out returned. The decision was made to change the lighting cycle to match the school day which resulted in the end of the acting out behavior. The problem of absenteeism also ended. The results of this study indicate the need for improved diagnostic and evaluation tools.

In discussion with care providers in other settings, it would appear that SAD may be more prevalent than suspected, with varying degrees of severity. In the last two weeks I have learned of the work of Dr. Oren et al. at the National Institute for Mental Health, where diagnostic and treatment methods are being developed. I will be sharing information with this group in an effort to explore these concepts. In addition I will be looking for evaluation methods that can be adapted to SAD.

I have recently developed a hydroponic therapy system that may be useful in the treatment of SAD. In addition to growing other crops, the system can be used to grow "Fast Plants," developed by the University of

Wisconsin. Fast Plants are a rapid cycling brassica that completes its life cycle in 35 days. The rapid growth promotes daily interaction with the patient, including pollination of blooms with "bee sticks" which are made from dried bees. The plants require 24 hour illumination with an intensity of 13000 lumens in the 4100 kelvin range. It is believed that exposure to these levels of light while daily working with the plants, may be useful in the treatment of SAD. It is hoped that diagnostic and evaluation methods are available to further study the system in the next fall-winter season.

As I began this study, I expected that all the information would be readily available, but have found that there is a need for additional research. I would encourage organizations such as American Horticulture Therapy Assc. and People-Plant Council to interact with medical research to further explore the diagnosis, treatment, and evaluation of SAD.

REFERENCES

Brown, Paul, Clinical Psychologist, Community Mental Health Services, P.O. Box 508, St. Clairsville, OH 43950.

Harris, Wes, Director, Lewis Wetzel Care Center, P.O. Box 428, 80 East Benjamin Dr., New Martinsville, WV 26155.

Obloy, John, Classroom Instructor, School of Hope, 330 Fox Shannon Place, St. Clairsville, OH 43950.

Oren, Dan A.; Jacobson, Fredrick M.; Wehr, Thomas A.; Cameron Christine L.; Rosenthal, Norman E.; National Institute of Mental Health, Building 10, Rm. 4S239, 9000 Rockville Pike, Bethesda, MD 20892.

Rosenthal, Norman E., M.D.; (1989) Light Therapy, Treatments of Psychiatric Disorders, Volume 3, American Psychiatric Assc. APA Press, 1989.

RESEARCH IMPLEMENTATION

Chapter 27

Historical Perspectives on the People-Plant Council

Diane Relf
Pete Madsen

BACKGROUND

Understanding and quantifying the values placed on plants by consumers of horticultural products and services and the impact these plants have on human behavior and well-being can have significant results. The most important of these is that such knowledge can enhance the quality of life for a wide range of people, from high-pressured executives to residents of low-income housing. Those of us involved in horticulture professionally recognize a side benefit of this research: knowledge of this kind can become an exceptionally valuable market development/planning tool for the horticulture industry and educators. This can have a very positive impact, particularly in urban areas, of increasing jobs and stimulating the economy.

As recently as ten years ago, exercise was seen primarily as something for athletes to do. Then, through research followed up by dissemination and promotion, exercise/fitness became recognized as something that everyone should be involved in, because it was shown to greatly improve physical and mental health. Incident to this change in lifestyle, the market for sportswear and equipment has exploded.

Plants also improve physical and mental health–often in more subtle ways, but with greater impact. You and I know this intuitively, but we don't have research to document that people who are concerned enough about their well-being to own a pair of running/walking shoes should also maintain plants in their home and work environments.

INTERDISCIPLINARY RESEARCH TEAM IN CONSUMER HORTICULTURE

Interdisciplinary research is imperative for conducting research on human issues in horticulture. At Virginia Tech, an Interdisciplinary Research Team in Consumer Horticulture (IRTCH) was established with the support of the Director of the Agricultural Research Station and the Dean of Research. It consists of Virginia Tech faculty, staff, and graduate students interested in horticulture's role in society; the roles of plants in history, culture, wellness, and behavior. This group recognized that although the aesthetic and stress-reducing values of plants are "known" to most of us, there is essentially no quantified documentation to establish the contribution of plants to human well-being or social development. The field is wide open for research of this type. The present growth of the horticulture industry makes this the time for such non-traditional research to be well accepted. Funding is becoming available, and the industry is fully supportive of our efforts in this area.

The mission of IRTCH is to quantify the effects of plants on human attitudes and behavior, to understand why humans respond to plants, and to disseminate this knowledge so it may be used to improve the quality of human life. Research projects are being conducted within several disciplines including psychology, sociology, and interior design which will develop data of value to the horticulture community.

Activities of IRTCH included a campus seminar and workshop on people/plant issues which featured Rachel and Stephen Kaplan, noted Environmental Psychologists from the University of Michigan; Charles Lewis, the most well-known advocate for PPI in the United States; and Charles Dunn, president of a prominent horticultural marketing firm. The University Provost opened the event by emphasizing the importance of interdisciplinary cooperation, especially in the study of people/plant interaction. The symposium was well-attended by faculty, staff, and students from many departments, and the afternoon discussion was dynamic and productive.

NATIONAL COMMITMENT

We asked several leaders of the U.S. horticulture industry to meet with us in November of 1988 in Washington, D.C. to discuss a plan for gaining documentation of the benefits plants provide for quality of life. This meeting resulted in the agreement that this is an idea whose time has come; that this is the kind of long-range planning the horticultural community needs. As a result of this meeting, the Horticultural Research Institute granted us seed money to begin working on this project.

SYMPOSIUM SPONSORED

To establish research initiatives in this area, the Department of Horticulture at Virginia Polytechnic Institute and State University, the American Society for Horticultural Science, the Association of American Botanic Gardens and Arboreta, and the American Horticultural Therapy Association co-sponsored a national symposium, "The Role of Horticulture in Human Well-Being and Social Development," in April 1990 in Arlington, Virginia. It was endorsed by most of the major horticultural associations including the American Association of Nurserymen, the Society of American Florists, the American Floral Endowment, the Associated Landscape Contractors of America, the American Society of Consulting Arborists, the Professional Grounds Management Association, and the U.S. Botanic Garden.

Organizers of the symposium recognized that an understanding of the psychological, physiological, and social responses of people to the plants in their environments can play a significant role in improved physical and mental health for individuals as well as communities. Information about these responses has significant implications for a wide variety of professionals, industries, and agencies including: urban planners; architects; office, mall, and space station designers; developers (particularly those involved in condominiums, planned communities, hotels/motels, etc.); insurance companies and hospitals (research indicates a view of plants may reduce the length of stay in the hospital); nursing homes; hospices;

horticultural therapists; rehabilitation facilities (concerned with quality of stay, treatment time, program design, and job training); and government and private employers of office workers (plants to reduce stress and increase productivity).

Current trends in horticultural research are inadequate for the future of horticulture in relation to the economic and social role it could play in this country. It is important that members of the horticulture community become pro-active in designing and conducting research cooperatively with social scientists to understand the interplay of horticulture with various aspects of human development. This symposium will provide a forum for such interdisciplinary discussions.

To address these issues, the organizing committee established the following focus and objectives for the symposium:

- to collect current information on the psychological, physiological, and social responses of people to plants and publish it in symposium proceedings
- to identify research priorities that will lead to an understanding of the relationships between people and plants, and to seek ways horticulturists can work with social scientists and others to more fully understand and utilize these relationships
- to initiate the development of a network for researchers, implementors of research findings, funding sources, and information dispersal systems.

The symposium brought together four groups: those currently conducting research; horticulturists and social scientists interested in doing future research; representatives from trade groups, public and private agencies, and educational institutions who can utilize this research; and representatives from public and private funding sources.

The program included oral presentations and discussion groups in the following areas:

1. Plants and human culture
2. Plants and the community: human interaction in communities as altered by plants in the environment
3. Plants and the individual: their influence on human behavior, physical health, and perceptions of comfort
4. Developing a conceptual framework: the psycho/physiological responses of people to vegetation in a man-made context
5. Exploring a specific application: horticultural therapy
6. Implementation: putting the research into action.

At the conclusion of the symposium, more than 30 prominent members of the horticultural community and allied fields met to discuss the future needs for research in Human Issues in Horticulture (HIH). Led by a facilitator, participants first reached a consensus that the time had arrived for cooperating to address these issues. They established an action plan with a two-fold approach to structure and organizing.

Working Within the Existing Associations

Encourage all horticultural associations to support Human Issues in Horticulture (HIH) activities within the context of their existing activities: Publish articles in newsletters, solicit research articles for journals, conduct workshops/presentations at conferences, support research through endowments, etc.

Hold meetings including officers, staff, and leaders at annual conferences of the existing associations to communicate the importance of HIH and encourage research and/or funding of research.

Examples of possible activities:

American Society for Horticultural Science has two publications *HortTechnology* and *HortScience* that could appropriately have Associate Editors for Human Issues in Horticulture.

Horticulture Research Institute and American Floral Endowment could be encouraged to expand their funding to HIH research.

American Association of Nurserymen could be encouraged to have an Associate Editor for HIH in the *J. of Env. Hort.*

Trade and professional associations can include this area in the speakers for conferences and articles in their publications.

Establish an HIH Consortium or Council

Organize or structure an HIH consortium/council for communication and research. An ad hoc meeting was held May 24, 1990 at AAN offices in Washington, D.C. and attended by Jim Swasey, AABGA; Chuck Richman, AHTA; Larry Scovotto, AAN; Skip McAfee, ASHS; Marvella Crabb, SAF; Earl Wells, FN and GA; Diane Relf, VPI and SU; Candice Shoemaker, VPI and SU; Pete Madsen, VPI and SU.

PEOPLE-PLANT COUNCIL FORMED

The People-Plant Council (PPC) was formed on May 24, 1990, as a direct result of the national interdisciplinary symposium, "The Role of Horticulture in Human Well-being and Social Development."

The *mission* of this Council is to document and communicate the effect that plants and flowers have on human well-being and improved life-quality. This mission is to be carried out through a five-part *strategy* focusing on the effects that plants have on human well-being through the psychological, sociological, physiological, economic, and environmental effects they produce:

communication–maintain an interdisciplinary network between researchers, funders, users, and Council affiliates; provide research-based information to the horticulture and social science communities, commercial and private users of plants, and the general public.

research–encourage cooperative efforts in identifying research priorities and establishing interdisciplinary research methodologies.

funding–establish a network to link researchers to funding sources including government agencies, public and private foundations, and co-operatives.

implementation–encourage the use of horticulture for enhanced life-quality based on research findings and provide consulting services to users to implement research data.

education–encourage public curriculum development to include people-plant interaction as an essential subject in kindergarten through adult/continuing education in many fields of study.

The PPC serves as a link between organizations representing all facets of both the horticulture and the social science communities. The Council is a network, designed by the representatives of interested associations to enhance and focus their efforts toward documenting the human benefits derived from horticulture.

PEOPLE-PLANT COUNCIL AFFILIATION

The PPC is not a membership organization, but rather a link or affiliation between organizations. Affiliation is open to all organizations within the horticulture and social science communities and allied or interested organizations to include, but not be limited to: academic and professional associations; trade and commercial associations; volunteer, civic, amateur and concerned groups. Affiliation with the PPC is established through contributions to maintain its operational expenses. Contributions are based on the size and scope of the affiliating organization. All contributions are handled through Virginia Polytechnic Institute and State University through their Industrial Affiliates Program in the office of Grants and Contracts.

Researchers, educators, and others use the services of the People-Plant Council, including a bi-annual newsletter, periodic update reports, access to computerized information, and conference/educational program registration with a cost-of-service fee. As this is not a membership organization, individuals can receive the PPC information by submitting their names to the mailing list.

PPC also accepts contributions to support its goals. Contributors include commercial horticulture businesses, public relations and consulting firms, foundations, endowments, and individuals who have a commitment to the mission of the Council and seek to support its goals and fund its operational strategies.

Benefits of Affiliation. By affiliation with the People-Plant Council, an organization will ensure:

a continuous source of information for their newsletters to keep membership informed on human issues in horticulture

an increase in the use of plants in all areas of modern life

expansion of the research basis for human benefits from plants and flowers

an increase of information for communicating the value of both active gardening and the presences of plants in public, commercial, and residential settings

access to researchers with knowledge about the value of people-plant interaction

the availability of consultants to help you increase the application of people-plant knowledge within your organization

improved quality of life and human well-being.

PEOPLE-PLANT COUNCIL MANAGEMENT AND PRIORITIES

At the March 15, 1991 meeting of the organizing committee, Dr. Diane Relf was elected coordinator of the People-Plant Council. Dr. Relf is an associate professor of horticulture at Virginia Tech University and chair of the symposium, "The Role of Horticulture on Human-Well Being and Social Development." Her responsibilities at Virginia Tech include serving as Extension Specialist in Consumer Horticulture, supervising the staff of the Office of Consumer Horticulture, and chairing an interdisciplinary group of researchers looking at the psychological and social relationships between people and plants.

At the spring meeting of the PPC, priorities were established for implementing the mission strategies. Emphasis has been placed on research and funding development. The first step is to establish a base-line of existing knowledge. The bibliography currently under development is to be combined with computer searches conducted by the USDA Ag. Library, the Library of Congress, and other sources. The citations identified will be evaluated for their relevance to research in people-plant interaction, and a comprehensive annotated bibliography will be developed and continually maintained.

To identify potential funding sources, several computer data banks are being accessed. Emphasis will be placed initially on horticultural and plant-related foundations and funding agencies. For example, the United States Golf Association has issued a call for proposals that includes research on "Psychological and physical well-being of people, and the importance of landscape aesthetics to humans due to the interaction between people and plants."

Another equally important priority for the PPC is to establish an organizational structure that provides resources and personnel to address its mission. This will be done through work with the existing and potential affiliates of PPC.

RESOURCES AVAILABLE THROUGH THE PEOPLE-PLANT COUNCIL

Proceedings of the 1990 Symposium, "*The Role of Horticulture in Human Well-Being and Social Development*," Edited by Diane Relf, Associate Professor of Horticulture, Virginia Polytechnic Institute and State University. ISBN 0-88192-209-0, 254 pp, 8 3/4″ × 11 1/4″, 8 color photos, published April 1992 by Timber Press.

The aim of this volume, and the symposium at its origin, is to bring the reader a survey of Human Issues in Horticulture. Through a multidisciplinary approach involving researchers in the fine arts, sociology, psychology, urban planning, forestry, environmental psychology, history, and the horticulture community, the authors develop an overview of how plants affect people and explore diverse opportunities for research and acquisition of knowledge. This wealth of interrelated material will interest all plant professionals and also amateurs, such as Master Gardeners, who through their horticultural activities contribute to society.

Participation by the horticulture community is essential for the growth of horticulture into new and nontraditional areas. As research documents the life-enhancing potential of plants, it is the role of commercial horticulturists to supply crops and services that will support the broader role of horticulture. It is the role of arboreta and botanic gardens to display these plants and educate their clientele in appropriate use. Growers, landscapers, educators, and communicators–all whose economic well-being depends on people's involvement and satisfaction with plants–will have a significant role in implementing the findings of research focused on horticulture and human well-being. In addition, horticulture faculties at universities will have a renewed mandate to teach, conduct research, and carry information on horticulture to the public. Only by going beyond questions of growing and maintaining plants and the search for basic scientific knowledge on plant growth can this mandate be fulfilled in today's society.

The Role of Horticulture in Human Well-Being is $50 per copy plus postage and handling. Postage and handling is $4 for the first book and $2 for each additional book. To order, send your check payable to the *Treasurer, Virginia Tech* to: Dr. Diane Relf, PPC Coordinator, Department of Horticulture, Virginia Tech, Blacksburg, VA 24061-0327. (Note: Checks must be in U.S. dollars, payable on a U.S. bank.)

People-Plant Council Research Bibliographies. Through funding from various horticultural associations (including ASHS, ALCA, SAF, WF&FSA, and HRI) the People-Plant Council has been able to develop a computerized bibliography that will be of great value to researchers in the area of *People-Plant Interaction* and a second bibliography specifically for the area of *Horticultural Therapy.* The combined PPI and HT bibliographies contain approximately 2000 citations, 25 percent of which include an abstract. Due to the size and length of each bibliography (over 450 pages of hard copy), they are available on diskette. This will facilitate users searching for keywords or specific articles and allow them to rearrange the material as needed.

The People-Plant Interaction bibliographic database and search system requires a 286/386 processor and at least 1 MB RAM for correct operation. It currently consists of 829 citations with over 400 abstracted. We hold hard copies of 345 of these articles in our files. Much of the material in this database is a result of the literature accumulated while preparing for the first symposium on the Role of Horticulture in Human Well-Being held in April 1990. To develop this database, we have searched VTLS, ERIC, PSYCHLIT, AGRICOLA, NEWSBANK and INFOTRAC for the keywords *horticultur-, plant-, tree, garden- and flower-*; cross-checked with keywords *psych-, socio-, people, well-being and stress*; and checked keywords *flora, botanical and environmental effects*. More than 1000 citations were pulled up. Of these, about 50 were determined to be relevant and added to the database. To further expand the bibliography, we have written to top researchers requesting that they cross-check our citations with their personal list of publications.

We have conducted a search of dissertation abstracts with only four papers found. To overcome problems of keywords here, we plan to write to departments of horticulture, landscape architecture, environmental psychology, forestry, geography, and others as identified requesting copies of relevant theses and dissertations. Additional plans include direct searches of back issues of architecture, landscape architecture, planning, urban design, interior design, and other relevant periodicals as we have determined that computer searches do not identify many of the articles because the plant aspect of the research is not included in the keywords. A final step in making the bibliography as comprehensive as possible will be to cross-check literature citations in theses and publications on hand against existing bibliographies for new references.

As it currently stands, the bibliographic database will be extremely valuable both to researchers and educators to gain insight into people-plant interactions and determine research priorities, methodologies, and applications. Citations include research on people-plant interaction, related research and articles from the trade and popular press.

The Horticulture Therapy bibliographic database and search system also requires a 286/386 processor and at least 1 MB RAM for correct operation. It contains 1132 citations (many of these are newspaper articles and brochures from HT programs). Although most of these have been keyworded, they have not yet been abstracted. We have hard copies of 786 articles in our files. To establish this database, we have entered all citations from previously accumulated HT bibliographies, all citations from the resource material collected for HT classes and Extension lectures and all citations from issues of the *Journal of Therapeutic Horticulture* and the HT short courses at Kansas State University. We have not yet performed any CD-ROM database searches, nor conducted any of the more rigorous searches planned for the PPC bibliography. However, similar measures are planned as time and money permit. This bibliography should prove to be useful to anyone seeking a broad overview of HT or attempting to identify programming for a specific, special population. There is still relatively little research reported in the literature, and this lack is very evident in this bibliography.

The two bibliographies, People-Plant Interaction and Horticulture Therapy, are available on 3.5-inch, DS/HD diskettes containing all the citations in WordPerfect 5.0 for $15 each. If you do not have access to WordPerfect 5.0, the material can be ordered on a 3.5-inch diskette in DOS. To order, send a check payable to "Treasurer, Virginia Tech" to the Office of Consumer Horticulture, Virginia Tech, Blacksburg, VA 24061-0327.

Computerized List of Researchers and Users (Under Development)

A form is available to establish a list of researchers and users of research information to be easily posted and identified by specialty. If you would like your name included on this list, please send a self-addressed, stamped envelope to PPC for an application form.

1990 Symposium Abstracts and 1992 Symposium Abstracts. $5 each.

Industry Article Packet: Better Business through People-Plant Research. $5.

Videotapes: (1) Role of Horticulture in Human Well-Being and Social Development–reflections of Jules Janick, Charles Lewis, Roger Ulrich, Russ Parson and Diane Relf; (2) The Art of Rhonda Roland Shearer. $15 each. (All prices include shipping and handling.)

HORTICULTURE INDUSTRY INVOLVEMENT

Since the symposium in April 1990, horticulture trade publications have printed five major articles addressing the value of research into people-plant interaction and/or the People-Plant Council. In addition, state trade associations have published related articles.

The *Associated Landscape Contractors of America* have taken a leadership role in supporting PPC as its first affiliate. Their support of research and communication in this area is invaluable.

The *Society of American Florists* (*SAF*) has been very active in accessing its communications network to the floriculture industry and keeping members informed of PPC activities. SAF and the *American Floral Endowment* have supported research in this area and sponsored presentations at their 1990 meetings. SAF has become a charter affiliate of PPC.

In the summer and fall of 1990, the *Texas State Florists Association* and the *Texas Association of Nurserymen* had presentations at their state conferences.

The *American Association of Nurserymen* and the *North American Horticultural Suppliers Association* arranged for presentations to be given at their annual conference to allow their membership to better understand the importance of this type of research and how to apply it.

PROFESSIONAL ASSOCIATION INVOLVEMENT

American Society for Horticultural Science. At the ASHS conference in Tucson, the Sociohorticulture Working Group sponsored a workshop to inform the members of the society of the impact of the spring symposium and to involve them in expanding the research in this area. Over 100 ASHS members attended this workshop and the response was very positive. With a name change to Human Issues in Horticulture Working Group, this group conducted the workshop "Social Science Research Methods" at the 1991 ASHS annual meetings at Penn State. The key speaker, Rex Warland of the Department of Ag. Economics and Rural Sociology at Penn State, gave a comprehensive overview of methodologies. Candice Shoemaker of Berry College addressed the use of focus groups, and Virginia Lohr of Washington State University discussed using questionnaires. Publication of these presentations is being explored. For more information on research methodologies, write to Dr. Shoemaker at the Department of Agriculture, Berry College, 5003 Mount Berry Station, Rome, GA 30149-5003.

The board of directors of ASHS approved affiliation with the People-Plant Council and the appointment of an associate editor for *HortTechnology* in the area of sociohorticulture. The editorial board of *HortTechnology* selected human issues in horticulture as a major theme for the summer 1992 issue.

International Society of Horticultural Science. At the international conference in Florence, Italy, the social science side of horticulture was much in evidence. Plenary lectures included *The visual arts and the science of horticulture in Tuscany from the 16th to the 18th century* and *The biblio-*

metric analysis of world horticulture literature published during the past two decades, while the scientific program included four sessions on economics, two sessions on history and art, and two sessions on urban horticulture with most of these papers addressing some aspect of the people-plant interaction.

The next ISHS conference was held in Kyoto, Japan in August of 1994, and was of particular interest to those of us interested in the role of plants in life-quality. The title for this conference was *The Beautification of Life and Its Environment through Horticultural Science* and it included a special symposium on human issues in horticulture.

American Association of Botanical Gardens and Arboreta. During the June 1990 AABGA conference in Seattle, there was significant attendee interest in human issues in horticulture. In other words, how gardens and gardening affect people, why people visit gardens and their perceptions of them, how plants can be used in rehabilitative environments, etc. In Seattle, the AABGA board approved the establishment of a working group to explore this interest and ultimately, if the potential exists, to elevate this group to full committee status. This group will explore human issues research opportunities within both AABGA and allied organizations. A list is being developed of those within AABGA who may be/are interested in human issues research and programs.

In addition, the board authorized the AABGA newsletter to act as a clearinghouse for information on human issues. At the AABGA meetings, a session, "Human Issues in Horticulture," chaired by Charles Lewis, presented three talks: *Role of Horticulture in Human Well-Being and Social Development* by Charles Lewis, *Methodologies for Human Issues Research* by Dr. Candice Shoemaker, and *Using Human Issues Research to Understand and Expand Your Audience* by Dr. Diane Relf.

The AABGA Human Issues in Horticulture working group held its first business meeting at the 1991 Annual Conference. The meeting was attended by representatives from 11 different arboreta, botanic gardens, and related facilities. The group identified seven categories of diverse interests among its members and the 37 people who responded to an earlier survey. Coordinators were selected to clarify and articulate common goals among people in each interest area. An educational session was scheduled for the 1992 Annual Conference in Columbus, Ohio, to address "Human Issues Research at Arboreta and Botanic Gardens." In the two-hour session, environmental psychologists discussed the basis for human issues research in botanic gardens and arboreta. For more information contact: Gene Rothert, Chicago Botanic Garden, P.O. Box 400, Glencoe, IL 60022.

American Horticultural Therapy Association. The focus of this association is on the response of special populations to therapeutic and rehabilitative programs based on horticulture. In the last year, AHTA has provided increased leadership in strengthening research in horticultural therapy. The Association's 18th Annual Conference was held in Albuquerque, New Mexico, August 5-7, 1992. Its theme was "Protecting Health and Habitat in the 90s," and 150 attendees made it the best attended AHTA event to date. Speakers included Charles Lewis of the Morton Arboretum and Dr. Diane Relf, HTM. Immediately following the Conference, AHTA conducted a full-day training program entitled "Research and Writing for Publication," presented by Dr. Bob M. Gassaway of the University of New Mexico. It was so well received that an extensive article was commissioned for the AHTA just-published Volume V of the *Journal of Therapeutic Horticulture.*

AHTA's 1991 Annual Conference, "Advancing the Practice of Horticulture as Therapy," brought together 171 horticultural therapists and interested individuals for three full days of educational sessions, tours, and an awards banquet at Longwood Gardens. This record attendance, during hard economic times, underscores the growth of the people-plant connection in recent years.

Affiliates of PPC

American Society for Horticultural Science
Associated Landscape Contractors of America
Wholesale Florists and Florist Suppliers of America
American Horticultural Therapy Association
Society of American Florists

Contributors to PPC

Horticulture Research Institute
Florida Nurserymen and Growers Association
Bailey Nurseries, Inc.

Correspondence should be addressed: People-Plant Council, Department of Horticulture, Virginia Tech University, Blacksburg, VA 24061-0327. Telephone: (703) 231-6254; FAX (703)321-3083.

WHAT YOU CAN DO TO HELP PPC REACH ITS GOALS

There are several things you can do to support the goals of the PPC:

- Conduct and publish research in this area. Look at ways your current research can be modified to provide answers to human issue questions.
- Work through your professional organization to encourage others to conduct relevant research. You can also encourage your professional organization to affiliate with PPC.
- Send me copies of any relevant journal articles to include in our files and annotated bibliography. These articles need not be ones you have written, but can be ones you have found to be valuable in your work.
- Establish interdisciplinary research teams on your university campus or within your community.
- Add your name to the computerized list of researchers.
- If you are not a researcher, provide financial support, materials, and supplies to conduct research.
- Encourage your organization to become an affiliate of the People-Plant Council.

Chapter 28

Experimental Approaches to the Study of People-Plant Relationships

Russ Parsons
Roger S. Ulrich
Louis G. Tassinary

SUMMARY. People have cultivated plants and other natural elements in urban environments virtually since the advent of the earliest human cities. Recent research in the social sciences suggests that passive interactions with large scale natural environments may have health benefits. These early findings with large scale natural environments are consistent with enduring beliefs about the beneficial effects of natural environments, which in turn echo beliefs about the positive effects of horticultural pursuits. This paper describes two general classes of research methods that have been used to investigate the potential health effects of passive exposures to natural environments, methods that have direct applicability to the investigation of similar issues in horticultural environments. The first set of methods have been used in the field of environmental aesthetics to measure perceived environmental quality. Several of these specific methods are described and their applicability to horticultural research is highlighted. The second set of methods are taken from the field of psychophysiology, and have direct relevance for the study of potential health effects of environmental exposures. The emphasis in this section is on the issues involved in drawing inferences about psychological states from the measurement of physiological events.

INTRODUCTION

The cultivation and maintenance of natural elements in urban environments is an ancient practice. Trees, green grass and other vegetation were valued by the Greeks and Romans as aesthetic buffers to crowded city scenes (Glacken, 1967). Gardens, in particular, have been a part of human environments since the earliest civilizations (Shepard, 1967). Gardens provide both active enjoyment for the gardener and passive pleasures for those who appreciate the gardener's work. Beyond these simple, self-evident pleasures, however, recent research in the social sciences suggests that horticultural pursuits may provide more substantial benefits as well. Psychiatrists and other mental health workers have been using horticulture as a therapeutic tool for over one hundred years (see McCurry, 1963; O'Connor, 1958), with benefits cited including intellectual and emotional growth, as well as improved social and motor skills (Hefley, 1973). Additionally, beliefs in the curative effects of gardening for the mentally ill are at least several hundred years old (Watson & Burlingame, 1960). Unfortunately, documented evidence of the therapeutic benefits of gardening is equivocal at best (Burgess, 1990; also, see Horne, 1974 and Train, 1976 for reports of null and limited effects of horticulture therapy, respectively), and virtually no research has been reported concerning the effects of such passive interactions as merely viewing plants.

This state of affairs is unfortunate for several reasons. First, developing accurate knowledge of how people respond to plants should be of prime concern to horticulturalists. Gardens are for human consumption, both literally and figuratively. Thus, a well-designed and cultivated garden is not simply one that produces a bountiful yield, but one that also produces "collateral" benefits for the gardener and the visitor. Without knowledge of how people respond to different horticultural stimuli, it is impossible to assess the full range of potential benefits associated with horticulture. Second, documented evidence of collateral benefits, especially health-related collateral benefits, could prove to be a powerful political tool for horticulturalists faced with policy makers and decision makers who may undervalue the contribution of horticulture to human well-being. Third, there are theoretical perspectives, methods, and some empirical evidence in related fields, such as environmental psychology, that are relevant, and many of these theories and methods could be applied directly, or easily modified, to suit the needs of those interested in person-plant research.

The purpose of this paper is to describe some of the relevant research methods in environmental psychology and psychophysiology, and to suggest ways in which these methods might be used to investigate a range of horticultural issues from perceived preferences for flowers and other plants to the potential health effects of horticultural stimuli. The emphasis is on inferential issues that impact the choice among methods, rather than listing and describing all potentially relevant methods. As was implied above, a distinction can be made between effects associated with active and passive interactions with plants. In this paper we will focus on passive interactions with plants, such as sitting in a garden or viewing a flower bed from a window. We will not be concerned with horticulture or gardening as therapeutic tools *per se*. The first section provides a brief overview of some of the relevant methods used in environmental psychology to assess preferences for passive viewing of large-scale, natural environments and environmental surrogates. The next section focuses on psychophysiological methods that may help to determine the potential health effects of passive interactions with plants. The focus on preferences and psychophysiological methods reflects both the research on the beliefs about the benefits of contact with natural environments and theory in environmental perception.

People have often expressed visual preferences for natural environments (Parsons, 1991a). People also have long believed in the beneficial health effects of exposures to natural environments (Ulrich & Parsons, 1992). Typical beliefs about these effects associate time spent in natural environments (ranging from urban parks to backcountry wilderness) with psychological and physical well-being. Such beliefs have influenced urban planning (Walker & Duffield, 1983), urban park design (Olmsted, 1865), and laws affecting the preservation of natural areas (NEPA, 1969), yet very little empirical research has been done to test the accuracy of those beliefs, or their relationship to visual preferences. Thus far, the research suggests tentative support for the stress-reducing qualities of natural environments (Ulrich et al., 1991), but the findings are neither very extensive nor wholly consistent (Parsons, 1991b). In part, this lack of consistency in early research indicates that any relationships between passive environmental exposures and human health and well-being are bound to be complex, multifarious phenomena. The logic underlying beliefs about and early research concerning environments and health, however, has tended toward simplicity, as indicated by Figure 1. The seemingly straightforward reasoning that exposure to preferred environments leads to pleasant feelings, which in turn lead to improved psychological and physical health belies levels of complexity involving theories of environmental perception and aesthetics, the relationship between perceived environmen-

FIGURE 1. Logic underlying hypothesized relationship between environments and health

tal quality and affective responses, the independent psychological dimensions of affect and arousal,[1] the multiple feedback/feedforward relationships that exist between affective/arousal states and physiological homeostasis, and the types of systemic effects (cardiovascular system, immune system, etc.) that modulations in homeostasis can have on health and behavior.

Though it is by no means comprehensive, Figure 2 lists several classes of research methods that could be used to investigate the basic relationship outlined in Figure 1, taking into account some of these factors.[2] Listed under each methodological type are potential dependent variables pertinent to that basic relationship. Figure 2 also provides a context for the two classes of methods described below. We begin by focusing on different methods used to assess environmental preferences and visual quality, because a reliable means of establishing preferences is required before relationships among preferences, affective/arousal states and health effects can be explored. We then turn our attention to psychophysiological methods that have direct relevance for drawing inferences about affective/arousal states and stress reduction that may be associated with horticultural environments.

ENVIRONMENTAL RESEARCH METHODS

A. Perceived Environmental Quality

One of the more basic assessments of human responses to environments is that for visual preferences, or more broadly, environmental aesthetics. Extensive research has been conducted on visual preferences for outdoor environments, and methods used to assess environmental quality have been reviewed and categorized by a number of researchers. Daniel and Vining (1983) and Zube, Sell and Taylor (1982) have independently developed very similar classification schemes for such methods, suggesting some degree of consensus in the field. Though we will not discuss all of

FIGURE 2. Potential methods for investigating human-plant interactions

SELF-REPORT	BEHAVIOR	PSYCHOPHYSIOLOGY	ARCHIVES
Preferences	Performance	*Electromyography	Frequency/Severity of:
Moods/Feelings	*Problem Solving	*Blood Pressure	*Colds
Arousal	*Creativity	*Heart Rate	*Viruses
	*Perceptual	*Skin Conductance	*Cardiovascular Disease
		*Respiration	*Immune System Disorders
	Emotional	*Endocrines	
	*Irritation	*Immunoglobulins	
	*Jitters		
	*Helping		

the classes of landscape assessment methods described by these researchers, several distinctions among methods are relevant for people-plant research. These distinctions can be illustrated by mentioning three classes of methods, formal aesthetic methods, psychophysical methods and psychological methods.

A1. Formal Aesthetic Methods

One particularly relevant distinction made by both classification schemes is between methods that require expert judges to rate the visual quality of environments versus those employing non-expert judges, or naive observers. An example of the former type of method is the formal aesthetic model of landscape evaluation, some version of which is widely used by landscape architects in federal government agencies, such as the U.S. Forest Service and the Bureau of Land Management.

Adherents to formal aesthetic methods of landscape assessment assume that the interplay among abstract qualities, such as forms, lines, colors, and textures, comprises the aesthetic value of landscapes (Daniel & Vining, 1983). High quality landscapes are those that are highly diverse or varied, and yet maintain a degree of unity or integrity in their diversity. An example of a formal aesthetic method is the Visual Management System (VMS), routinely used by landscape architects with government agencies (Daniel & Vining, 1983). A VMS assessment begins with the categorization of landscapes according to character type: gorge lands, steep mountains, foothills or rolling plateaus. Within character types, landscapes are then classified according to *variety* and *sensitivity*. The former classification refers to the variety of and contrast among the abstract aesthetic qualities mentioned above (forms, lines, colors, etc.), and constitutes the primary emphasis of landscape categorization. The latter classification refers to the visual importance of the landscape as a scenic resource, and takes into account such factors as visual accessibility from different viewpoints and the likelihood that a typical visitor will encounter the landscape.

Because abstract qualities are the focus of assessment, formal aesthetic methods can be used with any landscape type, from deserts through grasslands, hardwood and coniferous forests, to oceans, and including human-made and dominated environments. This suggests that these methods could be adapted for use with gardens and other small-scale horticultural environments. However, because the application of these methods requires formal training, they are limited to use by experts. This limitation is, *a priori*, neither good nor bad, but it does raise a host of issues associated with the question of who is the appropriate arbiter of aesthetic value. This is a question of philosophy, not science, but the manner in which it is answered will have implications for the use of formal aesthetic methods in the evaluation of horticultural environments. If experts are chosen over the users of the environment, or more broadly, over the general public, then it is encumbent upon researchers in the field to establish whether there is in fact a difference between expert and non-expert assessments of the aesthetic quality of horticultural environments. There is some evidence on this point with respect to large scale natural environments, suggesting that trained experts' assessments of the aesthetic quality of an environment differ from naive observers when the experts use formal aesthetic methods, though their judgments are very similar to non-experts when both groups make a simple perceptual rating (e.g., scenic beauty; see Hetherington, 1992). If such a difference were established for experts judging horticultural environments, it would still be necessary to determine the reliability and validity of experts' assessments before formal aesthetic methods could be usefully applied. For instance, the validity of the formal aesthetic approach would be called into question if considerable agreement among experts using the method on similar horticultural environments could not be established.

A2. Psychophysical Methods

Among the commonly used methods for assessing landscape quality that are not based on expert judgments is one derived from classical psychophysics. Psychophysical methods represent attempts to mathematically relate measurable physical characteristics of a given environment to human perceptual responses. In the case of landscapes, the physical characteristics measured can range from the very specific, such as the number of ponderosa pines per acre greater than 20″ in diameter, to more general categories, such as the predominant vegetation type or tree species in a given area. The perceptual responses measured are typically preference ratings, but psychophysical methods have also been used to relate physical characteristics of environments to ratings of visual air quality (Parsons & Daniel, 1988), perceived naturalness (Daniel et al., 1973), and the fittingness of developments to their surroundings (Wohlwill, 1979). Psychophysical methods have been used in designed environments as well, in cognitive mapping studies and in gauging the perception of potentially harmful chemical odorants in studies of "sick buildings" (see Baird & Bergland, 1989). Because perceptual responses are obtained from relatively large observer groups (20-30 raters) who rate multiple environments, ratings are usually made off-site by using environmental surrogates (color slides or photographs). Studies comparing preference ratings of on-site and off-site observers have generally found good correspondence between the two types of ratings (e.g., Daniel & Boster, 1976; Jackson, Hudman & England, 1978; Malm et al., 1981; Shafer & Richards, 1974; Zube, Pitt & Anderson, 1975), and a similar correspondence between on- and off-site ratings for horticultural elements would have to be established to use psychophysical methods in people-plant research.

Though preferences can be expressed in a number of different ways (numerical rating scales, paired comparisons, Q-sorts, rank orders, etc.) across studies, perceptual responses theoretically should not (Daniel & Vining, 1983) and empirically do not (Daniel & Boster, 1976; Hull & Buyoff, 1981; Pitt & Zube, 1979) differ as a function of the manner in which preferences are solicited. Thus, the library of potential perceptual metrics available to the person-plant researcher interested in applying psychophysical methods to the question of horticultural preferences is fairly broad. On the other hand, the task of measuring the relevant physical characteristics in an environment, at first glance, appears daunting. This task has been greatly facilitated within landscape aesthetics by two circumstances. First, the physical characteristics of large-scale environments are sampled, rather than measured *in toto*; second, many relevant physical characteristics are routinely inventoried by Forest Service, Park Service and other government personnel who manage federal and state lands. In the case of horticultural psychophysical research, the manner in which physical characteristics are measured would depend on the size of the environment in question. If it were a small garden, for instance, it is conceivable that all of the relevant physical characteristics could be measured (e.g., number, size, color and spatial density of constituent plant species). For larger horticultural environments, either on-site sampling of physical characteristics, or the use of photo-based physical measurements would be appropriate.

One variant of this latter technique involves measuring relevant physical characteristics from photographs of the environment, and then expressing the physical variables in terms of the percentage of the scene each variable occupies. This method has been successfully used in the assessment of "vista" scenes, that is, panoramic scenes where the dominant features are at a great distance from the observer (Buyoff & Leuschner, 1978; Buyoff & Riesenman, 1979; Buyoff, Wellman & Daniel, 1982). An example of findings from this research suggests that sharp mountains are a positive feature of vista scenes up to a point; if the mountains comprise too much of the scene, however, preference ratings decline (Buyoff & Wellman, 1980). This photo-based procedure seems easily adaptable to a large garden where on-site physical measurement would be difficult, expensive and time consuming. As an example, Figure 3 shows a garden scene that includes rather diverse elements, only some of which are horticultural. The photo-based psychophysical procedure could be used here to help determine the relative contribution of horticultural elements to the perceived scenic beauty of the depicted environment. Figure 4 shows the same scene with a grid overlay from which one could determine the percentage of the scene comprised of architectural, sculptural and horticultural elements. The percentages associated with each type of element would constitute the "physical" variables to be related to perceptual responses, such as scenic beauty judgments, in the mathematical model.

A3. Psychological Methods

The final category of methods for gauging perceived environmental quality to be mentioned has been referred to as the psychological model (Daniel & Vining, 1983). As the name implies, psychological methods represent attempts to assess environmental quality by associating various psychological qualities or characteristics with landscapes. These characteristics are both cognitive and affective in nature, and typically include

FIGURE 3. Courtyard of Trammel Crow Building, Dallas

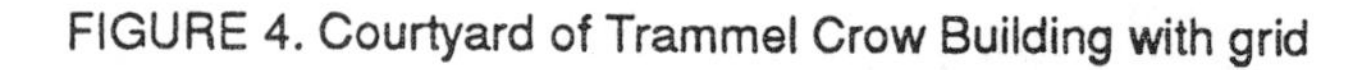

FIGURE 4. Courtyard of Trammel Crow Building with grid

such qualities as warmth, relaxation, interest, cheerfulness, etc., for high quality environments, and tension, gloom, fear, stress, etc., for low quality environments. Often, perceptual characteristics are of interest as well, and observers are asked to rate environments for qualities such as openness or depth, size, information rate, complexity, etc. As with the psychophysical methods, psychological assessment procedures usually involve the use of environmental surrogates (e.g., color slides) shown to groups of observers. Unlike the psychophysical methods, however, there is no attempt to relate ratings of psychological qualities to specific physical characteristics of the environment. Instead, ratings across a large battery of psychological qualities are reduced (through factor analysis or multidimensional scaling, for example) to a small number of factors or dimensions (3-5), which can then be used to describe environmental quality. Researchers occasionally relate scores on these higher order dimensions to independently gathered preference ratings for the same environments. A good example of work that relates psychological variables to preferences is research by the Kaplans and their colleagues (e.g., R. Kaplan, 1975; S. Kaplan, 1975; S. Kaplan, R. Kaplan, & Wendt, 1972). These researchers have consistently found that two higher order psychological qualities, which they have named mystery and legibility, are associated with visual preferences for environments. An environment has mystery if one perceives that further exploration of the environment holds the promise of more information; legibility refers to the relative ease with which an environment can be comprehended.

Adapting the general framework of the psychological model of methodology for use with horticultural environments would be a relatively straightforward task. Indeed, many of the same psychological qualities (e.g., warmth, relaxation, cheerfulness) used to describe large-scale environments would likely be appropriate in horticultural settings, given their obvious face validity for flowering plants. It is less likely that specific higher order dimensions, such as the Kaplans' mystery and legibility variables, could be used directly with horticultural stimuli, though that is an empirical question. One important caveat should be recognized, however, before any attempt is made to use psychological methods with horticultural stimuli. As was mentioned, when psychological methods are used, there is no attempt to relate the psychological qualities of environments to specific, measurable physical characteristics. Thus, if a given environment elicits very positive responses across a broad range of psychological variables, there would be no indication which specific elements in that environment (trees, grass, water, other vegetation, amount of blue sky, etc.) were responsible for the positive responses. Similarly, a garden that elicits positive responses from observers would have little utility as a design model without some knowledge of the specific elements in the garden scene and how they are related to the positive psychological responses. This is especially important for ornamental gardens, which are explicitly designed to have a visual effect, but it is important for other gardens as well, which are also likely to have visual effects.

A4. Summary

All three methods outlined above for the assessment of perceived environmental quality seem well-suited for application to horticultural settings. The primary distinction among them is the reliance on expert judges in the formal aesthetic model, while the psychophysical and psychological models employ naive observers. As was mentioned, there is no compelling scientific reason to prefer one type of judge over the other for the assessment of aesthetic quality. However, if the research question involves the relationship among preferred environments, stress-reduction and health, as suggested in Figure 1, then a case can be made for the use of naive observers, at least at first. One would expect the generalizability of findings from formal aesthetic assessments of environmental quality to be limited by the extent to which they differ from naive observers preferences for environments. This is because in the relationship shown in Figure 1, preferred environments are thought to be related to stress-reduction and health because they produce positive affective/arousal states; and the preferred environments believed to elicit those states are those of the typical users of the environments. Thus, an important empirical question regards the extent to which formal aesthetic assessments of environmental quality overlap with assessments of environmental quality derived from analyses of preferred environments and restorative environments.

B. Psychophysiology and Potential Health Effects

Though there is no generally agreed upon definition of psychophysiology, a recent definition offered by Cacioppo and Tassinary (1990) is especially appropriate for studying responses to horticultural environments. They define psychophysiology as ". . . the scientific study of social, psychological and behavioral phenomena as related to and revealed through physiological principles and events." This definition is especially appropriate because it implies a focus of study on organism-environment transactions, as distinct from more narrowly defined studies of physiology and anatomy. It is difficult, therefore, to overemphasize the importance of psychophysiological methods to the study of the health effects of environmental exposures.

We turn now to the consideration of methodological issues in psychophysiology relevant to the association of health effects with environmental exposures. A brief section on the important advantages of psychophysiological measures is followed by a description of what is necessary to draw strong inferences about psychological states from physiological data. A final section provides some illustrative examples from psychophysiology and environmental psychology, focusing on affect and arousal states, and suggests some possible directions for horticultural psychophysiology.

B1. The Importance of Psychophysiological Measures

Considering the investment in time, equipment and expert training required to conduct psychophysiological research, it is worthwhile to state explicitly some of the advantages of psychophysiological methods. First, the inclusion of psychophysiological methods in a research study can potentially validate self-report and behavioral indicators of affect and arousal. When several self-report or behavioral indicators suggest that a given horticultural environment produces positive moods and low arousal, for instance, one can more confidently conclude that the environment actually does have that effect if psychophysiological data are included. Second, psychophysiological measurements can be collected continuously during exposure to an environment, allowing a more fine-grained analysis of responses to that environment than would be possible with discrete behavioral observations or administrations of a mood adjective checklist or other self-report instrument. A third advantage of psychophysiological measurements is that they can provide information about affective and arousal states that cannot be had from behavioral and self-report measures. For example, it might be possible to measure the occurrence of subtle psychological states that are not easily articulated or accurately captured by questionnaires and behavioral observations. In addition, many research situations may not elicit complete candor on the part of research participants, who, for instance, may respond simply to please the researcher. Finally, an important advantage of psychophysiological measures is that they can provide data that are directly relevant to the potential health effects of exposures to horticultural stimuli. If passive exposure to horticultural stimuli is stress reducing, as it has been hypothesized to be for active interactions with plants, then the large body of literature relating stress to cardiovascular and immune diseases is germane to the question of the potential benefits of horticulture.

B2. Inferences in Psychophysiology

The implicit causal chain of inference indicated in Figure 1 suggests that psychological states (positive feelings, lowered arousal, stress reduction) are associated with physiological events (blood pressure, heart rate, EMG activity, immunoglobulin levels, etc.) that are directly related to organismic integrity and health. In considering which specific methods should be used to establish such relationships, it will be helpful to briefly examine a taxonomy of possible psychophysiological relationships suggested by Cacioppo and Tassinary (1990).

Figure 5 shows the possible relationships that can exist between elements in the psychological and physiological domains (indicated by the Greek letters Ψ and Φ, respectively, in the diagram). As Cacioppo and Tassinary note, only the first two of these relationships allow psychological elements to be specified as a function of physiological elements. This is significant, because physiological elements are often taken to be indicators of psychological events, both conceptually (as seen in Figures 1 and 2) and as experimental operationalizations (examples of which will be presented below). Thus, in the many-to-one and many-to-many psychophysiological relationships shown in Figure 5, it is not clear which psychological element is implicated by a given physiological event when

FIGURE 5

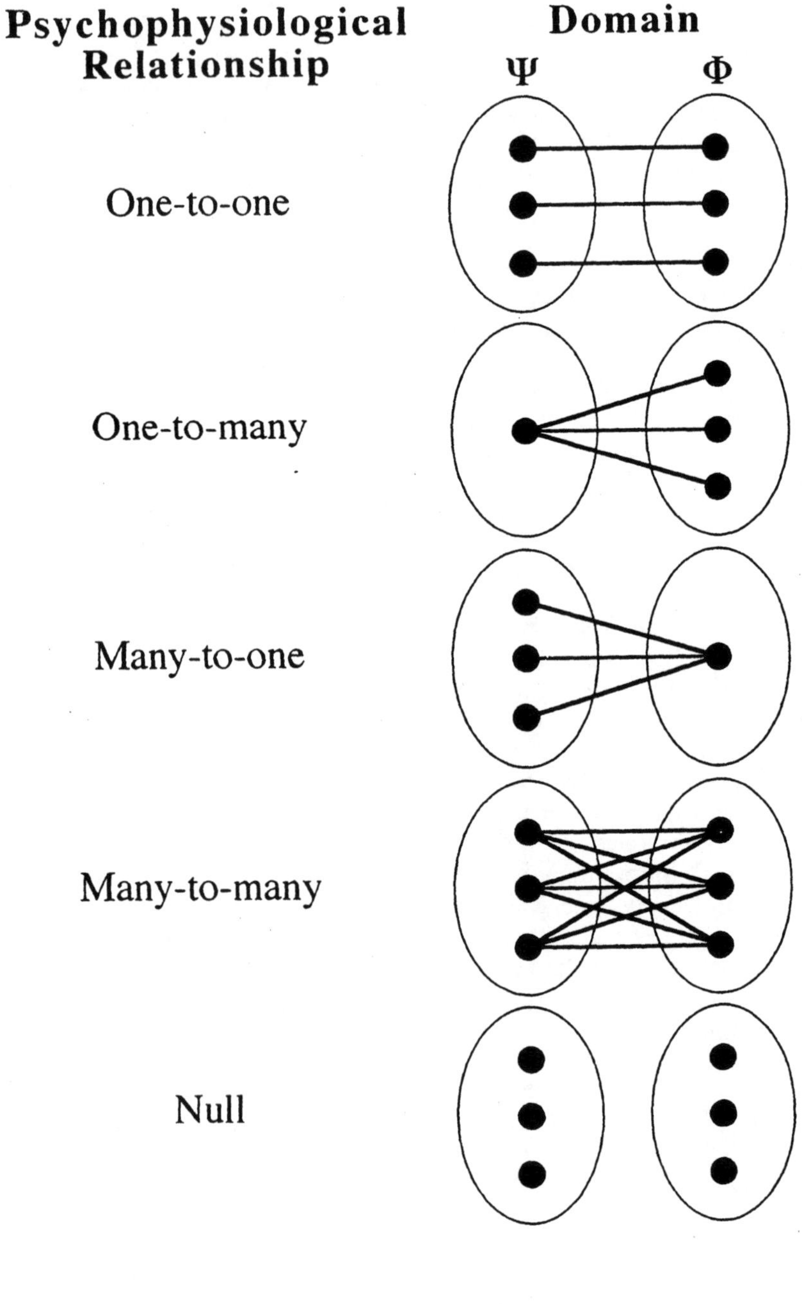

Psychophysiological Relationship

Domain

Ψ Φ

One-to-one

One-to-many

Many-to-one

Many-to-many

Null

there are multiple psychological events associated with it. Such ambiguous relationships can often be reduced to one-to-one relationships when spatial and temporal patterns of physiological responding are considered. Figure 6 presents an example of how this can be done. Three elements are listed in the psychological domain, orienting, startle and defense responses. The elements in the physiological domain are phasic skin conductance responding (a measure of moment to moment differences in sweat gland activity; SCR) and heart rate (HR). As the first panel shows, all three of the psychological responses are associated with skin conductance and heart rate responding, a many-to-many relationship. This relationship is reduced in the second panel by the introduction of a second class of physiological element, Φ', which represents a *pattern* of physiological responding. The pattern for the orienting response, increased SCR and decelerating HR, is distinct from that for the startle and defense responses, which are characterized by increased SCR and accelerating HR. Thus, the Ψ/Φ relationship for the orienting response has been reduced to a one-to-one relationship, with the physiological element consisting of a pattern of responding. In the third panel of Figure 6, the remaining many-to-one relationship among the psychological elements of startle and defense, and the Φ' element of increases in SCR and HR is further reduced by the introduction of a temporal differentiation, represented by Φ''. In this instance, the *rate* of HR acceleration helps to distinguish between startle and defense responses, being abrupt in the former and lingering in the latter, and allowing one-to-one relationships with physiological elements (Φ' or Φ'' response patterns) to emerge for each of the psychological elements.

B3. Some Examples

This taxonomy of potential Ψ/Φ relations, and the example used to illustrate how some of the more complex relationships can be reduced to allow reasonable inferences about psychological states, suggest that extreme care must be exercised when designing psychophysiological studies. This point can be further illustrated with examples that are more directly relevant to the potential health effects of interactions with horticultural environments. As has been mentioned, the two primary benefits associated with exposure to horticultural and other natural environments are positive emotions or moods, and reduced arousal. One approach to drawing inferences about these psychological states from physiological data would include a set of measures associated with emotional states and a set of measures associated with arousal. The former set would be especially useful in helping to establish the nature of the relationships between pref-

FIGURE 6

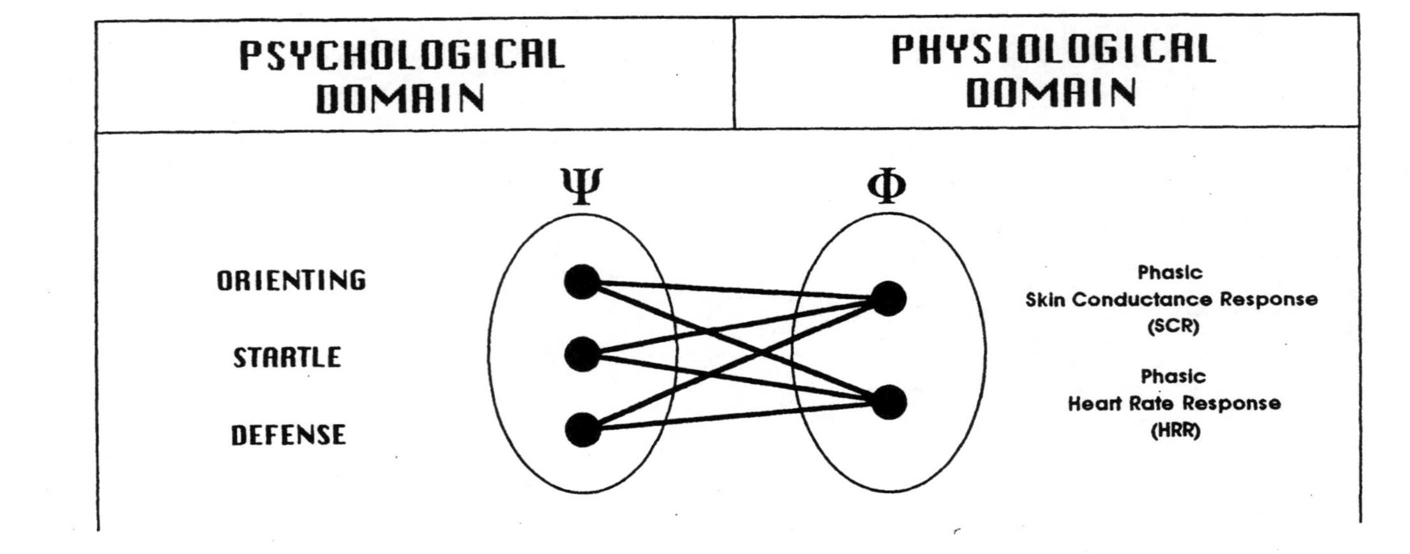
PSYCHOLOGICAL DOMAIN
PHYSIOLOGICAL DOMAIN
Ψ
Φ
ORIENTING
STARTLE
DEFENSE
Phasic
Skin Conductance Response
(SCR)
Phasic
Heart Rate Response
(HRR)

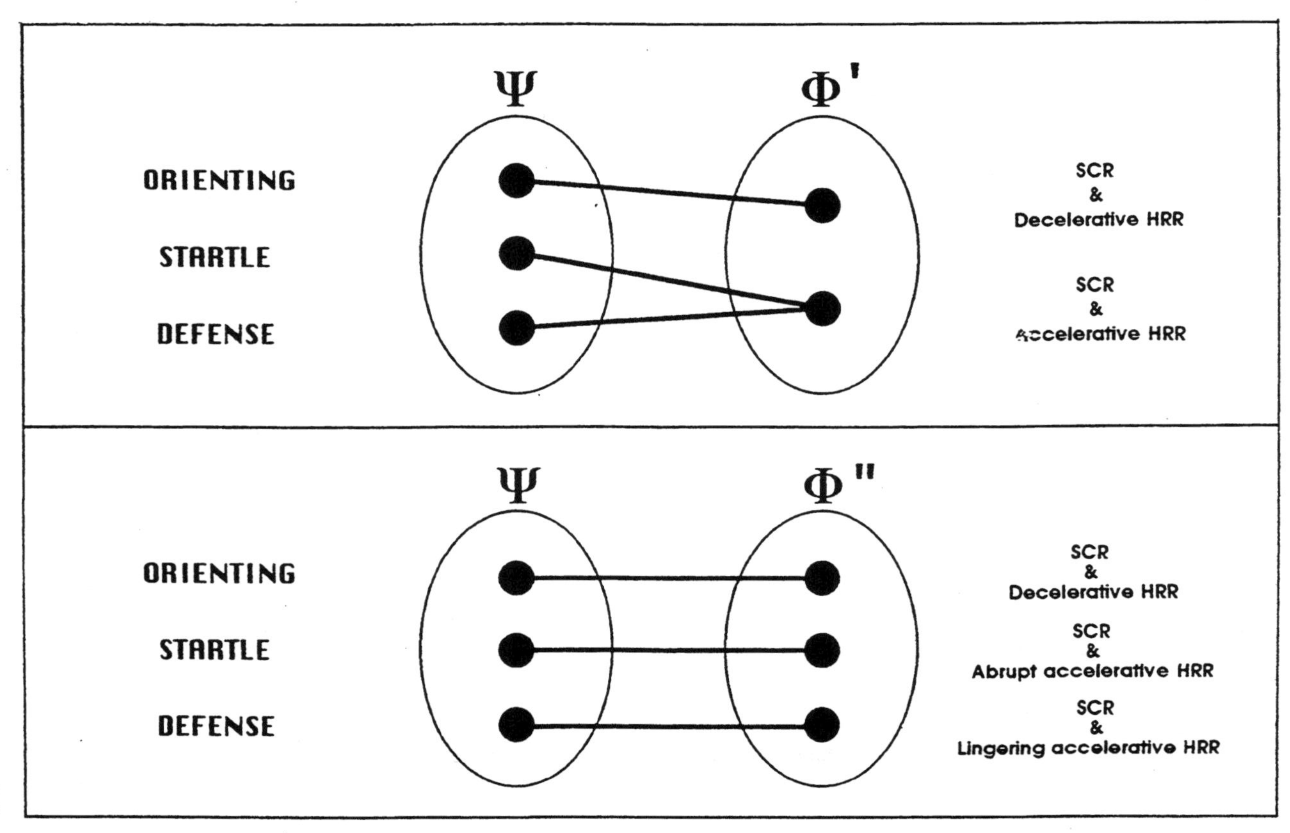
Ψ
Φ'
ORIENTING
STARTLE
DEFENSE
SCR & Decelerative HRR
SCR & Accelerative HRR
Ψ
Φ''
ORIENTING
STARTLE
DEFENSE
SCR & Decelerative HRR
SCR & Abrupt accelerative HRR
SCR & Lingering accelerative HRR

erences and other emotional responses to environments, an important theoretical issue, while the latter set may be more directly related to potential health effects. For instance, it may be that preferences are more closely related to the affective dimension of affect/arousal states, varying as a function of emotional valence, while being relatively independent of arousing qualities. For each dimension, however, the set of physiological measures is to be viewed as a pattern, or response profile, allowing inferences to be made about one-to-one relationships between the physiological response profiles and psychological states (mood and arousal). Below we present some examples of physiological response profiles that might be measured to help draw inferences about psychological states. We begin with a brief description of the response profiles that may be associated with positive and negative emotional responding. We then present a more lengthy description of responses that are indicative of arousal states. The measurement of arousal is emphasized because changes in arousal states have been hypothesized to have more direct relationships to health effects than changes in emotion or mood states.

One physiological response profile to be associated with emotional responding might include electromyographic[3] (EMG) measurement of three facial muscles, *corrugator supercilii*, *zygomaticus major* and *orbicularis oris*. Activity in the former two facial muscles has been associated with negative and positive emotional responding, respectively. The *corrugator* muscle is active in facial expressions that involve furrowing or knitting of the brow, and the *zygomaticus* is active in expressions and movements that draw the corners of the mouth up and back, as in a smile. The third muscle is associated with mouth movements and is useful as a check for artifactual activity recorded from the *zygomaticus* muscle placement (see Figure 7). This is necessary because activity can be recorded from a *zygomaticus* muscle placement in the absence of positive emotional responding (i.e., yawning, coughing, certain grimaces, clenching of the jaw, etc.). Among these three muscles, then, the patterned response profile that is most unambiguously associated with positive emotional responding is increased activity recorded from the *zygomaticus* muscle placement that is not associated with increased activity in either of the other two muscle placements (that is, in the absence of any negative emotional responding or nonemotional mouth movements).

For example, Cacioppo, Petty, Losch and Kim (1986) used these (and other) facial EMG placements to help discriminate subjects' responses to moderately unpleasant, mildly unpleasant, mildly pleasant and moderately pleasant scenes depicted by color slides. Subjects viewed each scene and then rated it for familiarity, arousal and liking. Liking was associated with

FIGURE 7

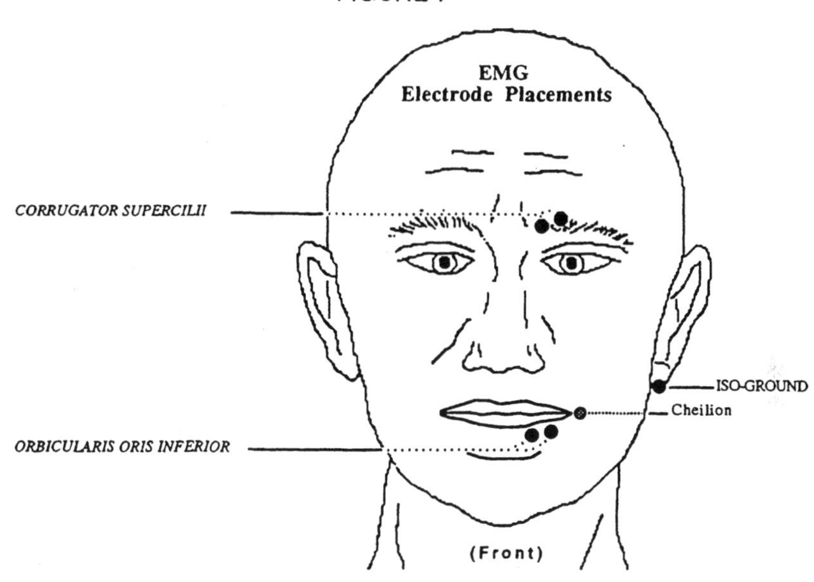
EMG
Electrode Placements
CORRUGATOR SUPERCILII
ISO-GROUND
Cheilion
ORBICULARIS ORIS INFERIOR
(Front)

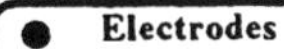
Electrodes
Anatomical Landmarks

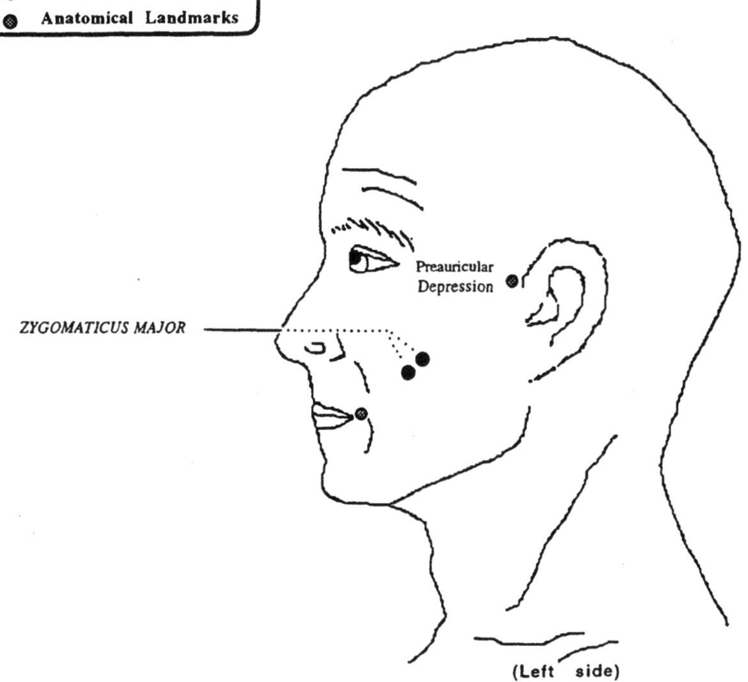
Preauricular
Depression
ZYGOMATICUS MAJOR
(Left side)

increased activity recorded over the *zygomaticus* (smile muscle) and decreased activity recorded over the *corrugator* (brow region), while self-reported negative responses produced the opposite pattern of recorded muscle activity–that is, increased activity recorded from the *corrugator* and decreased activity recorded from the *zygomaticus*. Importantly, activity recorded from the *orbicularis oris* region (lower lip) did not vary as a function of emotional valence, indicating that differences in activity recorded from the *zygomaticus* region were not due to extraneous, nonemotional mouth movements. In a study that used a similar slide presentation procedure, Dimberg and Ulrich (1992) recorded facial EMG responses to outdoor scenes dominated by green vegetation, but differing with respect to perceived openness or depth and aesthetic preference. They reported increased *zygomaticus* activity and relatively quiescent *corrugator* activity associated with the pleasant scenes, while the unpleasant scenes produced quiescent responding over both muscle regions. Though there was corroborating self-report data for the *zygomaticus* finding, activity was not recorded from the *orbicularis oris* muscle area, and the inference of positive emotional responding is not as strong as it might have been.

The response profile used to indicate changes in arousal depends upon one's definition of arousal, among other things. In psychophysiology, the term arousal is often used interchangeably with activation, which is typically defined in terms of increases in sympathetic nervous system (SNS) activity (Johnson & Anderson, 1990). The SNS is the division of the autonomic nervous system that energizes an organism physiologically in fight or flight situations, as well as in potentially threatening situations when fighting or fleeing are not possible or appropriate. Activation of portions of the SNS precipitates the release of catecholamines and corticosteroids, "stress hormones" that help to orchestrate the changes in physiological responding that support action in stressful situations. These changes often include (but are not limited to) increases in blood pressure, heart rate, respiration rate, electrodermal activity (e.g., skin conductance responding; EDA), and muscle tension. Which of these response systems is measured depends in part on the research question, the experimental procedure, and the available equipment and expertise. The two most frequently measured responses are EDA and heart rate (Dawson, Schell & Filion, 1990). Both responses are relatively easily acquired and relatively noninvasive, and both responses show changes soon after exposure to stressful stimuli or events. If a choice must be made between the two, electrodermal activity is often chosen over heart rate, for several reasons.

First, EDA is considered a more direct measure of SNS activation, because the relevant sweat glands are innervated almost exclusively by sympathetic nerves, whereas the heart receives both sympathetic and parasympathetic innervation. This is important because parasympathetic nervous system (PNS) activation generally has effects that are opposed to SNS activation–decreasing blood pressure, heart rate, respiration rate, etc.–serving restorative, anabolic functions of the body during nonstressful episodes. Thus, observed changes in heart rate (increases or decreases) could be due to changes in activation of either or both the parasympathetic and sympathetic nervous systems, undermining inferences about arousal. A second reason that EDA is often preferred over heart rate as an indicator of arousal is that EDA is relatively immune to somatic influences on heart rate that are independent of arousal. For example, heart rate oscillates in rhythm with respiration (though not perfectly), decelerating with inhalations and accelerating with exhalations. This means that inferences about arousal from heart rate data are strengthened by knowledge of the concomitant respiratory cycle, especially if the experimental protocol requires relatively short observation periods that coincide with inhalations or exhalations. A third reason to favor EDA over heart rate as an indicator of arousal is evidence that one's heart rate may decelerate during outwardly focused processing and accelerate during inwardly focused processing (Lacey & Lacey, 1974), though both types of processing may produce arousal. If both types of processing occur during a given recording period, then heart rate may well show no arousal effect at all; in this case, a second arousal measure would be needed to help disambiguate the heart rate data. EDA does not discriminate between arousal produced by inwardly and outwardly focused processing.

Though these reasons seem to argue strongly for the measurement of EDA over heart rate as an indicator of arousal, there are limitations to the inferences about sympathetic activation that can be drawn from EDA data. Because EDA is a measure of sweat gland activity, it is affected by differences in skin temperature, and by ambient temperature and humidity. These characteristics require that the measure be taken in a relatively stable environment, both within and across individuals. Also, while EDA is a relatively "clean" measure of sympathetic activation, there are more direct methods. For instance, microneurography allows direct measurement from sympathetic nerve endings through the insertion of fine wire recording electrodes into the skin or sympathetically innervated muscles. In addition to being a direct measure of SNS activation, muscular microneurographic insertions provide information about skeletal muscle blood flow that cannot be gleaned from surface electrode placements (Johnson & Anderson, 1990). Finally, because of its significant advantages, such as ease of acquisition, the relative directness of the measure, and its independence from other non-arousal somatic influences, there is the temptation to make EDA the sole measure of arousal. This is problematic for two reasons. First, a single measure of SNS activation ignores distinctions that have been made among cortical, SNS and behavioral arousal (Lacey, 1967; Stern & Sison, 1990). In the absence of any practical or theoretical reasons to focus on only one form of arousal, additional measures can provide information on the extent and character of arousal. A second reason to include multiple arousal measures if possible is the variability across individuals in electrodermal responsivity. More generally, the notion of individual response stereotypy has gained acceptance in the field of psychophysiology (Stern & Sison, 1990). Individual response stereotypy refers to differences in typical physiological response patterns across individuals to a variety of stimuli. Thus, while some individuals may discriminate among stimuli with varying degrees of electrodermal responding, others may show relatively stable EDA to the same stimuli, but respond differentially with respect to heart rate, blood pressure or muscle tension responses. The use of multiple response measures maximizes the likelihood of being able to detect changes in arousal across individuals.

In a recent study, Ulrich et al. (1991) recorded multiple arousal measures and used a recovery-from-stress experimental procedure in an attempt to gauge differential responses to natural and urban environmental surrogates. Participants in this research viewed a brief, mild videotaped stressor, followed by a brief videotaped presentation of an outdoor environment. Among the physiological measures continuously recorded were electrodermal activity and pulse transit time (PTT), which is inversely related to systolic blood pressure. As PTT increases, systolic blood pressure decreases. Ulrich and his colleagues found that those subjects who viewed natural environments had fewer SCRs and longer PTTs (lower blood pressure) during the recovery phase of the experiment relative to those who viewed urban environments. Both of these findings indicate lowered arousal, and the inference is a relatively strong one given the agreement between measures. If only one of these indicators had suggested lowered arousal, then the inference would be considerably weakened; and, if one of them had been in the opposite direction (faster PTTs, for instance, suggesting increasing blood pressure), unambiguous inferences about arousal would have been unwarranted.

CONCLUDING REMARKS

Taken together, these two disciplines, environmental aesthetics and psychophysiology, have developed methods that can be extremely useful for those interested in the potential salutary effects of passive exposures to horticultural environments. Specifically, they offer the means by which assumptions underlying the relationship outlined in Figure 1 can be systematically tested. Each of the two groups of methods described here is well-suited to address issues in that relationship. Those methods borrowed from environmental aesthetics provide for the assessment of the perceived visual quality of horticultural environments, while the methods drawn from psychophysiology allow researchers to test the restorative potential of horticulture with measures and procedures that have some cachet in the medical community. Thus, both types of methods can generate information that can be applied to the design of horticultural environments, and may ultimately garner greater community support for horticultural pursuits. Moreover, information generated by both classes of methods would be pertinent to theoretical issues that have yet to be reconciled in environmental psychology, such as establishing the extent to which preferred and restorative environments overlap.

Though the potential for theoretically and practically useful information is great with these methods, their use can be expensive and may require specialized training, especially the psychophysiological methods. On the other hand, few environmental psychologists or psychophysiologists are trained horticulturalists, and thus they are ill-equipped to apply these methods to horticultural environments. This situation suggests that collaborative efforts among these disciplines could be beneficial to all involved, and we wholeheartedly advocate such cross-pollination.

ACKNOWLEDGEMENT

We would like to thank Terry Daniel and John Hetherington at the University of Arizona for help with the section on psychophysical methods.

NOTES

1. Here and throughout, "affect" is used interchangeably with "emotion." Also, a distinction will be maintained between psychological and physiological arousal, the former referring to a perceived deviation from one's ordinary waking tranquility, while the latter refers to sympathetic nervous system activity (see section on psychophysiological methods, below).

2. At an even broader level, to fully understand the potential health effects of passive environmental exposures, one would also need to include factors relating to individual differences (coping strategies, diet and exercise habits, genetic predispositions for coronary disease and immunocompetence, etc.) and the socio-

physical environment (various pollution levels, residential crowding and safety, availability and quality of local health care systems, social cohesion and sense of community, etc.; the interested reader should see Stokols, 1992, for a more complete list).

3. Electromyography refers to "... a continuous record of the intrinsic electrical activity associated with muscle contraction by means of surface or needle electrodes" (Cacioppo & Tassinary, 1990).

REFERENCES

Baird, J.C. & Bergland, B. (1989). Thesis for environmental psychophysics. *Journal of Environmental Psychology, 9*, 345-356.

Burgess, C.W. (1990). Horticulture and its application to the institutionalized elderly. *Activities, Adaptation & Aging, 14* (3), 51-61.

Buyoff, G.J. & Leuschner, W.A. (1978). Estimating psychological disutility from damaged forest stands. *Forest Science, 24*, 424-432.

Buyoff, G.J. & Riesenman, M.F. (1979). Experimental manipulation of dimensionality in landscape preference judgments: A quantitative validation. *Leisure Sciences*, 2, 221-238.

Buyoff, G.J. & Wellman, J.D. (1980). The specification of a non-linear psychophysical function for visual landscape dimensions. *Journal of Leisure Research, 12*, 257-272.

Buyoff, G.J., Wellman, J.D. & Daniel, T.C. (1982). Predicting scenic quality for mountain pine beetle and western spruce budworm damaged vistas. *Forest Science, 28*, 827-838.

Cacioppo, J.T. & Tassinary, L.G. (1990). Psychophysiology and psychophysiological inference. In J.T. Cacioppo & L.G. Tassinary, Eds., *Principles of psychophysiology: Physical, social and inferential elements.* Cambridge: Cambridge University Press, 3-32.

Cacioppo, J.T., Petty, R.E., Losch, M.E. & Kim, H.S. (1986). Electromyographic activity over facial muscle regions can differentiate the valence and intensity of affective reactions. *Journal of Personality and Social Psychology, 50*, 260-268.

Daniel, T.C. & Boster, R.S. (1976). *Measuring Landscape Esthetics: The Scenic Beauty Estimation Method.* (USDA Forest Service Research Paper RM 167) Ft. Collins, CO: Rocky Mountain Forest and Range Experiment Station.

Daniel, T.C. & Vining, J. (1983). Methodological issues in the assessment of landscape quality. In I. Altman & J. Wohlwill, Eds., *Human behavior and environment (Vol. VI).*

Daniel, T.C., Wheeler, L., Boster, R.S. & Best, P. (1973). Quantitative evaluation of landscapes: An application of signal detection analysis to forest management alterations. *Man-Environment Systems, 3*, 330-344.

Dawson, M.E., Schell, A.M. & Filion, D.L. (1990). The electrodermal system. In J.T. Cacioppo & L.G. Tassinary, Eds., *Principles of psychophysiology: Physical, social and inferential elements.* Cambridge: Cambridge University Press, 295-324.

Dimberg, U. & Ulrich, R.S. (1992?). Facial electromyography can discriminate among emotional responses to everyday physical environments. Unpublished manuscript, Department of Psychology, Uppsala University, Sweden, and College of Architecture, Texas A&M University, College Station, Texas.

Glacken, C.J. (1967). *Traces on the Rhodian shore: Nature and culture in western thought from ancient times to the end of the eighteenth century.* Berkeley: University of California Press.

Hefley, P.D. (1973). Horticulture: A therapeutic tool. *Journal of Rehabilitation, 39* (1), 27-29.

Hetherington, J. (1992). *Toward an integrative theory of environmental aesthetics.* Ph.D. Dissertation, University of Arizona.

Horne, D.C. (1974). *An evaluation of the effectiveness of horticultural therapy on the life satisfaction level of aged persons confined to a rest care facility.* M.S. Thesis, Clemson University. As cited in Burgess (1990).

Hull, R.B. & Buyoff, G.J. (1981). On the law of comparative judgment: Scaling with intransitive observers and multidimensional stimuli. *Educational and Psychological Measurement, 41*, 1083-1089.

Jackson, R.H., Hudman, L.E. & England, J.L. (1978). Assessment of the environmental impact of high voltage power transmission lines. *Journal of Environmental Management, 6*, 153-170.

Johnson, A.K. & Anderson, E.A. (1990). Stress and arousal. In J.T. Cacioppo & L.G. Tassinary, Eds., *Principles of psychophysiology: Physical, social and inferential elements.* Cambridge: Cambridge University Press, 216-252.

Kaplan, R. (1975). Some methods and strategies in the prediction of preference. In E.H. Zube, R.O. Brush & J.A. Fabos, Eds., *Landscape assessment: Values, perceptions and resources.* Stroudsburg, PA: Dowden, Hutchinson & Ross, 118-119.

Kaplan, S. (1975). An informal model for the prediction of preference. In E.H. Zube, R.O. Brush & J.A. Fabos, Eds., *Landscape assessment: Values, perceptions and resources.* Stroudsburg, PA: Dowden, Hutchinson & Ross, 92-101.

Kaplan, S., Kaplan, R. & Wendt, J.S. (1972). Rated preference and complexity for natural and urban visual material. *Perception and Psychophysics, 12*, 354-356.

Lacey, J.I. (1967). Somatic response patterning and stress: Some revisions of activation theory. In M.H. Appley & R.H. Trumbull, Eds., *Psychological stress: Issues in research.* New York: Appleton-Century-Crofts.

Lacey, B.C. & Lacey, J.I. (1974). Studies of heart rate and other bodily processes during sensorimotor behavior. In P.A. Obrist, A.H. Black, J. Brener & L.V. DiCara, Eds., *Cardiovascular psychophysiology.* Chicago: Aldine, 538-564.

Malm, W., Kelly, K., Molenar, J. & Daniel, T. (1981). Human perception of visual air quality (uniform haze). *Atmospheric Environment, 15*, 1875-1890.

McCurry, E. (1963). "Flowers and gardens–Therapy unlimited." Pontiac: Pontiac State Hospital. As cited in Hefley (1973).

O'Connor, A.H. (1958). *Horticulture as a curative.* Cornell Plantation, Ithaca: 14(3): 42. As cited in Hefley (1973).

Olmsted, F.L. (1865). The value and care of parks. Report to the Congress of the State of California. Reprinted in *Landscape Architecture, 17*, 20-23.

Parsons, R. (1991a). The potential influences of environmental perception on human health. *Journal of Environmental Psychology, 11*, 1-23.

Parsons, R. (1991b). *Recovery from stress during exposure to videotaped outdoor environments.* Ph.D. Dissertation, University of Arizona.

Parsons, R. & Daniel, T.C. (1988). Assessing visibility impairment in Class I Parks and Wilderness Areas. *Society and Natural Resources, 1*, 227-240.

Pitt, D.G. & Zube, E.H. (1979). The Q-sort method: Use in landscape assessment research and landscape planning. In *Our National Landscape* (USDA Forest Service Technical Report PSW-35). Berkeley, CA: Pacific Southwest Forest and Range Experiment Station.

Shafer, E.L. & Richards, T.A. (1974). *A Comparison of Viewer Reactions to Outdoor Scenes and Photographs of Those Scenes* (USDA Forest Service Research Paper NE-302). Upper Darby, PA: Northeastern Forest Experiment Station.

Shepard, P. (1967). *Man in the landscape: A historic view of the esthetics of nature.* New York: Alfred A. Knopf.

Stern, R.M. & Sison, C.E. (1990). Response patterning. In J.T. Cacioppo & L.G. Tassinary, Eds., *Principles of psychophysiology: Physical, social and inferential elements.* Cambridge: Cambridge University Press, 193-215.

Stokols, D. (1992). Establishing and maintaining healthy environments: Toward a social ecology of health promotion. *American Psychologist, 47* (1), 6-22.

Train, R.L. (1976). *The effectiveness of horticultural therapy in maintaining the life satisfaction of geriatrics.* M.S. Thesis, Kansas State University. As cited in Burgess (1990).

Ulrich, R.S. & Parsons, R. (1992). The influences of passive experiences with plants on human well-being and health. In D. Relf, Ed., *The role of horticulture in human well-being and social development.* Portland, OR: Timber Press, 93-105.

Ulrich, R.S., Simons, R.F., Losito, B.D., Fiorito, E., Miles, M.A. & Zelson, M. (1991). Stress recovery during exposure to natural and urban environments. *Journal of Environmental Psychology, 11*, 201-230.

Walker, S.E. & Duffield, B.S. (1983). Urban parks and open spaces: An overview. *Landscape Research, 8*, 2-12.

Watson, D.P. & Burlingame, A.W. (1960). *Therapy Through Horticulture.* New York: The MacMillan Company.

Wohlwill, J.F. (1979). What belongs where: Research on fittingness of man-made structures in natural settings. In T.C. Daniel, E.H. Zube & B.L. Driver, Eds., *Assessing Amenity Resource Values* (USDA General Technical Report RM 68). Ft. Collins, CO: Rocky Mountain Forest and Range Experiment Station.

Zube, E.H., Pitt, D.D. & Anderson, T.W. (1975). *Perception and Measurement of scenic resources in the Southern Connecticut River Valley* (Institute of Man and His Environment Publication R-74-1). Amherst: University of Massachusetts.

Zube, E.H., Sell, J.L. & Taylor, J.G. (1982). Landscape perception: Research, application and theory. *Landscape Planning, 9*, 1-33.

Chapter 29

Nurturing People-Plant Relationships in Order to Foster Environmental and Community Stewardship: The Rutgers Environmental and Community Stewardship (R.E.A.C.S.) Program

William T. Hlubik
Harry Betros

SUMMARY. Environmental programs have traditionally focused on alleviating environmental problems with an isolated or fragmented approach, rather than addressing the interrelated issues that constitute such problems.

Success of environmental stewardship depends on research, establishing techniques for integrating the diverse elements that constitute environmental problems.

The Rutgers Environmental and Community Stewardship (R.E.A.C.S.) Program is a proposed application of interdisciplinary research. In addition, it offers the opportunity to collect quantified data to evaluate the efficacy of such programs. The methods used to evaluate the program could serve as a useful model for similar programs. Currently the program is in the initial stages of implementation.

> Scientific investigation into the value of people-plant relationships can be applied to environmental problem solving. The R.E.A.C.S. program will integrate people-plant relationships into all aspects of program delivery.

Humankind's intrinsic affinity for plants may provide the logical impetus and practical focal point for establishing effective community models for environmental and community stewardship. Understanding the need for plants is only the first step; one must then incorporate plants into personal and community planning processes with a desire to fulfill human needs and create a healthy environment.

ENVIRONMENTAL AND COMMUNITY STEWARDSHIP DEFINED

Common terminology can be quite subjective. The word "environment," for instance, may elicit a variety of responses from the listener, depending on the knowledge and experiences of the listener. The concept of environmental and community stewardship is the individual and collective acceptance of responsibility for the long term nurturing and preservation of the natural physical environment as well as a healthy community environment.

Unless one fosters the healthy development of local community environments, one can not hope to solve environmental problems on a global scale. There are numerous opportunities within one's home and community to begin practicing environmental stewardship.

Community Defined

According to Lewis (1992), "The term community refers to people who live in some spatial relationship to one another and who share interests and values. The community might be a neighborhood, housing project, school, prison, or other spatially defined relationship." Lewis (1992) points out that "Ideally, a community should have a clear sense of itself and seek cooperatively to improve its physical, economic and social conditions."

A study reported in the American Journal of Community Psychology (Brogan, Douglas 1980) examines the relationship of the physical environment to the psychosocial health of urban residents. The study gathered data from 100 city blocks in Atlanta, Georgia where 21 indices of deviant psychosocial behavior, 104 physical environmental indices (such as landscaping and land use), and 106 sociocultural environmental indices (such as population density and income) were measured. The results of the study revealed that the characteristics of the physical environment (landscaping) were as important as the characteristics of the sociocultural environment in relating differences in psychosocial health.

The physical condition of the natural or man-made environments can greatly influence how one chooses to interact with these environments and with one another. The physical condition of our surroundings can be changed in subtle, yet significant and meaningful ways. The concept of "nearby nature" (Kaplan 1992) demonstrates the subtle, yet significant impact that plants have on the way people perceive their physical environment.

Nearby Nature Concept

Kaplan (1992) defines "nearby nature" as "one plant or many plants and also the place created by them." The significance of the nearby nature concept is that it relates the importance of a person's exposure to nature or the natural world in terms of their personal experiences to their immediate environment. That environment may be limited to a single houseplant, a garden or simply a view of plants from a window. Kaplan disagrees with commonly accepted definitions of nature by some professional groups as "something reserved for wilder places that does not exist in the urban context." Kaplan goes on to say that "nature is not merely an amenity, luxury, frill or decoration. The availability of nearby nature meets an essential human need; fortunately, it is a need that is relatively easy to meet. A garden patch, some trees nearby, and a chance to see them can all be provided at minimal cost and for enormous benefits" (Kaplan 1992).

In describing the variety of important psychological and cultural effects that plants can have on people, Schroeder (1988) explains that the mere visual exposure to natural vegetation can promote relaxation, recovery from stress, and recuperation from surgery. Trees not only enhance our perceived quality of different environments but provide a symbolic link between human cultures and their underlying spiritual values. The importance of planting and maintaining a garden near one's home is that it may help individuals achieve a sense of self, place, and temporal identity (Schroeder 1988).

The physical appearance of a community can have an enormous impact on how individual community members feel about themselves (Lewis 1990). The natural and man-made physical environments represent the fabric or the landscape upon which the human community interacts with one another and, in turn, impacts their local and global environments.

Lewis (1979) discusses humans' innate need for nature and the natural settings within which man evolved. He indicates that part of the stress encountered in urban environments may be related to the lack of natural settings or plants in these environments. As part of a solution the author suggests that planning and establishing urban activities and settings involving nature would help to create life enhancing and tranquil settings for residents. He describes successful gardening projects in low income areas of New York, Philadelphia, Chicago and Vancouver which "provide human benefits and satisfactions of enhanced self-esteem, increased sociability, reduction in vandalism, cleaner streets, painted buildings, and revitalized neighborhoods."

THE ENVIRONMENTAL AND COMMUNITY STEWARDSHIP PROGRAM

Resolving the needs and problems of local community environments must be considered as the starting point in order to resolve environmental concerns on a global scale. The ability to resolve environmental degradation depends on the collective action of individuals. The fate of the earth hinges on the large scale adoption of environmental stewardship. The success of environmental stewardship depends on individual human beings taking responsibility for environmental degradation and making the necessary changes in their everyday lifestyles.

In order to help individuals come to terms with accepting responsibility and taking positive action for environmental stewardship, the Rutgers Environmental and Community Stewardship (R.E.A.C.S.) program was formed. The program is an action oriented community environmental and education program that utilizes an intricate network of volunteers at all levels of program development and implementation. Although the program is still in its infancy, the volunteer network is growing and preparing for program implementation.

The objectives of the R.E.A.C.S. program are to provide practical community projects that empower people to take an active role in saving their local and global environments. The program involves individual, family and community participation in the reuse, recycling and reduction of natural resources and the development of hands-on planting projects. The planting projects include the establishment of community and school flower and vegetable gardens, tree planting projects for parks and streets, and the cleanup and rejuvenation of abandoned lots.

Community projects and environmental education will be guided by a core group of youth and adult environmental stewards, who are volunteer leaders. The core leaders will be trained by a network of staff and volunteers through the Rutgers University Cooperative Extension Program. The teachers include county extension staff, trained volunteers, college staff and other competent professionals. Leaders will receive formal classroom training in leadership development, problem solving, ecology, and horticulture. In return for this training, core leaders will be required to donate 100 hours of community volunteer service through the R.E.A.C.S. program. Trained R.E.A.C.S. leaders will share their newly developed knowledge and skills with their fellow community members through educational sessions and hands-on projects. R.E.A.C.S. core leaders will help develop and maintain an extensive volunteer network to implement programs throughout the community.

In order to achieve success, it is vitally important that community members be involved in the planning and implementation process of community programs or projects. If community members feel that they are an integral part of the program, then they are more likely to get involved and stay active. According to Ellis and Ray (1988) "Achieving loyalty, participation, and involvement on the part of staff, volunteers or other organizational members is essentially a function of each person's sense of ownership or proprietorship of the organization."

The R.E.A.C.S. program will be guided by recognized community leaders who genuinely care about improving the community environment. Program success will depend on continued, active participation of community members and recognized leaders. Tapping into the diverse knowledge and experience of concerned participants will strengthen the program and help participants develop a needed sense of ownership.

The challenge lies in eliciting a sense of individual responsibility for environmental stewardship. The collective historical actions of humanity have equated the definition of civilization or culture with the ability to dominate and control the environment, rather than develop mutual coexistence (Janick 1992).

The Oakland, California urban tree planting program is a successful example of how to use plants to bring people together and inspire community activism. Ames (1980) states that the program "promotes a community of interest, social organization and participation in activities consistent with American values and success." Ames stresses the importance of community participation in the tree planting process in order to enhance the sense of community among participants and ensure survival of the program.

PEOPLE-PLANT INTERACTIONS

People-plant interactions can provide the stimulus for positive people-people interactions within a community. The collective interactions of individual community members with one another define the characteristics and health of a particular community. Community gardens bring people together in a natural and nonthreatening way that allows for positive communication and interaction.

In relation to the context of this paper, an appropriate definition for people-plant interactions is given by Relf (1992) as follows: "People-Plant interaction–the wide array of human responses (mental, physical, and social) that occur as a result of both active and passive participation with plants. Because of the human benefits plants engender, this interaction implies stewardship of plants that are susceptible to man's impact on the environment."

The Universal Attraction of People to Plants

The universal attraction of people to plants, and the positive power that plants exert over our psyche, provides an excellent nucleus from which to evolve an environmental stewardship program. Such programs require support from a diverse group of people commonly found within a community.

According to Lewis (1992), plants have intrinsic qualities that encourage responses from all human beings. The world is a fast-paced, competitive and threatening place for many people. Plants are nonthreatening and nondiscriminating organisms, that respond to the proper care of any individual, regardless of their race, intellectual or physical capacities.

Humans' dependency on plants transcends the basic survival needs of food and shelter and involves psychological benefits that provide "a sense of beauty, peacefulness, joy and excitement" (Burde 1991). According to Lees (1973) "It is obvious that the concept of a garden goes beyond an aesthetic composition; it also signifies the necessity of men to live intimately with nature." In the same paper, Lees theorizes that "man needs to have some part of the forest (from which he came) around him" (Lees 1973).

BENEFITS OF COMMUNITY GARDENING

A study by Clark (1980) discussed the importance of community gardens for social, economic and personal benefits. Clark states that "two million people in the United States participate in community gardening programs and another 6 million express an interest if more programs were available" (Clark 1980).

Gardening is an example of an active people-plant interaction that calls for the care and nurturing of a dependent living creature outside of ourselves. The ability of plants to evoke nurturing characteristics instinctive to all human beings has important implications for the healing of a wounded society.

Gardening provides a means by which to empower people to take positive action to improve the community environments they live in. The fruits of their gardening activities provide a very powerful, visual reinforcement to the gardener and to those who experience it. Gardens offer a unique opportunity to bring the young and old together to talk, to work, to play, and to enjoy one another and share the values and wisdom of the past and present.

The Newark urban gardening program (Patel 1992) is an excellent example of how a successful community gardening program can become the foundation for community development. Program participants convert vacant lots strewn with garbage into beautiful vegetable and flower gardens. Over $450,000 worth of vegetables were produced in 905 community gardens in 1989. According to Patel (1992), "For the neediest members of society, community gardening provided more than money, nutrition, and open space. It provided a sense of self-pride and self-worth in producing food on their own and for themselves. Community gardening is an activity that sells itself."

According to Lewis (1992), "It is our humanity, not our economic or social status, that qualifies us to benefit from gardening." Gardening naturally lends itself to group activities, and the social interaction that often occurs in a community garden setting can provide a "catalyst for social intervention" (Lewis 1992). The community garden provides a nonthreatening common ground where people begin to communicate with their neighbors about gardening and other affairs that effect one another and their entire community. The development of a sense of community depends upon individual community members coming together to communicate and break down the walls of isolation that separate them. The separation can lead to a sense of powerlessness and apathy. The key is to present practical opportunities for people to change their immediate environments.

In order to solve environmental problems, society needs to establish ecological values. The prioritization of environmental education can help the next generation attain the knowledge and experience necessary to create an environmental ethic. The ability of individuals to take personal responsibility for the future of the world they live in will depend on the quality of their environmental education. This educational process should not be limited to science classes. Environmental education needs to be integrated through the entire curriculum in order to demonstrate the dynamics of environmental problems. The concept of environmental and community stewardship is long overdue.

CONCLUSION

Responsible stewardship will happen only if people can be re-awakened to a childlike sense of wonder. Individuals must first develop an awareness and excitement about the natural environment. This must be followed by an understanding of the interaction of plants, animals and people in natural and man-made communities. With understanding comes a concern for the present state of the environment and an exploration of ways to reverse negative trends. Finally, activism can initiate positive changes. Unless people's environmental growth is properly developed, activism will probably not be sustained over long periods.

According to Janick (1992), "Horticulture must increase in importance in schools, in homes, and in communities, to underscore the inter-connectedness of the living world and to improve the beauty and quality of life here on earth."

Almost every human being is capable of making significant contributions towards the survival of "spaceship earth." The intent of the R.E.A.C.S. program is to help individuals understand and witness how they can affect positive changes in their local community environments. This newly acquired knowledge is then shared with others. The future of the earth rests in the collective actions of concerned individuals and communities to foster an environmental ethic of global proportions.

REFERENCES

Ames, R.G. 1980. Urban tree planting programs: A sociological perspective. Hort Science 15(2):135-137.

Brogan, D.R. and J.L. Douglas 1980. Physical environment correlates of psychosocial health among urban residents. American Journal of Community Psychology 1980 Oct. Vol. 8(5):507-522.

Burde, L. 1991. Why people love trees. National Woodlands 14(3):12.

Clark, G.A. 1980. Selected factors influencing the success of a community garden. M.S. Thesis, Kansas State University.

Ellis, J. and J. Ray. Winter 1987-1988. Organizational Proprietorship: A Participation Model. The Journal of Volunteer Administration. VI(2):12-19.
Janick, J. 1992. Horticulture and Human Culture. In *The Role of Horticulture in Human Well-Being and Social Development*, edited by Diane Relf. Portland, Oregon: Timber Press: 19-27.
Kaplan, R. 1992. The Psychological Benefits of Nearby Nature. In *The Role of Horticulture in Human Well-Being and Social Development*, edited by Diane Relf. Portland, Oregon: Timber Press: 125-133.
Lees, C.B. 1973. Why men make gardens. Longwood graduate program seminars. 5:14-19.
Lewis, C.A. 1979. Healing in the urban environment: A person/plant viewpoint. Journal of the American Planning Association 45: 330-338.
Lewis, C.A. 1992. Effects of Plants and Gardening in Creating Interpersonal and Community Well-Being. In *The Role of Horticulture in Human Well-Being and Social Development*, edited by Diane Relf. Portland, Oregon: Timber Press: 55-65.
Patel, I.C. 1992. Socio-Economic Impact of Community Gardening in an Urban Setting. In *The Role of Horticulture in Human Well-Being and Social Development*, edited by Diane Relf. Portland, Oregon: Timber Press: 84-87.
Relf, D. 1992. Conducting the Research and Putting It Into Action. In *The Role of Horticulture in Human Well-Being and Social Development*, edited by Diane Relf. Portland, Oregon: Timber Press: 193-206.
Schroeder, H.W. 1988. Psychological and cultural effects of forests on people. In *Proceedings of the Society of American Foresters National Convention*. Rochester, New York, October 17, 1988.

Chapter 30

A Socio-Economic Impact of New Crops Production on Diverse Groups of People: A Case Study in Northwest Missouri

Alejandro Ching
Duane Jewell

SUMMARY. Some professional workers and business people in cities with populations ranging from 10,000 to over 100,000, often purchase a parcel of land to live on and be in contact with plants. Besides providing positive psychological and social benefits to their well-being, the economic factor will affect these positive benefits, creating a socio-economic pattern that eventually will be conducive to fostering their relationship with the land. Implementing a continuing adult education system will help promote and improve their socio-economic condition. As part of the Northwest Missouri State University's Alternative Crops Program, any interested individual can retrieve information on plant culture from a computer data base program. Furthermore, experimental plot demonstrations at the experiment station, off-campus farm demonstrations, periodic workshops and technical meetings will alleviate people's anxiety and will help foster a positive attitude that eventually will impact their socio-economic condition. Above all, this creates a trusting relationship to work on the land. A feedback system can also be developed with university input, people and plant/environment response and vice versa.

INTRODUCTION

Historically, farming was established by people migrating from the eastern seaboard to the midwest. A lot of the pioneers and homesteaders came with different backgrounds and occupational trades. There are many reasons for the wave of people moving to farm the open prairies of the Midwest, with utopias of building a future, love for the land, food production to feed the new colonies, and the flight of the unemployed from a collapsing economy of the 1930's. In the decade of the 1980's, a farm crisis forced farm foreclosure due to debts acquired by farmers, and low market prices for soybeans and corn in the midwest. This crisis created "a farm for sale" condition. As a result of this condition, a significant number of urban people with and without a farming background have been purchasing these farms. Most of these persons hold jobs in cities and nearby towns. Most of these land holdings tend to be small, and are planted with grasses and leguminous species. Most of these people have a desire to be close to nature and establish a relationship with plants. Often times, some have notions of what they want to grow, but others do not know what plants to raise.

This paper intends to provide information and strategies for extension personnel, researchers, students of agriculture, and potential urban farmers. The focus will be on issues concerning institutional support and motivational cues on the relationships of plants and people, especially people coming to live in rural areas.

THE BASIS OF THIS STUDY

In the 1980's, a number of people living in cities ranging from 10,000 to over 100,000 in population bought small farms in the United States of America, and Northwest Missouri was no exception. According to Lewis et al. (1978), many persons think of relocating from the city to rural areas because of the "back to nature" considerations, while others due to preconceived desires. Many of the would-be farmers consider growing non-traditional crops on a few acres of land. A segment of the population with different professional backgrounds, distance from the place of work, and size of their operation was studied, as shown on Table 1.

Their relationship to plants was mostly with vegetable plants while a small number of persons are working with fruits, mushrooms and ornamental plants. Distances from the places they work to the farm site ranges from 4 to 30 miles. Some of these persons have moved to their farm site by now. They appear to cope very well with the new environment. Previous experience has shown that many people that have moved to rural areas become dissatisfied with their decision, lost time and money, and are compelled to return to their city or urban living (Lewis and Glade, 1978). An important factor to consider in people's decision to stay in such a rural setting is the size of the farm and the kind of plants they become involved with. In our study, the size of the farm ranges from 0.5 acre to as large as 40 acres. Interest in the land, environment, and plant interaction are tied to tangible effects such as the economic aspect of this plant relationship. Pietsch (1978) indicated that since the part-time farmer is likely to spend less time operating the farm, location relative to primary employment may be an important factor. However, we found that if the economic side tends to be alluring enough to dedicate more time to being in the farm, this economic factor becomes the primary reason for allocating more time to growing plants.

SOCIO-ECONOMIC CONSIDERATIONS

A large majority of individuals and families moving to rural areas after purchasing a small acreage of land tend to leave the land to pasture. One contributing reason for not cultivating the land to economic plants is the lack of knowledge. The "grow for profit" idea is an excellent incentive to get people closer to plants. The large majority of people moving to small farms in Northwest Missouri who have no farming background receive technical and advising services from the Alternative Crops Research Center at Northwest Missouri State University. This service creates an atmosphere conducive to eliminating uncertainty in growing a crop for profit. Dyer (1990) indicates, for would-be farmers to produce vegetable crops for profit, individuals must make a decision to what types of vegetable crops to produce. Market opportunity is then critical, especially in South Carolina. For the Northwest Missouri area, these people must use the "window of opportunity" to marketing crops that would bring a satisfactory income. Typically, the major questions often asked by people interacting with plants in Northwest Missouri are as follows:

1. Types of crops or plants they can grow in the area.
2. Cost of production.

3. Investments on infrastructures, loan sources, etc.
4. Where to obtain plant materials or seeds, etc.
5. Disease and insect prevention and control.
6. Environmental concerns on the use of chemicals, possibility of producing organic vegetable crops.
7. Market outlets, etc.

From our study of 13 individuals, eight of them have responded with a likelihood of increasing their farm operation, four of them will not likely increase their operation and one individual expressed that depending on how much money is being made in the operation, an increase decision will follow on his vegetable and fruit tree production (Table 2). None of the four individuals who do not plan to expand intend to either reduce or eliminate their current plant operation. As the percent of their total income increases, so will his/her commitment to increase the operation and time on the farm as well.

A survey was made to ascertain whether the reasons for wanting to work with plants or the land was purely economic or was because of being close to nature. In Table 3, one can observe that for individuals coming from strong farming backgrounds, 80% stated their reasons as purely economic, while 20% stated their reason was to be close to nature. For people with some farming background, 50% do so because of economics, while 25% desire being close to nature. It was found that of people having no farming background, 50% expressed their reasons for working with plants as wanting to be close to nature. However, lately these individuals are expressing a desire to explore the possibility to raise vegetable crops for profit. This tendency to profit from their relationship with plants can be explained in part by the state of the economy in the United States, which forces people to supplement their total income.

Individuals that were involved in this study had a varied preference on the type of market size. This choice of market is actually dependent on the types of crops being considered, and whether a specific "window of opportunity" exists, or if requests are being made by the population at large in cities close to their farms. Funt (1977) made an economic study comparing various small fruit species produced in Maryland; selection to grow a particular small fruit crop species was based on yield and ease of harvest. In close-by cities, U-pick operations are a popular choice for would-be farmers. Crop selection is, then, critical to achieve the socio-economic level sought by these individuals. Technical personnel from the Alternative Crops Research Center recommend sound planning before a particular individual initiates a production plan for a specific vegetable

TABLE 1. People's background and their plant relationship, location and size of the operation involved

Backgrounds	Plant Relationship	Distance from Place of Work (miles)	Size of Operation (acres)
Hospital Adm./Youth Counsel.	Vegetables/Fruits	12	5.0
Retired Soil Conserv. Off.	Vegetables/Small Fruits	7	1.5
Postal Carrier	Fruits/Vegetables	20	2.0
Store Owner	Small Fruits/Vegetables	4	5.0*
Accountant	Chipping Potatoes	10	40.0
Airport/Airline	Greenhouse Plants	30	1.0
Horses/Blacksmithing	Christmas Trees/Vegetables	6	5.0
Printer	Vegetables/Small Fruits	5	8.0
Mechanic	Shiitake Mushroom	12	0.5
College Professor	Vegetables	6	0.5
Insurance Representative	Vegetables	5	5.0
Extension Worker	Vegetables	30	2.0
Trucker	Vegetables	4	8.0

* Planning stages.

TABLE 2. Socio-economic consideration and the impact on people's plant relationship

		Socio-Economic Considerations		
Backgrounds	Likely to Increase Operation	Percent of Total Income/Revenue	Market Distance	Market Size*
Hospital Adm./Youth Counsl.	Yes	60	Short	Small/Large
Retired Soil Conserv. Off.	Yes	20	Short	Small
Postal Carrier	Yes/No	5	Short	Small
Store Owner	Yes	–	–	–
Accountant	No	60	Long	Large
Airport/Airline	No	50	Short	Small/Medium
Horses/Blacksmithing	Yes	50	Short	Small/Medium
Printer	Yes	70	Short/Long	Medium
Mechanic	No	–	–	–
College Professor	No	–	–	–
Insurance Representative	Yes	10	Short/Long	Medium
Extension Worker	Yes	20	Long	Medium**
Trucker	Yes	80	Short/Long	Small/Large

* Small Market = Farmers' Market, U-Pick Operation; Medium = Local Markets; Small/Medium and Small/Large = Farmers' Market, Local Markets, Brokers, Supermarket Chains and Associations; Large = Under Contract to Processors, Shippers to Long Terminals.

** Pooling with other farmers to a marketing association.

TABLE 3. Socio-origins and reasons for wanting to work with plants or the land

	Reasons		
Socio-Origins	Purely Economic	Being Close to Nature	Plant Relationship/Economic
Farming Background	80%	20%	0%
Some Farming Background	50%	25%	25%
Non-Farming Background	0%	50%	50%

crop. One particular occurrence in northwest Missouri is the lack of hand labor. Hence, a reason to establish continuous harvests of a single crop that can be achieved by successive planting at regular intervals of time limits the necessity of hired hand labor. Furthermore, the use of plastic mulches and drip irrigation reduces the use of chemical herbicides, hand labor and other cultural practices, thus reducing the cost of production. The response by these would-be farmers in adopting these recommended production practices has been very positive.

PERSONAL AND FAMILY INVOLVEMENT

A number of people from cities and large towns that have bought small farms and have become active in farming, do so to complement their total income for the family. Brooks and Kelbacher (1990) indicate that rural residence farms are individually operated, and many of them serve primarily as a residence for their family. For a young family, the wife and/or children get involved, as is the case with some of the residential farms in northwest Missouri. It was found that sometimes this arrangement does not always work, especially when children are in school session and activities will not permit them to help with a particular chore. In some other cases, the family does establish a weekly schedule with children to perform certain duties. All children surveyed about working with plants in the farm, responded that it was a fun activity; however, emphasizing that they would not do it for themselves when they grow older. Many of the studied families that sell their produce at roadside and farmer's markets tend to have a positive response with respect to family involvement. Some of the main socio-psychological reasons for being involved with plants can be summarized as:

1. Relationships with other people, when selling their plants or produce on their farm or at the farmer's market, added a new dimension into their lives.
2. Positive attitude and good mood toward their own family members when all members interact.
3. Feeling good physically and spiritually.

Allen (1990) found that with this type of relationship, a typical family becomes a close-knit unit, with dedication to their business, and especially to their plant produce. Their success can be traced to work ethic, togetherness and a willingness to follow recommendations from University and Extension Service Specialists relating to their production and marketing techniques.

COMMUNICATION PROCESS AND PEOPLE-PLANT RELATIONSHIPS IN NORTHWEST MISSOURI

In recent decades, the large commercial farms benefited from using the results of university research, but small and residential farms received less attention. Since many small farms and residential farms are operated by city dwellers, they are in need of very basic information, as is the case in the northwest Missouri region.

In 1988, Northwest Missouri State University established a Center to investigate alternative crop possibilities, and offer new economic opportunities for farmers and urban would-be farmers alike. Now in its fourth year of operation, research has begun to demonstrate the economic benefits of alternative crops, and the importance in the life of those people owning rural land and working in the city, to dedicate more time to the farm. The most popular alternative crops with rural residence farm operation are vegetables and fruits in northwest Missouri (Table 1). Since a limited acreage is being used by residential farmers in northwest Missouri, investment on infrastructures are not critical at this stage of the program. People being served by the Alternative Crops Research and Development Center at Northwest Missouri State University have a variety of communication alternatives to obtain the necessary information for the activities of growing plants. The actual communication process being used is illustrated in Figure 1.

This communication process has proven to be critical for the decision and for the commencement of farming operations by these would-be farmers. More often, city people with farming interests visit the office of the Alternative Crops Research Center to discuss crop possibilities to suit their objectives. This experience is also explained by Frutchey (1966) in the Cooperative Extension Service learning and teaching process. People who

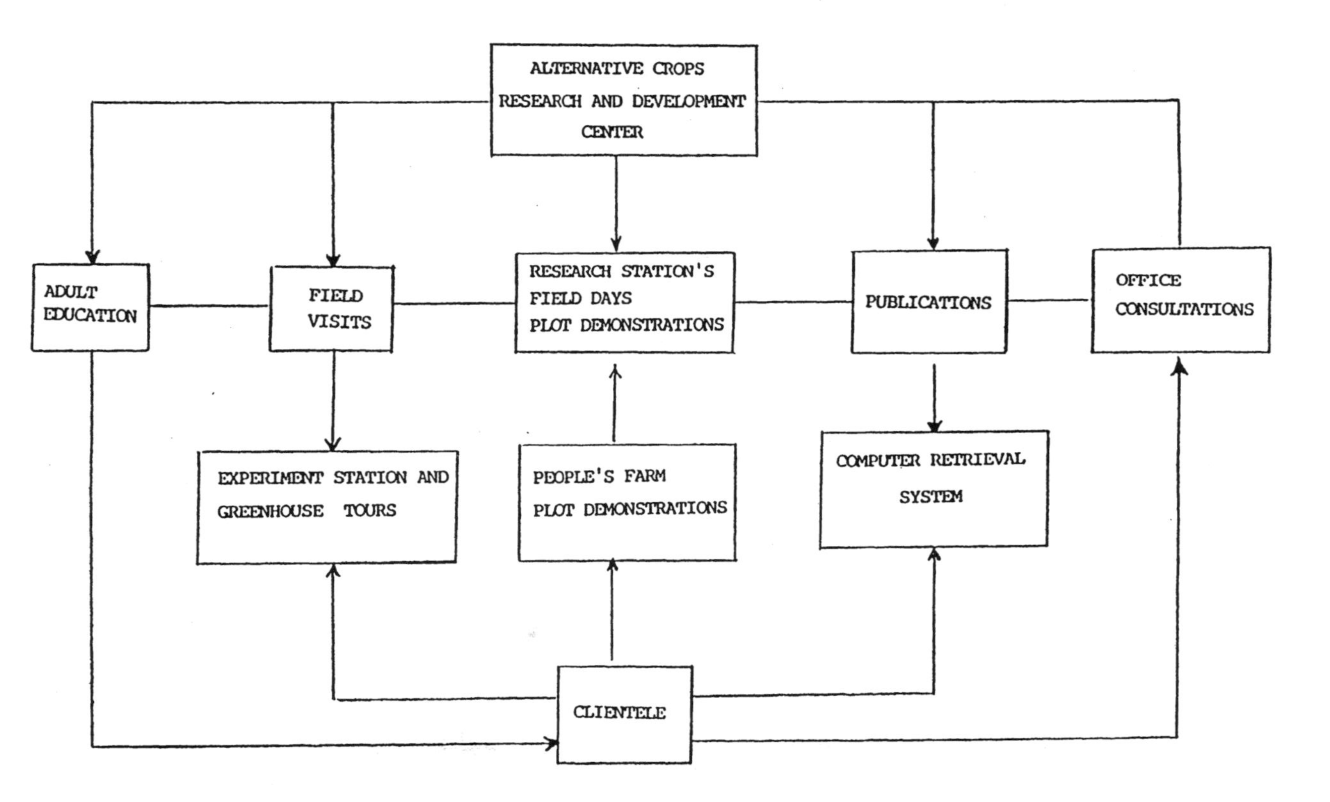

Figure 1. Actual communication process between Northwest Missouri State University's Alternative Crops Research and Development Center and People Plant-relationship in Northwest Missouri.

are beyond the first year with the Alternative Crops Program know that there is a wide range of media where information can be obtained or retrieved by personal computer. At present, all the means of supplying information and demonstration of plant growing techniques, harvesting and marketing strategies, have shown positive effects, and have become the pivotal arm in the development of these types of rural residential farms. Furthermore, it furnishes a positive psychological state of mind on the individual to break the fear of not having technical support, and provides assurance to produce a crop alien to them. Field visits reinforce this psychological response on people. Lately, these rural residential farmers have been encouraged to read the Alternative Crops Research Center's publications, and recommended reading; to listen to speakers invited to campus or to a specific community, and to visit each other's farm and other farms. To keep abreast of new developments in growing vegetables and fruits, they are encouraged to become members of the Missouri Vegetable Growers Association and The Missouri Horticultural Society (Fruits), and attend the joint annual meeting. The interaction between the Alternative Crops Research Center and the rural residential farmers has created positive impacts, giving high expectations to future would-be rural residential farmers, and a sense of security to get involved with plants, production and marketing. Moreover, the economic incentives tend to validate the concept that the plant/people relationship creates a wholeness in themselves and their environment.

CONCLUSION

Clearly, the development and formation of emerging new rural residential farmers are a fact of life in rural northwest Missouri. Professional workers and business people from large cities who purchase and settle rural land tend to seek a close relationship that brings about a socio-economic pattern; one that provides positive psychological and social benefits to their well-being. To carry out this endeavor, people in this situation need moral and technical support, and communication to these people can not be overlooked. Implementing a continuing adult education system will help promote and improve their socio-economic condition. Field visits and plot demonstrations provide solid encouragement to these rural residential farmers. It is essential to keep monitoring and studying the number of people with nonfarming backgrounds who purchase rural land to live on, and the types of plant crops they would like to grow. For the survival of these individuals, marketing studies are critical for further growth and development.

BIBLIOGRAPHY

Allen, A.L. 1990. The Carpenter Family: Farming vegetables and fruit keeps their dream alive. USDA Yearbook of Agriculture. pp. 53-54.

Brooks, N.L. and J.Z. Kalbacher. 1990. Profiting the diversity of America's farm. USDA Yearbook of Agriculture. pp. 18-22.

Dyer, J. 1990. Marketing, a window to success. USDA Yearbook of Agriculture, pp. 174-175.

Frutchey, F. 1966. The learning teaching process. In: The Cooperative Extension Service. Ed. H.C. Sanders. Prentice-Hall. Englewood Cliff, NY pp. 51-68.

Funt, R.C. 1977. An economic comparison of several small fruit production and harvest systems in Maryland. Maryland Agric. Exp. Sta. Miscellaneous Publication 922.

Lewis, J., E. Glade and G. Gustafson. 1978. Changing to a new lifestyle: The little things *do* add up. USDA Yearbook of Agriculture, pp. 25-31.

Pietsch, W.H. 1978. Selecting a region, community, site. USDA Yearbook of Agriculture. pp. 38-45.

Chapter 31

Encouraging Nurturing Behavior of Two to Seven Year Olds by Introducing Plants and Flowers

Karen Green

SUMMARY. The purpose of this project was to design a curriculum which encourages nurturing behavior of two to seven year olds by introducing plants and flowers and to design an observation instrument to be used by the teachers involved in the implementation of the curriculum.

A six week curriculum comprising thirty lessons was designed to use materials to aid children in the development of nine nurturing behaviors. Each week's lessons included sensory activities that were meant to foster practical life experiences. Books, music, outdoor activities, cooking and fantasy play were incorporated into the curriculum for at least one hour each day. The children were encouraged to learn positive social skills.

Out of sixteen evaluations by a range of professionals teaching varied populations of children, all thought that the nine goals had merit. They varied in their responses as to what ages would best be reached. They thought the six week curriculum could be easily implemented with not too much cost. They predicted that the children might enjoy working with the materials, but that some activities would have to be adapted for the various age levels.

Every day we are told from one source or another that many of our children are living in home and school situations which are unacceptable–

that they are succumbing to societal pressures including drugs and other forms of violence and are not achieving their unique potential. These feelings of helplessness and frustration experienced by us, as educators and parents, result from our perceived inability to battle these mounting societal pressures. There is no one answer that will resolve the situation. All we can do–and what we must do, is take steps and make changes, both in and outside the classroom, that positively effect the development of our youth. I have prepared a "plant play" curriculum entitled *Tools to Grow Curriculum for Early Childhood* that I feel will help to develop nurturing behavior and social skills and give children the feeling of empowerment that they need to thrive in our rapidly changing society (see Appendix I).

The curriculum that I have developed relies upon activities and experiences designed to encourage nine specific elements of nurturing behavior (see Appendix II).This overhead, which identifies the nine behaviors, is in the form of an observation/evaluation worksheet for classroom use. The nine elements of nurturing behavior can be broken down into three general categories. The first is the developmental stage of 2- to 7-year-olds–both stereotypical and unique behavior. I will include past experiences, their need for concrete experiences and the development of sensory awareness. The second area is the need to teach elements of practical life and the third area combines the child's unique differences with his or her need for autonomy.

Now I will set forth the highlights of my research that support my conclusions, and conclude with a discussion of a possible method for implementing such a curriculum in an early childhood education environment.

Vast research suggests that education in the garden aids in the development of many skills and behaviors that will, when acquired by children, lead to the nine nurturing behaviors.

One of our foremost childhood educators, Maria Montessori, believed that "children had amazing qualities of mental concentration, love of repetition, love for order, need for freedom of choice, no need for rewards or punishment, a love of silence, spontaneous self-discipline and a sense of dignity" (Ellicott, 1989, p. 15). She also believed that "the young child is a creature with special powers (or sensitivities) unlike any found elsewhere in nature." Montessori stressed early training of social skills to maintain a peaceful world and was the first educator to introduce gardening into her curriculum.

For a child to be appropriately educated, his education must have philosophical as well as practical implications. Many researchers support this view. Piaget states it this way:

> The principal goal of education is to create men who are capable of doing new things, not simply repeating what other generations have done–men who are creative, inventive and discoverers. The second goal of education is to form minds which can be critical, can verify and not accept everything that is offered. So we need pupils who are active, to learn early to find out by themselves, partly by their own spontaneous activity and partly through material we set up for them; who learn early to tell what is verifiable and what is simply the first idea to come to them. (Katz, 1990, p. 79)

Former Secretary of Education, William Bennett, believes that education's aim is to produce children who score high on tests of achievement; that they should "know how mountains are made, and that for most reactions there is an equal and opposite reaction. They should know where the Amazon flows and where Ethiopia is" (Katz, 1990, p. 79).

Both Piaget and Bennett have stated goals that are necessary and viable in education and can be reached partly through the techniques learned through plant/play. A plant/play curriculum connects with such areas as art, music, literature, science, fantasy play and lays a concrete basis for continuing interest in all or some of these areas.

Another aspect of education is that of "efficacy expectations," or Bandura's theory which states that:

> . . . goals that are specific, moderately difficult, and seen as reachable in the not too distant future are most likely to stimulate persistent effort and to lead to increases in efficacy expectations if reached successfully.

A plant/play curriculum allows a child to view and assist with a plant's development from seed–to sprout–to flower. The child is able to see goals reached within a relatively short period of time.

The notion of interest is a necessary element of practical life, for without it there is no attention paid to what the child needs to learn.

> If he becomes deeply enough absorbed in something to pursue it over time, with sufficient commitment to accept the routine as well as novel aspects of work, growth will occur. (Katz, 1990, p. 48)

Children can learn the elements of practical life through a curriculum of plant/play, as well as developing an appreciation of the beauty, wonder and perpetuation of life. Being surrounded by plants and flowers on a daily basis, children see the care that goes into keeping them healthy and strong–much like his/her caregivers take of him. Outdoors, "the youngest child will watch the movement of the leaves and revel in holding a leaf or branch in his hand. Water, soil and sand play delights most every child" (Hahn, 1968, p. 8).

Defining stages, as Piaget does, we are able to more easily discuss the developmental process. As we see changes in nature, in plants and flowers, we see there is no sharp break between stages (seasons) and there are no completely new beginnings.

> With children the main stages follow one another in an order that is held to be the same for all children. But this is not because they are 'preprogrammed.' It is because each stage builds on the one before it. Thus the earlier construction is necessary for the later one. While the order of stages is the same for all children, the speed of movement is certainly not. (Donaldson, 1978, p. 144)

Through plant/play children learn their unique as well as age appropriate behavior which allows them to begin to work independently and in groups.

It is important that early on, a child is aware of what makes him or her feel good and bad, stressed or unstressed, frustrated or not. Through observation of the child, a teacher can help the child to verbalize his or her needs and learn about how he works. Naming things, developing sensory awareness through observing, listening and feeling through smelling and tasting materials that are living and natural broadens his/her range in language development. The child is then able to communicate with others and begin to develop long and lasting caring and sharing relationships.

Equilibrium is the term that Piaget uses to define the "self-regulating process to correct or compensate for any disturbance to the system. The disturbance may be compensated for if the child understands that the movement can be balanced or reversed by a movement in the opposite direction" (Donaldson, 1978, p. 156). A child may pull a plant out with its roots and be taught to replant it in a bigger pot or back in the same spot. However, he must be instructed at the proper time that if he tears a flower from its stem or a leaf from a tree, he cannot put them back or reverse the action. In other words, through plant/play he begins to learn that there are consequences to his actions.

Learning to care for and help one another naturally feels good and must be encouraged. Hahn states: "In a world full of destruction we cannot get a child early enough to learn how to protect life and how to support the weak. In gardening we teach that damaged plants get a bandage to heal and weak ones a support."

Evelyn Weber from Columbia University and head of Children's Defense Council believes that: "If during this period of preschool the child can get some sense of the various roles and functions that he can perform as an adult, he will be ready to progress joyfully to the next stage."

> To 'mess about' is a term often used in early childhood educational literature to describe exploration-experimentation behavior. The behavior can be promoted in playgrounds. The first may be referred to as representational modes. As children play with materials, they often recreate representations of experiences or their conception of the world. (Cornell, 1984, p. 611-612)

[For example, in the garden they may become: Caregivers, architects, farmers, scientists, carpenters, to name a few.]

In an urban environment, particularly in high density inner city school districts, if we want ultimately for children to care for each other and their environment, they must be nurtured by the beauties of those elements as well.

If we can help them to create an environment that can nourish and provide beauty, we are giving them tools to keep their spirits high.

Many educators today are in favor of peer help and cooperative play and state that working in small groups productively is as important as playing alone. Plant/play gives the opportunity for this to take place both inside and outside the classroom environment.

In order to develop memory, another important element of practical life, Vygotsky states that:

> The possibility of combining elements of the past and present visual fields (for instance, tool and goal) in one field of attention leads in turn to a basic reconstruction of another vital function, memory. (Vygotsky, 1978, p. 36)

Plant/play at a young age affords children "hands on" or "concrete" experiences which aid in his acquiring memory skills.

We are constantly reminded that the United States lags behind other countries in the field of science. In order to raise the standard, one of the areas of the NYS science curriculum aims to teach children nurturing skills by setting down the following goals:

- Appreciating the natural world as an essential resource in fulfilling human needs, both physical and aesthetic.
- Respecting life in the natural world. Valuing the use of what is known about the natural world for the benefit of future generations.
- Enjoying work with living and nonliving materials.
- Valuing resourcefulness and innovativeness in the solution of practical problems.
- Caring about the needs and problems of others.
- Valuing the benefits that technological change may have on people and the environment.

The curriculum further states: Like people, living plants and animals live and thrive when their needs are met.

That each plant has properties that enable the plant to meet needs:

- That the parts of plants have functions that help the plant to live and thrive.
- That parts of some plants change and meet the immediate needs of the plant.
- That the parts of some plants undergo seasonal change that enable the plant to grow.
- That each kind of plant goes through its own stages of growth and development during the life span of the plant (NYS Syllabus, 1990, pp. 22-23).

The above stated goals are metaphors for the human experience and lay the basis for future exploration into other fields of science. Given these objectives, a plant/play curriculum in any school environment also helps all involved to learn what Weber and Tuan have stated eloquently:

> Idiosyncracies of the individual are the greatest blessing of nature and must be respected to the highest degree. (Weber, 1984, p. 15)

> No two persons see the same reality. With good will one person can enter into the world of another despite differences in age, temperament and culture. (Tuan, 1974, p. 5)

In a multicultural environment like we have in most of our major cities, we must recognize these truths. We must build on children's past experiences, their likes and dislikes, whether it be plants, flowers or foods from their native countries, and allow them to share this knowledge with their new friends. Parental participation should be encouraged.

There are many benefits of educating through a plant/flower curriculum. But how does one go about designing a curriculum to build on children's natural sensitivities and needs? Indeed, we must recognize, even at a young age, the child's innate need for autonomy and self-determination. In order to foster this type of growth, we must present a wide variety of plants, flowers and materials. Some children prefer using their tactile or olfactory sense more often than the other senses. Children must often be allowed to choose their own work/play experience. Some might just be observers at the beginning. Some may want to be alone near the plants in silence and others will want to share the experience with teachers or peers.

And how will materials be selected? Catherine Eberbach of the Children's Garden Program at the New York Botanical Garden, writes: "Plant selection should generally favor those with ornamental qualities. Children have a sensitivity to beauty. Their perception of beauty may be very different from that of an adult. However, bright bold colors attract and amaze children (red, orange, yellow, purple should be used freely)." It is interesting to note in her research that colors play an important part in human emotions, and may constitute man's earliest symbols. She concurs with Tuan that, "Primary colors designate strong emotions. Young children appear to have little interest in mixed or impure colors because they denote ambiguities that lie outside their experience. Among chromatic colors, red is the most dominant and its meaning is the most widely shared by peoples of different cultures."

This could prove to be still another vehicle to draw out children's feelings about themselves and others which might help them still further to recognize each others' similarities and differences.

The initiation of a plant and flower curriculum may prove useful since "human development in all its forms is seen as the primary purpose of education" (D'Emidio, 1987, p. 13). "Plant/play helps children to experience through all the senses the beauty and strength and fragility of other living things. Montessori expressed the danger of personal growth at the expense of others" (D'Emidio, 1987, p. 13). By this she meant that the achievement or the distance one goes does not matter if it hurts another; that fairness and cooperation are most important. Plant/play enables children to see that each of us is unique, that we grow in different ways at different times and that the most important thing is that we continue to grow while not losing sight of the love and respect we must always have for one another.

"It is important to utilize the elements in the natural environment in conjunction with other avenues of learning to encourage behavior that in the long run will benefit society. The education experience is designed to provide gradually a reciprocal relation between individual fulfillment and socialized experience. Only a certain kind of curriculum where there is opportunity for more collective than competitive enterprise, supports that process" (Weber, 1984, p. 189).

A plant curriculum may encourage that kind of collective play. However, when a child finally chooses a plant to nurture or simply observe and enjoy, he/she begins to realize too that he/she *can make a difference* both in *productivity* and *attitude*. This is the beginnings of the child's feelings of *empowerment*.

Children are using their senses to learn at every moment.

> Using herbs along with vegetables and flowers can be a most effective way of getting children truly involved with plants and gardening by awakening a sensory appreciation for the plant world. (Dunks, 1976, p. 27)

Children want to discover a world that is beautiful, gentle and kind to their senses. Many have careful love and nurturing from home, let them find a lovely environment that they will want to perpetuate at school. This quote from Maria Montessori sums up what I believe this kind of curriculum has the potential to achieve:

> For children no matter how young, this appeal will form a wonderful basis for an introduction to the pleasures and later on the necessity for nurturing our living things. Through plant nurturing children are initiated into the virtues of patience and into confident expectation which is a form of faith and philosophy of life. (Montessori, 1964, p. 159)

REFERENCES

Aksakov, Sergei. *The Scarlet Flower.* A Russian Folk Tale.

Aliki. *Corn is Maize, The Gift of the Indians*, Harper Trophy, N.Y., 1976. A Let's Read and Find Out Book.

Althea. *Flowers.* Life-Cycle Books, 1977. Beginning Botany.

Bjork, Christina and Anderson, Lena. *Linnea's Almanac*, R & S Books. Linnea is the name of a flower. An almanac and time. March–the first spring flowers. April–garden of herbs gets full. May–garland for your hair. June–fruit drinks. Sept.–City trees–housing, climbing, resting under, cleaning the air, absorbing noise, showing the seasons, looking at. P. 45 Birch, Maple, Aspen, Elm, Rowan, Beech, Lime, Oak, Poplar, Ash, Horse Chestnut, Alder, Willow.

Bridgehampton Works and Days, 1974c. Handbook, almanac and recipes of Eastern, L.I., written by its residents.

Burnett, Frances, Hodgson. *The Secret Garden*, David R. Todine, Boston, 1986. First published in 1911. Can be edited for any age. Can be told or read in part or in its entirety over a period of time.

Butning, Eve. *The Wall*, 1990. Deals with war (Vietnam) in a very tender way.

Chaucer, Geoffrey. *Canterbury Tales*. Selected, translated and adapted by Barbara Cohen, Lothrop, Lee & Shepard Books, N.Y., 1988. The foibles of people of all generations and classes. Must be edited and adapted. Mostly for older children. Good basis for discussion of personality traits, excesses and historical perspective.

Ehlert, Lois. *Growing Vegetable Soup*, Harcourt Brace Jovanovich, 1987. Garden tools (shovel, rake, hoe), seeds, sprouts, water and sun, grow, weeds, picking.

Ehlert, Lois. *Eating the Alphabet*, 1989.

Garden catalogs–(Burpee, Michigan Bulb)

Gile, John. *The First Forest*, Worzalla, 1978. Story about beauty, greed and love.

Growing Things

HBJ Science–Nova Edition–Yellow and Blue, Harcourt Brace Jovanovich Publishers, Orlando, 1989. Children's Edition. For school use with teacher's guide. Lovely illustrations and questions for discussion. Flashcards included with suggested activities.

Hennessy, B.G. *The Dinosaur Who Lived in My Backyard*, 1988.

Ichikawa, Satomi. *Rosy's Garden, A Child's Keepsake of Flowers*, Philomel Books, 1990. Story of love and romance. Also history for older children. Printouts.

Johnny Appleseed–An American Classic, Starlight Editions.

Ketchum, Robert Glenn. *The Hudson River and the Highlands*, 1985. Essay and Photographs. Children from N.E. will know the Hudson River.

Kid's Cooking, a Very Slightly Messy Manual, 1987. Rules and Recipes.

Krauss, Ruth. *The Carrot Seed*, Harper & Row, 1945. "Patience and quiet expectation."

Kuhn, Dwight. *More Than Just a Vegetable Garden*, Silver Press, 1990. Vegetables and insects (seeds and roots).

Linnea Visits Monet's Garden, Bjork & Anderson, 1985.

Little Hiawatha, Walt Disney, 1990.

Lobel, Arnold. *The Rose in My Garden*, Greenwillow Books, 1984. Full color $1.00 to: Greenwillow, Dept. RM6, 105 Madison, NY 10016.

Look Inside a Tree–A Poke and Look Learning Book by Gina Ingoglia, 1987.

Lyon, George Ella. *Cecil's Story*, 1991. A Civil War story for very young children.

Markman, Erika. *Grow It! An Indoor/Outdoor Gardening Guide for Kids*, Random House, 1991.

Miller, Lenore (P.S. 54R). *Using the Outdoors to Enrich Learning*, NYC Teacher Centers Consortium. Month by month activities.

Oda, Mayumi. *Happy Veggies*. Japanese author.

Pallotta, Jerry. *The Flower Alphabet Book*, Quinlan Press, 1988. Pictures of flowers with poems about each one.

Paul, Aileen. *Kid's Gardening*, Doubleday, 1972. How-to's of indoor gardening. Information book.

The Kid's Gardening Book

The Please Touch Cookbook, 1990. From Please Touch Museum, PA.
Watson, Clyde. *Father Fox's Feats of Songs*, Philomel Books, 1983. Songs, p. 10, "Oh My Goodness"; p. 11, "Huckleberry, Gooseberry."

APPENDIX I

Tools to Grow Curriculum for Early Childhood

Karen R. Green

© May 1991

TOOLS TO GROW CURRICULUM FOR EARLY CHILDHOOD–SAMPLE METHODOLOGY FOR IMPLEMENTATION

Subjects

The subjects may be two to seven year olds in any early childhood setting.

Required Materials

Outdoor environment:	park with trees, moss, etc. boxes to plant bulbs 2 garden plots, 10′ × 4′ (built with parents' help)
Indoor environment:	cut flowers (new bunch each week) (dead flowers can be dried) piggyback plant spider plant with babies African violets bulbs in bright colors herb plants (parsley, basil, thyme, oregano, rosemary, sage) peppermint geranium plant dried herbs and spices (peppermint, spearmint, sassafras, lavender, lemon, cloves, cinnamon, whole nutmeg, vanilla beans) soil, sand and water children's size garden tools books (related to curriculum) plant and flower catalogs fruits and vegetables, beans and seeds art materials (magic markers, crayons, paints and natural materials) musical tapes

Procedure

The major divisions of the designed curriculum are the following:

1. Everything Has A Name
2. Size and Shape of Things
3. Smell of Things
4. Taste of Things
5. Touch of Things
6. Contrast and Comparison (likes and dislikes of children)
   (Allen and Allen, 1969, pp. 41, 51, 55, 75, 83, 87, 91, 105, 107, 109, 115)

APPENDIX I (continued)

It is planned that each division will be covered over a one week period; 5 hours, 1 period each day (but can be expanded). There will be an ongoing project of creating a book for each child which will be added to each week and may include pictures drawn by the child, words to songs or poems, pages with pasted in leaves of the various spices, herbs and flowers. It might include cutouts from garden catalogues of favorite colors of flowers, favorite fruits and vegetables or those that the children do not like. Class project will be a bulletin board based on the design of a tree and how it changes. Children who would like to begin collections might do so (seeds, dried leaves).

On *Monday* the main activity will be centered around a book. It might be a story book, an information book or a book of poetry or art within the plant and flower curriculum area.

On *Tuesday* the activity will center around music. We will listen to music or learn a song about nature and do some movement game or dance.

On *Wednesday* the main activity will be centered around an outdoor play activity–a field trip or some planned activities in the playground.

On *Thursday* the main activity will be cooking and eating.

On *Friday* we will combine all these things in dramatic and fantasy play activities.

APPENDIX II

OBSERVATION INSTRUMENT

Based on the six curriculum divisions this instrument is designed to measure which activities and materials most encourage the development of the nine elements of nurturing behavior for each child.

Child's Name:

Weekly Topic
Choice of: A. Naming Things; B. Size and Shape of Things; C. Smell of Things; D. Taste of Things; E. Touch of Things; F. Contrast and Comparison.

Appearance of Nurturing Behavior (+)

Nurturing Behavior	M	T	W	T	F
1. Being able to share attention of others, including the teacher					
2. Developing eye to eye contact with adults and peers					
3. Learning to smile at others					
4. Being helpful to others					
5. Learning to gain attention from others in positive ways					
6. Developing responsible behavior					
7. Learning to compliment rather than criticize others					
8. Showing tolerance for others and their differences					
9. Being able to express sorrow when actions or words have hurt another					

Key (Day): M Books; T Music; W Outdoors; T Cooking; F Dramatic Play

Weekly Summary Description:

Chapter 32

Risk Communication Methods for Newspaper Gardening Columns

Clare S. Liptak

SUMMARY. Educators writing about gardening and environmental risk can be more successful by writing text readable at the ninth grade level, choosing topics that reflect readers' concerns, and using risk communication methods. Such newspaper articles offer an overview of the plant or pest problem, its solutions, and include factors for the reader to consider in reaching a decision. These methods increase the effectiveness of mass media to disseminate information. They do not usually generate questions on the topic of the article for the Extension professional, and may increase credibility for the Cooperative Extension office.

My weekly newspaper column, GROWING, is published by Gannett and mailed free to 81,000 households in Somerset County. Forty-eight percent of the readers are between the ages of 25 and 54; most readers are affluent and well-educated. The articles provide information on plant science, entomology, and environmental risk.

The editor wants friendly, conversational, personal articles written for the novice gardener. An editor's request that an article be conversational is significant because our spoken language is more informal than our written language. Since sentence fragments are common in conversation, using them in writing makes the text more conversational for the reader. Also, the editor inserts fragments for effect.

The topics for articles can be challenging but the language is not. The text is usually at the ninth grade level and I avoid technical terms or explain them. (See "Yew saves lives" for a way to draw the reader's attention to a new word such as "aril.") When the topic is complex, I write as if I were talking to an "intelligent but uninformed friend" (Hance et al., 1988).

In her book *Trashing the Planet* (Regnery Gateway, 1990), Dr. Dixie Lee Ray suggests that science illiteracy exists because our educational system has failed to teach science to the 80% of students who do not prepare to be scientists. As adults these people get news about science and technology from TV, newspapers, radio, and news magazines. Assuming this is true, my readers would probably be receptive to easy-to-read articles on plant science. In addition, the information they gain could enable them to make better decisions. The sentence "Imagine harnessing the Blob to work for mankind." (from "Yew saves lives") is an attempt to make an unfamiliar laboratory procedure more understandable to the reader.

The article on taxol was well received but one reader was concerned that children are likely to be poisoned by yews. The second article, "Poisonous plants perspective," explained the reasons yews are not a common cause of childhood illness and mentions another problem parents may not know about. Children frequently put seeds in their noses.

Articles on seasonal gardening topics interest readers, but I write these focusing on the science involved rather than from an aesthetic or recreational point of view. I would write about new columbine introductions, for example, by explaining how plant breeders develop hybrids or the value of hybrid vigor.

Readers' concerns often determine my choice of topics. An important principle of risk communication theory is "Perception is Reality." According to Dr. Max Pfeffer, Research Professor in Human Ecology at Cook College, Rutgers University: "If people think a danger is real, they act as if it is, and consequences develop as if it really is real" (Personal communication).

GROWING articles include information on alternative methods to resolve a problem and consequences of those alternatives. Some readers may not want the most efficient, or cheapest solution to a problem, especially if it involves the use of a pesticide. The articles conclude with a statement that interested readers can call the Rutgers Cooperative Extension office to ask for the publication on the topic. Since the majority of these callers do not ask to speak to an agent, the articles are an efficient way to reach large numbers of people.

People at risk commonly ask personal questions. It's helpful to answer these questions before explaining the basis for the risk because it alleviates some concern in the person's mind. This is contrary to the instincts of most Cooperative Extension System personnel who see the explanation of the origin of the problem as a logical beginning.

Research suggests that government officials, scientists, and educators should consider speaking personally about a risk and practice an answer that acknowledges options and provides perspective. The following example is the response a scientist, researcher or Extension person might use in a situation where he or she is speaking with an alarmed consumer about a near-negligible amount of a toxin in drinking water.

Consumer asks the scientist: "Would you drink the water?"

The scientist responds: "Yes, I would drink the water. But my sister, who eats natural foods, would drink bottled water, even though bottled water isn't regulated" (Hance et al., 1988).

It probably takes years to build credibility personally and for an agency. My assumption is that by writing frequently about non-chemical means to manage pests, people will consider or accept my information when I write about using a pesticide for the same purpose.

I believe that scientists, researchers and educators involved in controversial issues will find it productive to use risk communication methods and write readable text that acknowledges the concerns and values of those affected by the controversy. The importance of these techniques may reflect a change in the type of person who is effective in the role of expert. In 1974, Karrass wrote: "Those (experts) with the highest credibility are people who do original and controlled experiments. Below that are those who do independent analysis. Those who put information in categories and fool around with semantics are at the lower rung of the expert ladder" (Karrass 1974).

As a county agricultural agent who does applied research, I think of myself at the lower rung of the expert ladder. In 1992, though, the expert who can translate complex information into plain English may be more credible and get more attention because citizens trust government less than they did in 1974.

I find that risk communication methods are often difficult for some of my agricultural colleagues in the Cooperative Extension System to accept. First, I believe that too often we assume that readers have our values–that is, the values of a person trained in science. Second, we also assume that the public doesn't care or isn't smart enough to learn the "facts." Third, too often we use technical terms, acronyms and jargon to explain the topic or establish ourselves as experts. We talk to the public the same way we talk to each other. And finally, we sometimes identify with a defensive farm community. We may be perceived as advocates rather than as impartial educators.

* * *

YEW SAVES LIVES

Taxol. The National Cancer Institute calls it "one of the most important cancer drugs discovered in the past decade." The drug has effectively treated ovarian cancer, which kills more than 12,000 women annually.

Taxol is derived from the bark of the Pacific yew, a native plant on the West Coast found from British Columbia to California.

The Pacific yew doesn't grow in New Jersey but its relatives do. One of them–the Hicks yew–is a columnar form with upright branches. Its needles have a fair amount of taxol, but not nearly as much as the bark of the Pacific yew.

Poachers, who strip bark from old trees in the West, are a source of taxol these days. Drug companies, plant breeders and the nursery industry also are working to find other sources because the drug is a complex molecule difficult to synthesize from scratch. The options they are considering include the following:

- Harvest twigs and foliage of different yews, not just the Pacific type. This method has the advantage of being a renewable resource. A disadvantage, though, is the variable quality of the taxol, because other yews don't contain as much as the Pacific yew.
- Obtain an intermediate form of the drug, probably from a large-scale harvest, which is then converted to taxol.
- Grow masses of cells from yew bark, roots or needles placed in a "soup" of nutrients and energy sources. Then harvest the taxol from the cells. Imagine harnessing the Blob to work for mankind.

I have new respect for the yew plant, whose only major problem in the landscape is that it doesn't tolerate wet soil.

We get this wonder drug from a common ornamental plant that is poisonous to children and livestock who eat it. Taxine, a poisonous alkaloid, is produced in the foliage, bark and seeds. (The botanical Latin name for all yews is Taxus, so you can see how the drug and the poison were named.) The seeds are enclosed in a non-poisonous, pulpy red covering that birds find very tasty. This way Mother Nature insures that birds eat the fruits and distribute yew seeds to new areas.

Veterinarians say that horses and cows are affected by relatively small amounts of yew foliage. It would take even less to affect a toddler. Keep a child away if he or she is at the stage where everything they touch goes in their mouths.

The offices of Rutgers Cooperative Extension of Somerset County are in the 4-H Center on Milltown Road in Bridgewater. Call 526-6293 for more information.

Clare S. Liptak

* * *

POISONOUS PLANTS PERSPECTIVE

It's Poison Preventive Week and my chance to clear up confusion about a GROWING column I wrote recently on yews, an ornamental plant that produces an anti-cancer drug, taxol.

In the article, I said that yews contain a toxic alkaloid, called taxine, found in the plant's bark, foliage and seeds.

One reader had questions about how much a child would have to eat to get sick.

The records of the National Clearinghouse for Poison Control Centers show that there were 2,456 incidents involving yews from 1971 to 1978. Five percent (123 people) were sick and .8 percent (20 people) were sick enough to be hospitalized. The clearinghouse also reports that plants account for 7.5 percent of the cases involving sick and hospitalized children under 5 years old. Medicines, cleaning and polishing agents, and cosmetics accounted for 68 percent.

The few children who eat yew berries are probably attracted by the bright fleshy red covering around the seed.

The covering is called an aril, which rhymes with barrel. (Now, I know that it's not exactly a term you'll want to remember but it saves me from trying to think of other words for covering.) Each aril contains one seed, which ripens in late summer or fall. The seeds contain taxine; the arils don't. In fact, one nurseryman told me that during the food shortages that were common in World War II, Europeans ate the arils and just spat out the seeds.

The bottom line is actually a book title–one of my references on toxicology: "The Dose Makes The Poison" by Dr. Alice Ottoboni.

Spelled out, this means: If you eat too much of something, even something harmless or nutritious, you get sick. Factor into this a common rule of thumb: the amount that makes a person sick depends on that person's weight. Anyone who consumes about .1 percent of their body weight in yew foliage would be sick enough to need hospitalization.

Assuming that seeds have as much taxine as foliage, this is how it works out for a child weighing 25 pounds: One tenth of 1 percent of 25 pounds is .4 ounces.

The U.S. Department of Agriculture publishes the "Woody Plant Seed Manual" for use by nursery operators. I checked, and it says that one pound of Japanese yew seeds, with their arils removed, averages 7,700 seeds.

A child weighing 25 pounds would have to eat .4 ounces of yew seeds to ingest a hazardous dose of taxine.

That's about 192 seeds. With the arils adding extra bulk, it's very unlikely that a young child would eat enough to be sick.

On the other hand, I've read about another problem that's more common and yet doesn't seem to be as well-known: Parents who garden with their kids should know that children will put seeds in their noses. Too often, the seeds must be removed in a hospital emergency room.

We don't have publications on poisonous plants to send you, but we will answer your questions using our references. Just call Rutgers Cooperative Extension of Somerset County at 526-6293.

Clare S. Liptak

LITERATURE CITED

1. Hance, B.J., Chess, C., and Sandman, P.M., Improving Dialogue with Communities: A Risk Communication Manual for Government. Environmental Communication Research Program, New Jersey Agricultural Experiment Station, Rutgers University, New Brunswick, NJ.

2. Karrass, C.L. 1974. "GIVE & TAKE: The Complete Guide to Negotiating Strategies and Tactics." p. 63, Thomas Y. Crowell, New York.

Chapter 33

Creation of A LIVING LIBRARY™: A Planetary Network of Interactive LIFE FRAMES™

Bonnie Sherk

SUMMARY. Since the early 70's the author has been developing environmental LIFE FRAMES–indoor/outdoor environments integrated with interactive programs and curricula. Beginning in 1981, the LIFE FRAMES have focussed on the creation of A LIVING LIBRARY, a network of international culture-ecology parks which promote horticulture, agriculture, and ecological awareness and involvement–locally and globally. The network of LIFE FRAMES made of individual branch LIVING LIBRARIES, is to be linked electronically using a variety of new communications technologies, forming an interactive planetary LIVING LIBRARY of diversity–ecological, cultural, and horticultural. The technology is meant to balance and counterpoint the non-mechanized forms of nature and to link us together around the globe promoting greater understanding and sensitivity.

A healthy future for cities, communities, and institutions may include the creation of unique environmental LIFE FRAMES™–indoor/outdoor culture-ecology parks and gardens, each designed in a site and situation-specific manner and integrated with programs and curricula. The LIFE FRAME synergistically links the resources of the community so that they work better together: human, ecological, economic, historic, aesthetic, and technological. As such, these LIFE FRAMES reflect the special characteristics of each place and its people showing the diversity of its culture, ecology, and built and horticultural environments. Each LIFE FRAME can serve the local populace and visitors as an integral community magnet through its participatory processes of creation, use, maintenance, and communication. With the use of various new communications technologies, the individual LIFE FRAMES can also be connected one with another throughout the world. The international network of the linked LIFE FRAMES can be thought of as A LIVING LIBRARY that brings to life the diversity of life showing us the many different styles, methodologies, and endeavors of human culture, including parks, gardens, and ecologies from around the world. An individual LIFE FRAME can also be thought of as A LIVING LIBRARY–a living embodiment of the local diversity.

A LIVING LIBRARY in a community promotes exciting involvement for all ages and also develops in people a planetary ecological consciousness. A.L.L. encourages sharing resources and information and understanding and appreciating differences and similarities.

An individual LIVING LIBRARY can be thought of as a public living, learning laboratory or a school of the future in which the park or garden serves as the vehicle to integrate seemingly disparate resources, information, and human consciousness. A.L.L. promotes an understanding of the relationships between biological, cultural, and technological systems through experiential involvement in nature and technology.

Each of these LIFE FRAMES, which is site and situation-specific in design, is interdisciplinary and can bring to life the humanities, sciences, social sciences, and arts. A.L.L. demonstrates the interconnectedness of phenomena and endeavor through plants and other living forms, the built and ecological environments, all the arts, programs of lectures, demonstrations, workshops, research institute, and state-of-the-art communications technologies. Each will use the environment (built and ecological) as part of its content and programs, and connect the patterns between subjects, endeavors, and systems–biological, cultural, and technological (see Figures 1, 2, and 3).

A LIVING LIBRARY results in intellectually and visually exciting interactive learning/creating environments that stimulate and support creativity and choice, and that motivate people to want to learn–so that they can learn, giving them practical as well as cognitive and sensory experience.

Since each environment with its integrated programs will be unique as it relates to the indigenous resources and characteristics of its locale and its people–when linked together, these individual LIFE FRAMES will create an extraordinary international network of local diversities–A LIVING LIBRARY.

FIGURE 1. A LIVING LIBRARY® is a school of the future and an indoor/outdoor culture-ecology park integrated with programs and curricula

Computer graphics by Diana Kline.

FIGURE 2. A LIVING LIBRARY® is a LIFE FRAME involving local people in creation, use, maintenance and communication

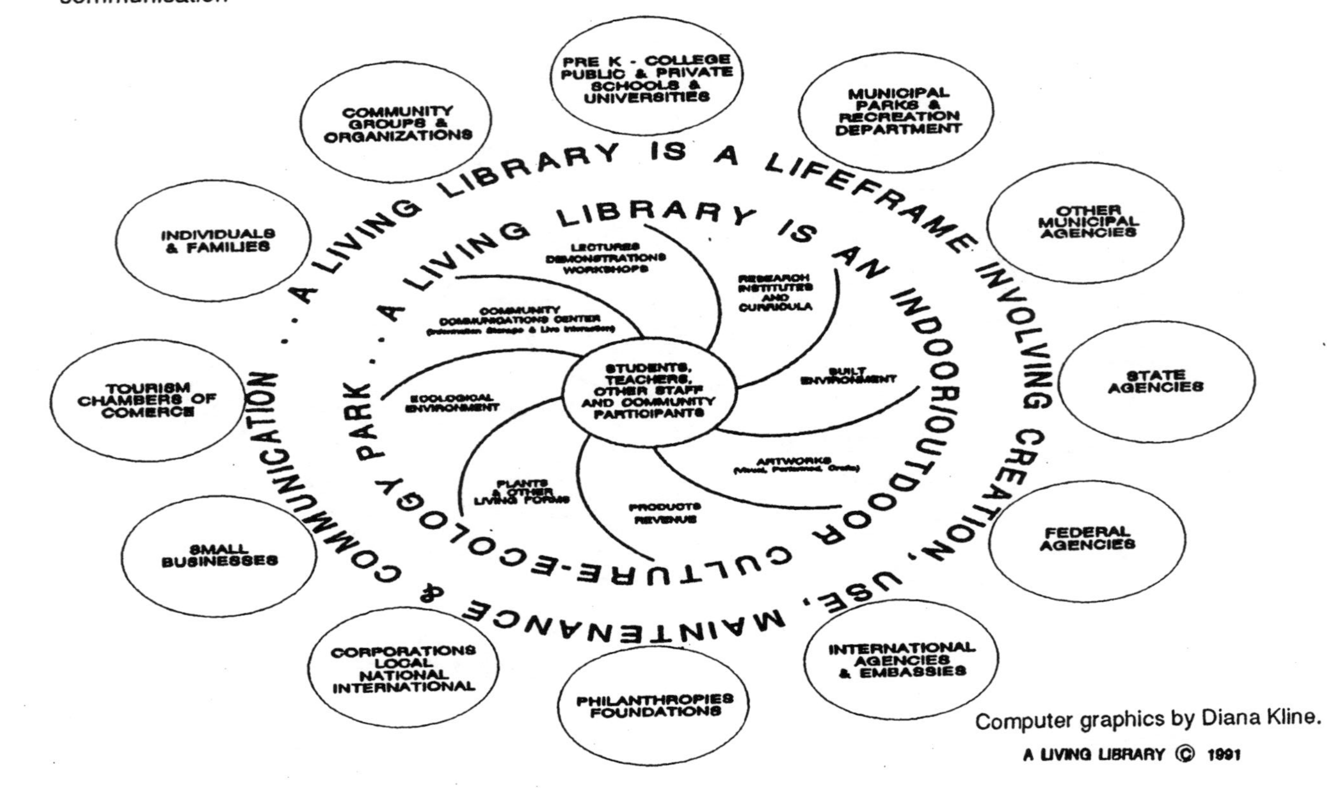

Computer graphics by Diana Kline.

A LIVING LIBRARY © 1991

FIGURE 3. A LIVING LIBRARY frame diversity

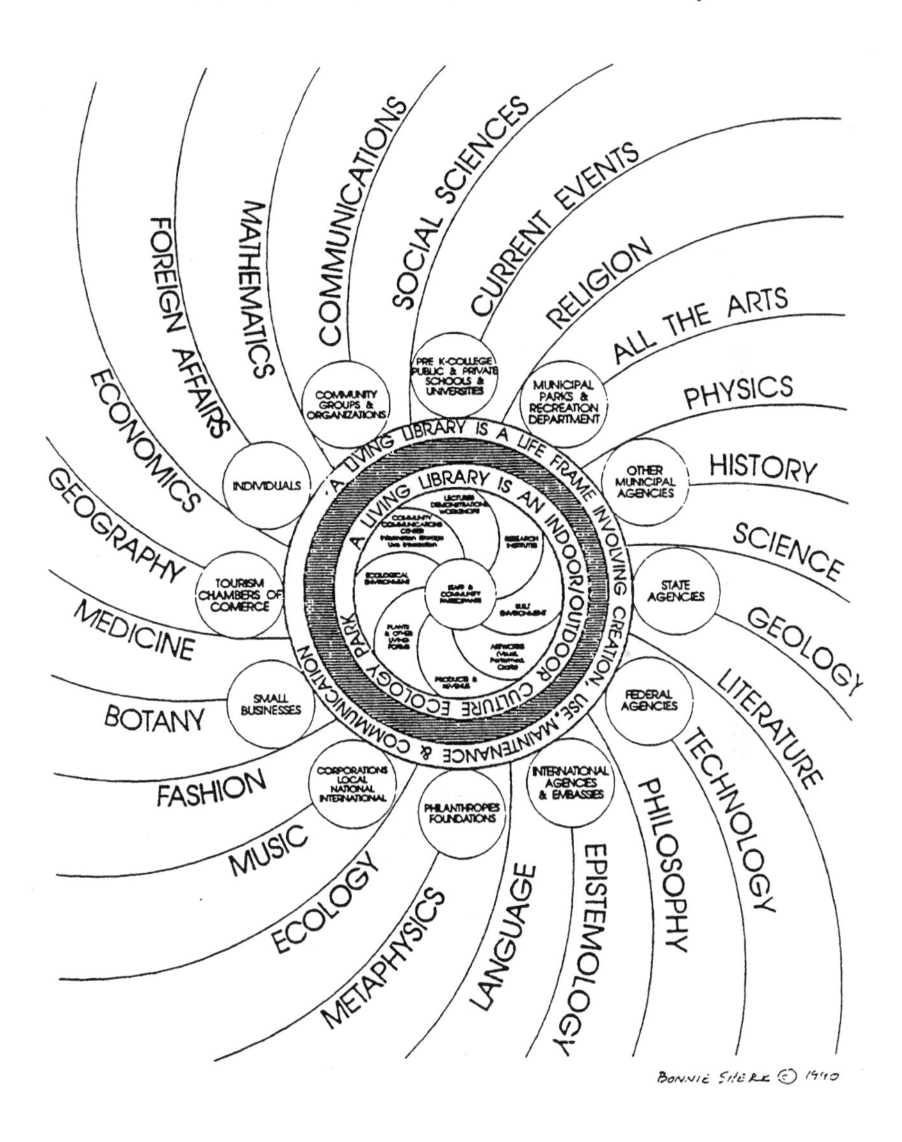

Computer graphics by Diana Kline.

A LIVING LIBRARY as it grows and matures will reflect the rich multicultural and ecological diversity that exists on the globe. And locally, each LIFE FRAME or LIVING LIBRARY will serve its locale bringing to life in the environment many subject areas and resources of the community while involving the public in the varied ongoing processes.

The park, plaza, museum, university campus, school yard, corporate headquarters, shopping mall, or other public or private indoor/outdoor open space will become a community magnet–bringing together many sectors of the community: students of all ages, teachers of all subjects, artists of all persuasions, historians, horticulturists, futurists, businesspeople, media technologists, scientists, environmentalists, families, corporations, senior citizens, foreign dignitaries, and others–all in a celebration of learning, creating, and maintaining the environment. The public open space will also become the site for live interactive broadcast allowing for unusual electronic interchange between the many LIFE FRAMES around the world.

At the local level, the resources will be integrated so that they work better together. For example, the curricula of the schools can be linked to the park or other environment and maintenance for the LIFE FRAME will become part of the ongoing program and learning experience for students and others. Businesses that wish to play a more active role in the larger community and also provide unique educational experiences for their employees and their families can also participate through sponsorship or program content development. A LIVING LIBRARY may also present an attractive and practical environment for professional artists as a new art venue–the gallery or museum of the future which functions to benefit the whole community.

At the national and international level, communities will exchange and share vital cultural and ecological information that promotes understanding, connectedness, and peace. The potential for positive global interaction is astounding !!

Shown here, the schematic drawing, "A LIVING LIBRARY–SKETCH OF INTERACTIVE LOCALES," illustrates two different potential locales for A LIVING LIBRARY. The sketch shows only a few elements of the proposed interactive program and environment for A LIVING LIBRARY. Each LIVING LIBRARY or LIFE FRAME will be designed in a site and situation-specific way. This illustration is meant merely to suggest a few interactive possibilities using new communications technologies in a garden setting (see Figure 4).

On the *left side of the picture*, we see an interactive park environment near an urban setting, possibly somewhere in Asia. On the *right side of the drawing*, we are in an interactive park setting near some mountains.

FIGURE 4. A LIVING LIBRARY–Sketch of interactive locales

The two sites are linked programmatically in a variety of ways through various multimedia computer and video technologies.

- In the locale on the right, watched by a group of people, we see a giant round electronic display showing images of the procession occurring on the left.
- Some of the people watching the live procession in the left-hand locale are also able to see simultaneously the performance occurring in the upper right locale on the "Video Parasol."
- The locale on the right shows interactive touch/sound/voice activated screens embedded in a "Video Computer Hedgerow" with a more detailed blow-up of hands on a screen in the foreground where the little girl is either accessing or inputting information from a local or international multimedia data base.
- The lower right-hand corner of the drawing shows a group of people gardening. Details of this activity appear in the "Tree Video Gate" framing the picture on the upper right.

A model for the urban park of the future, A LIVING LIBRARY is part of a larger concept of park and urban design that formulates creative cost-effective solutions to traditional problems plaguing not only public parks and gardens, but society-at-large, through integrating diverse human, technological, and ecological resources. Some issues that A LIVING LIBRARY addresses are:

- Healing the fragmentation of modern living and education and improving the quality of community life.
- Promoting a more profound understanding and appreciation of other cultures and ecologies around the world.
- Creating a sensitive balance between technology and non-mechanized nature.
- Proposing alternatives to the "business-as-usual" approach to environmental transformation, which is often merely cosmetic, unecological, overly expensive, and doesn't involve community participation and environmental education.
- Developing new approaches to civic management, park maintenance, and problems of vandalism.
- Creating innovative solutions for locating monies for the operation of public-oriented projects, such as parks.
- Providing accessible and up-to-the-minute in-depth information on issues of major significance to the globe, nation, or local region.

HISTORY OF A LIVING LIBRARY

Initially inspired and designed for a site (1981-83) in the middle of New York City, Bryant Park, adjacent to the Main Branch of the New York Public Library, A LIVING LIBRARY here would have gardens of knowledge arranged according to the Dewey Decimal System. There would be a Generalities Garden, Religion Garden, Philosophy Garden, Social Sciences Garden, Language Garden, Science Garden, Technology Garden, a Garden of the Arts, Literature Garden, and History and Geography Garden–ten in all (see Figure 5).

In each garden of knowledge, there would be live plants corresponding to each subject. Plants have different cultural meanings as well as botanical, medicinal, and economic values. There would also be visual artworks that relate to the subject, as well as a program of lectures, demonstrations, workshops, research institutes, and performing artworks–all connected thematically. In addition, there would be interactive computer/television/telecommunications capabilities giving more detailed information and linking the park electronically to other LIVING LIBRARIES or like projects throughout the world. There would also be International Garden Beds demonstrating styles and methodologies from different cultures around the world.

In the Bryant Park design there are three component parts to A LIVING LIBRARY, each involving a professional staff of experts and volunteers who are also educators. There's the implemented form–what it would look like. There's the way it's maintained, and that would be through programming, involving a multigenerational public in its planting, nurturing, and upkeep. And, thirdly, there's the way it's created. This is where the Research Institute component fits in.

The changing, seasonal themes of the park would be developed in conjunction with leading experts and educators in the area. In addition to involving professional artists, historians, ecologists, and media technologists, the programs of the park would relate to the curricula of schools–elementary, secondary, college. Students would be involved in the creation of programs and could choose their special area of concentration, whether it be literature, history, science, horticulture, or computer science. With expert staff guidance, students and other interested citizens would be involved in developing the indoor/outdoor environments and displays and the many elements within it such as theme gardens or local multimedia data bases. Libraries and other prime resources from the area would be used for creative research and development.

Suddenly, abstract learning could have a practical application! Think of this as it might affect the huge high school drop-out rate. With this kind of

FIGURE 5

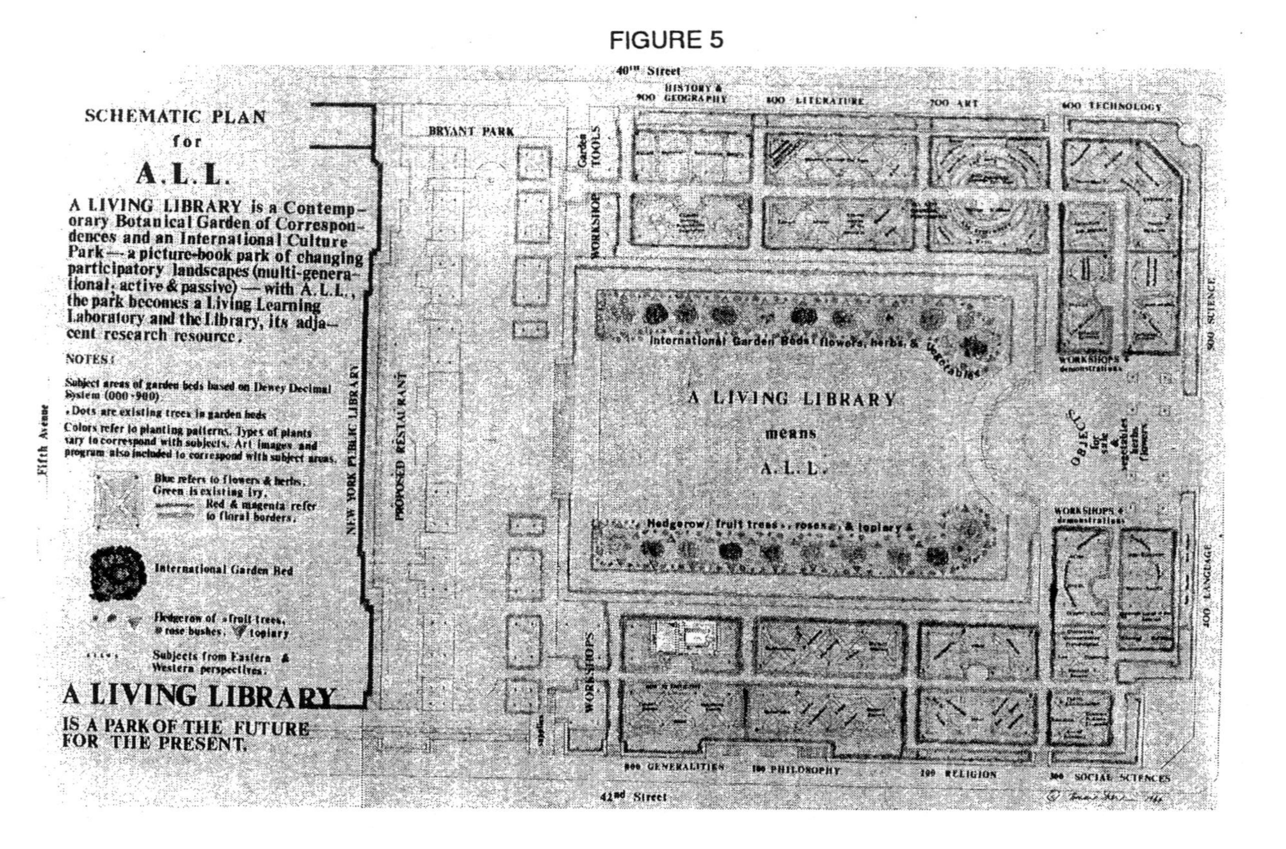
SCHEMATIC PLAN
for
A.L.L.
A LIVING LIBRARY is a Contemporary Botanical Garden of Correspondences and an International Culture Park — a picture-book park of changing participatory landscapes (multi-generational, active & passive) — with A.L.L., the park becomes a Living Learning Laboratory and the Library, its adjacent research resource.
NOTES:
Subject areas of garden beds based on Dewey Decimal System (000-900)
. Dots are existing trees in garden beds
Colors refer to planting patterns. Types of plants vary to correspond with subjects. Art images and program also included to correspond with subject areas.
Blue refers to flowers & herbs. Green is existing ivy. Red & magenta refer to floral borders.
International Garden Bed
Hedgerow of fruit trees, rose bushes, topiary
Subjects from Eastern & Western perspectives.
A LIVING LIBRARY
IS A PARK OF THE FUTURE FOR THE PRESENT.
Fifth Avenue
NEW YORK PUBLIC LIBRARY
PROPOSED RESTAURANT
BRYANT PARK
Garden TOOLS
WORKSHOP
WORKSHOPS
supplies
40th Street
900 HISTORY & GEOGRAPHY
800 LITERATURE
700 ART
600 TECHNOLOGY
500 SCIENCE
400 LANGUAGE
300 SOCIAL SCIENCES
200 RELIGION
100 PHILOSOPHY
000 GENERALITIES
42nd Street
International Garden Beds: flowers, herbs, & vegetables
A LIVING LIBRARY
means
A.L.L.
Hedgerow: fruit trees, roses, & topiary
WORKSHOPS & demonstrations
OBJECTS for sale & vegetables herbs flowers

a program, students might actually discover that they enjoy school and love learning!! The implications for the future of education are extraordinary.

In 1981 Bryant Park exhibited many urban problems typically found in public open spaces: the under-utilized, derelict, and vandalized environment inhabited primarily by drug dealers and drifters. What was envisioned to transform this place was an international culture park–A LIVING LIBRARY–which would relate to the Main Library, its location in the center of New York City, and New York City's international diversity and role as world leader of communications and culture. A LIVING LIBRARY here would create many reasons for people to come to the park, thus transforming the energy of the environment while preserving its history and original elegance.

Although built in 1934, Bryant Park is based on a seventeenth century French formal garden design, in which the garden was a place of social interaction and cultural creation. Even more interesting, is the fact (unknown to the author in 1981 when first conjuring this concept and design), that the Crystal Palace (1853-1858), stood on this very site, housing the first exhibit of its kind in America, of the Art and Industry of All Nations. So international culture has a magnificent precedence on this site.

But because New York's Bryant Park is probably one of the most prime pieces of real estate in the world, with all of its complex politics, it will be easier and faster to develop A LIVING LIBRARY elsewhere. And that is the current plan.

The author is currently developing plans and exploring possibilities for creating branch LIVING LIBRARIES in partnership with NASA Lewis Research Center in Ohio and with other communities throughout the United States, and in Canada, Japan, Denmark, Sweden, and France. If you, your community, garden club, or an affiliated institution would like to form A LIVING LIBRARY or LIFE FRAME, please contact:

Bonnie Sherk, Executive Director
A LIVING LIBRARY™
32 Cornelia Street
Suite 5C
New York, New York 10014
(212) 242-1700

West Coast Branch
A LIVING LIBRARY™
93 Mirabel
San Francisco, California 94110
(415) 206-9710

Index

Page numbers followed by f indicate figures; those followed by t indicate tables; and those followed by Ch indicate charts.